Soil Ecology and Land-Use Management

Soil Ecology and Land-Use Management

Edited by **Henry Wang**

SYRAWOOD
PUBLISHING HOUSE
New York

Published by Syrawood Publishing House,
750 Third Avenue, 9th Floor,
New York, NY 10017, USA
www.syrawoodpublishinghouse.com

Soil Ecology and Land-Use Management
Edited by Henry Wang

International Standard Book Number: 978-1-68286-221-6 (Hardback)

Printed in China.

Contents

Preface

Soil ecology and land-use management have become prominent fields of study in order to assess the damage caused to arable lands and soils because of pollution and industrialization. This book consists of chapters provided by experts from the field of soil ecology and land-use management, and includes concepts like soil science, soil nutrient management, crop yield, effects of chemical and organic fertilizers, irrigation practices, etc. It is bound to provide a detailed overview to the readers seeking comprehensive information in these fields.

The information contained in this book is the result of intensive hard work done by researchers in this field. All due efforts have been made to make this book serve as a complete guiding source for students and researchers. The topics in this book have been comprehensively explained to help readers understand the growing trends in the field.

I would like to thank the entire group of writers who made sincere efforts in this book and my family who supported me in my efforts of working on this book. I take this opportunity to thank all those who have been a guiding force throughout my life.

Editor

1

A Check List of Weeds in Rice Fields of Coastal Orissa, India

H. N. Subudhi[1], S. P. Panda[2], P. K. Behera[3] & Chitra Patnaik[4]

[1] Central Rice Research Institute, Cuttack, India

[2] P. G. Department of Botany, Utkal University, Bhubaneswar, India

[3] Panchayat College, Baragarh, Orissa, India

[4] P. N. College, Khurda, Orissa, India

Correspondence: H. N. Subudhi, Central Rice Research Institute, Cuttack 753006, India. E-mail: dr_hatanath_subudhi@yahoo.co.in

Abstract

A survey was undertaken to different rice ecologies to collect, identify and conserve the weed species as herbarium specimens in coastal districts of Orissa. In this programme, 201 angiosperm taxa and 3 pteridophytes under 146 genera belonging to 56 families were collected, identified and documented. Poaceae was the dominant family followed by Fabaceae and Cyperaceae. Families are arranged according to modified Bentham and Hooker's system of classification in the enumeration. Genus and species were arranged alphabetically under each family and genus. The number of weed species was more in upland followed by irrigated, semi deep water and deep water.Conservation strategies are also highlighted in this paper.

Keywords: exploration, weeds rice fields, Orissa, conservation

1. Introduction

Rice is widely grown cereal crop that feeds millions of people in world (Kumar et al., 2008). It is the most important cereal crop cultivated in wide range of habitat under varying temperature and water regime. Rice fields are one of the large and extensive fresh water ecosystem on earth.The total rice areas is about 44.6 Mha in India. These act as a ideal habitat for numerous species of aquatic, amphibious and terrestrial weeds that compete with rice crop for space and nutrients. With the dominance of rice crop, grasses, sedges and broad leaves plants are referred as weeds. Biodiversity is the key component for sustainable environment. Currently rice fields are under serious threat owing to rapid increase in human population and urbanization. This phenomenon sets the biodiversity of rice fields at stake making it necessary to take immediate step for conservation. Application of agrochemical and inorganic fertilizer is the major concern for biodiversity loss. Rice fields are the transitional ecosystem that acts as both aquatic for one part of the year and terrestrial the rest. This unique ecosystem acts as potential habitat for numerous species of biota that have adapted to the highly disturbed environment. In Orissa most of the rice fields are rainfed which means it depends on rainwater only.

During rainy season, the rice fields are full of rain water but in Rabi season the fields are completely dry, so the floral succession differs in both the season due to different water regime. Most of the aquatic plants did not appear during dry season. Rice fields are colonized by terrestrial, semiaquatic and aquatic species (Moody & Drost, 1983). Moody (1989) reported 2049 species of weeds from 15 south-east Asian countries. Till now, no systematic survey of weed species was made in Orissa in particular and eastern India in general except some sporadic reports. For effective weed management practice, study of weed phenology and their biology is highly essential. Some sporadic reports regarding the occurrence of weeds in Orissa were made by Patnaik (1956), Tripathy and Mishra (1988), Maiti et al. (1983). Mishra and Choudhury (1996) made extensive survey of aquatic plants in different areas of Bhubaneswar in Orissa state. Subudhi et al. (2002) and Subudhi and Choudhury (2000) reported deep water weed flora and floristic diversity of weeds in Cuttack district of Orissa. Besides these, Bahar and Bhat (2012) reported rice weeds of Kashmir. Pragada and Malliboyana (2010, 2012) and Singh and Rao (1973) reported the phytosociological study rice weeds of Andhra Pradesh respectively. Realizing these, extensive survey were undertaken in seven districts of coastal Orissa in different seasons. A total 201 Angiosperm taxa and three pteridophytes were collected identified and housed in as herbarium specimen in P. G. Department of Botany,

Utkal University Bhubaneswar. In the enumeration, families are arranged according to modified Bentham and Hooker's system of classification (1882-1883).

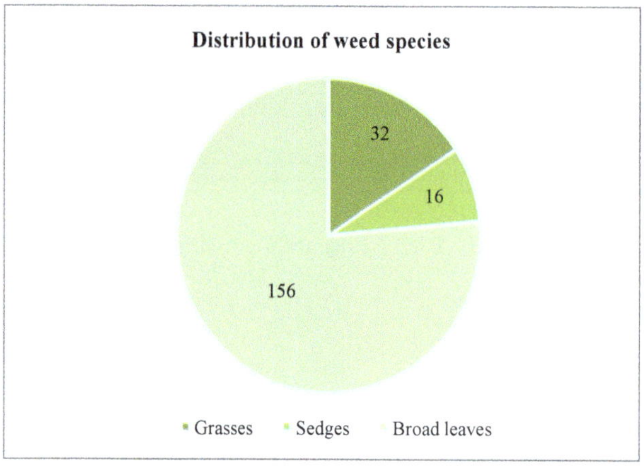

Figure 1. Distribution of weed species

2. Material and Methods

Field trips were undertaken regularly to different rice growing areas during Kharif and Rabi season. Weeds in upland are different from irrigated and lowlands. In irrigated, usually marshy plant are metwith. In lowlands, aquatic and submerged plants are commonly grown, in bunds, terrestrial plant are grown. The plant samples were collected with flowering and kept inside polythene bag. The detailed information regarding associate, ecology, other field characters are recorded in note book. After collection, the specimens were dried, poisoned and fixed in herbarium sheets (Jain & Rao, 1977). The samples are identified in consultation with regional and modern flora.Information regarding utility were collected from local people. Some unidentified specimen were confirmed after matching with the herbarium at CentralNational Herbarium, Calcutta and regional herbarium at Regional Research Laboratory, Bhubaneswar. Correct name was given to each plant in consultation with the flora of British India (Bentham & Hooker, 1882-1883; Saxena & Brahmam, 1994; Haines, 1921-1925) and modern floras.

2.1 Study Area

Orissa is one of the coastal district lies in coast of Bay of Bengal.It is situated in 17°49′ to 22°34′ N latitudes and 81°27′ to 87°29′ E longitudes. The coastal districts are Cuttack, Puri, Bhadrak, Balasore, kendrapara, Jagatsinghpur and khurda etc. (Figure 2). These districts are having maximum rainfed areas. High rainfall occurs due to south west monsoon during the month of July and August (Figure 3).

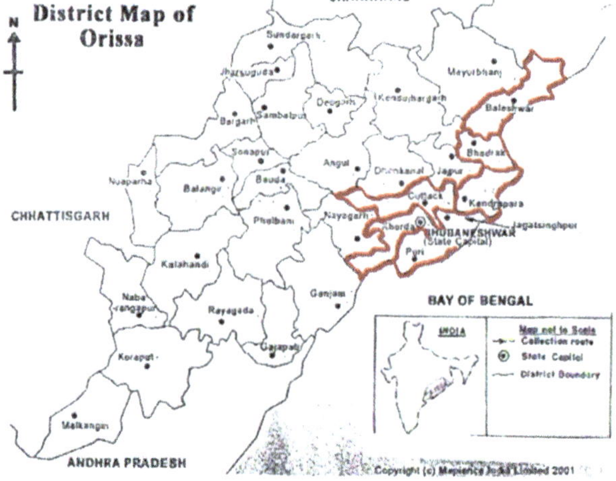

Figure 2. Study area

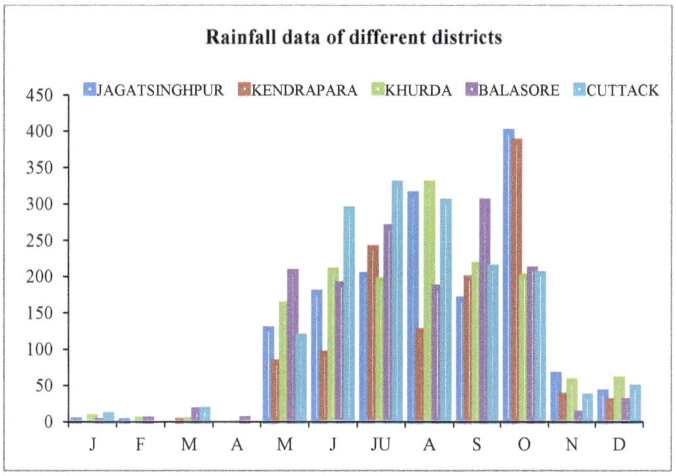

Figure 3. Rainfall data of different districts

2.2 Climate

Rice is the main crop in Orissa. It is grown in two seasons i.e. Kharif (June-December) and Rabi (January-May). South west monsoon is the principal source of rain contributing 80% of rainfall during kharif season. Kharif crop starts with onset of monsoon i.e. on June, the diversity of weeds is very rich in this season due high rainfall. During Rabi season, diversity is less due to less water. The average annual rainfall in Orissa is 1350.6 mm. Maximum rainfall occurs in the month of July-October. The temperature varies from 15-44 °C. December is the coldest month and May is the hottest month of the year.

2.3 Cultural Practices

In Orissa, most of the lands are rainfed because there is no irrigation facility and direct seeded in rainfed areas is the usually practice of the farmers. It is also economical and better suited for lowlands. After summer plough, in the month May/June, fields are broadcasted. Germination occurs after onset South west monsoon along with weeds. In July, there is heavy rainfall, which encourages the growth of weed species of different category much quicker than rice. After one month of germination, beusaning followed by laddering are the usually practices in Orissa which destroyed most of weed species. Water remains in fields up to October in lowlands, most of the hydrophytes grow in lowland as weeds, it remains up to December, there after, weed species dies. During Rabi season, other weed species appear and diversity is very less due to non availability of water. In irrigated field, the species richness is low but some dominant species occur very luxuriantly.

3. Results and Discussion

During exploration, 201 numbers of species under 145 genera belonging to 56 families were collected, identified and documented. The weed species are categorized 1) upland, 2) irrigated, 3) lowland, 4) saline ecology. The species are different from ecology to ecology. The number of weed species are more in upland (65) followed by irrigated (61), semi deep water (54) and deep water (21) (Figure 4). It is interesting to note that upland weeds can not beavailable in lowland/irrigated conditions. Upland weeds are having xeric adaptation due less water and flowers in August and September, whereas lowland weeds are hydrophytic adaption. These plants flower in October-December, the vegetative growth starts with onset of rainfall.

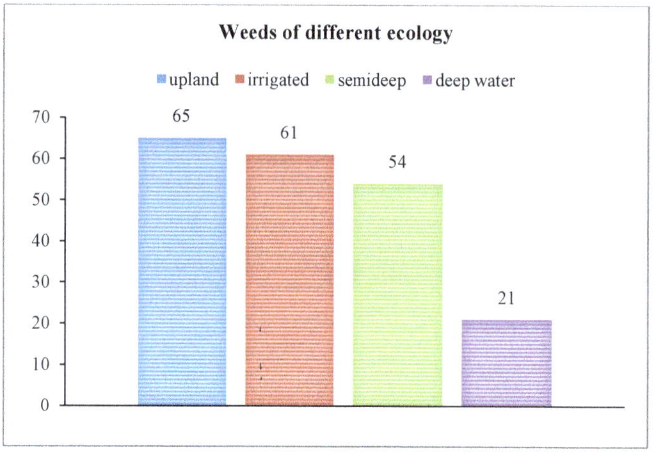

Figure 4. Weeds of different ecology

During the present survey programme a total of 201 Angiosperm species under 146 genera belonging to 56 families along with three pteridophytes. Out of 201 species, 123 species of dicots are distributed under 88 genera belonging to 40 families, while 78 monocot species spread over 57 genera under 16 families. It is interesting to note that out of 78 species of monocots 48 species are sedges and grasses, which constitute 64.97% of the total monocots. The approximate ratio of monocot and dicots species is 1:2.8. So it indicates that dicots are represented by nearly about three times than the monocots, which is represented in the Table 1. Out of 56 families, 20 families are represented by single species. Poaceae is the first dominant family where as it is in the 2nd and 3rd position in the flora of Bihar and Orissa and flora of British India respectively (Figure 5). On the other hand, Fabaceae and Cyperaceae are the 2nd and 3rd largest family respectively.

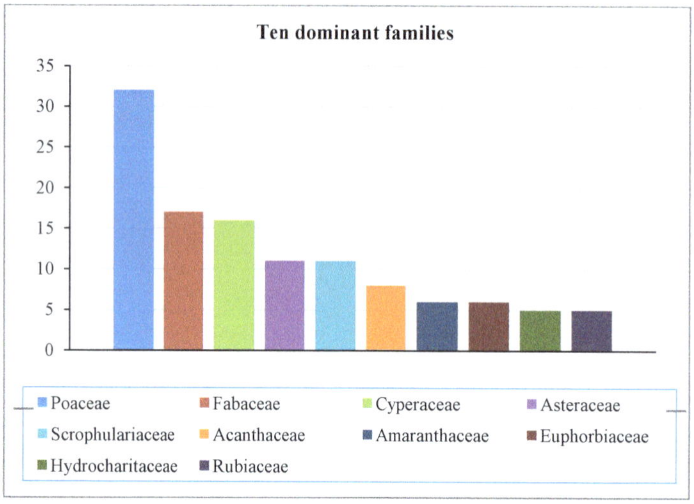

Figure 5. Ten dominant families

Table 1. Percentage of families, genera and species of dicots and monocots are represented below

Groups	Families		Genera		Species	
	No.	%	No.	%	No.	%
Dicotyledones	40	78.76	88	75.52	123	73.73
Monocotyledones	16	21.24	57	24.47	78	26.26

Due to population growth, operation of biotic factors like conversion of agriculture land to residential complexes, application chemical, pesticide and inorganic fertilizer, the biodiversity both flora and fauna in rice fields are declining at alarming rate. Again many weed species are having potential value for food, medicine and check soil erosion (Oudhia, 2001). So awareness should be given to the farmers/common people to conserve the species in general and weeds in particular for future use. There are four rice ecologies namely upland, irrigated, semideep water and deep water.

3.1 Upland Weeds

Rice fields, where there is no standing water are called as Uplands. The plant species in uplands are having some xeric adaptation and tolerance for drought. The species germinate immediately after rain in the month of mid June. Weeds remain in the field up to Octoberor till maturity. The weed diversity is very high in this ecology. The common species are *Cleome viscosa, Cleome monophylla, Argemone mexicana, Spermacoce articularis, Spermacoce pusilla, Allmania nudiflora, Celosia argentea, Aerva lanata, Sebastiana chaemelia, Scoparia dulcis.* The common sedges are *Bulbostylis barbata, Cyperus rotundus, Fimbristylis dichotoma, Fiurena ciliaris*, etc. Similarly The grasses are *Chloris barbata, Chrysopogon aciculatus, Cynodon dactylon, Dactyloctenium aegyptium, Dicanthium pertusum, Digitaria ciliaris, Eragrostis ciliaris, Imperata cylindrical, Oplismenus compositus, Chysopogon aciculatus, Perotis india.*

3.2 Irrigated Ecology

Water is irriged at regular interval to the field, so that there is no water stress to the plant. Sometimes the field is having standing water also which promotes luxuriant weed population. The diversity is very rich but in some cases two/three species are dominant in the fields, other species are rare. The common species are *Portulaca oleracea, Bergia ammaniodes, Crotalaria protrata, Desmodium triflorum, Rotala indica, Glinus oppositifolia, Dentella repens, Grangea maderaspatana, Bacopa monnieri, Oxalis corniculata, Marsilia minuta, Zornia gibbosa, Centella asiatica, Gnaphalium polycaulon, Mecadonia dianthera, Lindernia crustacea, Hygrophila polysperma, Commelina benghalensis*. The grasses are *Elusine indica, Elytrophorus spicatus, Leptochloa chinensis, Panicum repens, Paspalidum flavidum, Setaria intermedia, Echinochloa crus-galli, Echinochloa colona*, etc.

3.3 Semi-Deep Water Ecology

In this ecology, the standing water remain up to 50 cm throughout the crop season. So only aquatic/semiaquatic plants occur in this ecology. In this ecology, the paddy seeds were broadcasted before onset of monsoon, so after rain, weed seeds germinate quickly and grows faster than rice plants. But the diversity of species is very less. The common species in this ecology are *Bergia capensis, Oxalis corniculata, Hydrocera triflora, Aeshynomene aspera, Aeshynomene indica, Sesbania bispinosa, Myriophyllum tetrandrum, Ludwigia adscendens, Trapa bispinosa, Enydra fluctuans, Sphenoclea zeylanica, Ipomoea aquatica, Hydrolea zeylanica, Limnophila aquatica, Limnophila heterophylla, Polygononum hydropiper, Alternanthera philoxeroide, Alternanthera sessilis, Ottelia alsimoides, Monochoria hastata, Typha elephantiana, Typha angustata, Limnophyton trifolia, Limnophyton obtusifolia, Potamogetum pectnatus, Potamogetum nodosus.* The sedges and grasses are *Schoenoplectus articulatus, Cyperus iria, Cyperus platistylis, Cyperus corymbosus, Echinocloa crusgalli, Echinochloa stagnina, Leersia hexandra, Panicum paludosum, Sacciolepis indica, Eriochloa procera, Oryza rufipogon*, etc.

3.4 Deep Water Ecology

In deep water ecology, standing water remains above 50 cm. through out crop season. After the field was dried, boro rice crop was grown. The species diversity was still less than semi deepwater. The floating plant and emergent types were met with in this ecology. The common species are *Nymphaea nouchali, Nymphaea stellata, Nelumbo nucifera, Neptunia oleracea, Nymphoides indica, Nymphoides hydrophylla, Utricularia stellaris, Blyxa echinosperma, Ceratophyllum demersum, Nechamandra alternifolia, Vallisnaria spiralis. Pistia stratiotes, Spirodela polyrrhiza, Wolffia globosa, Azolla pinnata, Oryza rufipogon, Eichhornia crassipes, Leersia hexandra, Utricularia inflexavar stellaris.*

4. Conservation

Every plant has theraptical value so weeds are not exception. Weeds have many utilitarian values such as medicinal, food, fodder, ritual purpose and soil conservation, etc. The species viz., *Enydra fluctuans, Hydrolea zeylanica, Monochoria vaginalis, Limnophila indica, Nelumbo nucifera, Neptunia oleracea, Nymphaea nouchali, Pistia stratiotes, Trapa natans* var. *bispinosa*, etc. are having medicinal properties for different diseases. Oudhia (2001) reported many medicinal weeds from Chhatisgarh. The species such as *Alternanthera sessilis, Centella asiatica, Commelina benghalensis, Ipomoea aquatic, Portulaca oleracea, Mollugo pentaphylla* and *Marsilia minuta*, etc. are used as leafy vegetable but the rhizome of *Nelumbo nucifera, Trapa bispinosa* var. *bispinosa*, etc. are used as food. The grasses like *Panicum repens, Cynodon dactylon, Paspalum flavidum, Digitaria cliaris, Eriochloa procera, Chloris barbata*, etc. are used as fodder. *Azolla pinnata* is very good fodder for cattle. Similarly *Wolffia globosa* is used as feed of duck and fishes. The species like *Cynodon dactylon* is used in every ritual of local people. Rice ecologies have been destroyed due to habitat destruction, construction of houses and conversion of rice field for other purposes so the flora in general and economic plants in particular are on the verge of extinction. It is high time to conserve all the useful weed species as species cafeteria/weed garden so that awareness will be developed among the common people about the utilitarian value of weed species.

References

Bahar, S. W. N., & Bhat, G. A. (2012). Taxocoenosis and distribution of weed flora in the rice field monoculture of Kasmir valley (J and K) India. *Inter. Journ. Botany, 8*(2), 73-78. http://dx.doi.org/10.3923/ijb.2012.73.78

Bentham, G., & Hooker, J. D. (1882-1883). *Genera Plantarum* (3 volumes). London.

Haines, H. H. (1921-1925). *The Botany of Bihar and Orissa* (6-parts). London.

Jain, S. K., & Rao, R. R. (1977). *A handbook of field and herbarium methods.* Todays and Tomorrows Printers and Publishers, New Delhi.

Kunmar, G. S., Rajaratan, A., Thavaprakash, N., Babu, C., & Umasankar, R. (2008). Nitrogen use efficiency in rice (*Oryza sativa*) in systems of cultivation with varied N level under N tracer technique. *Asian Joun. Agric. Res., 2*, 37-40. http://dx.doi.org/10.3923/ajar.2008.37.40

Maiti, M. M., Mishra, G., & Ghosh, R. B. (1983). Some new records of weeds from crop fields of Orissa. *Journ. Econ. Tax. Bot., 4*, 308-310.

Mishra, S., & Choudhury, B. P. (1996). Rice fields flora of Bhubaneswar. *Bio. Sci. Res. Bull., 12*(2), 99-112.

Moody, K. (1989). *Weeds reported in rice fields in South and South East Asia.* International Rice Research Institute, Manila, Philippines.

Moody, K., & Drost, D. C. (1983). The role of cropping systems in weedsin rice. *Weed control in rice* (pp. 73-88). International Rice Research Institute, Manila, Philippines.

Oudhia, P. (2001). Common rice weeds used for first aid by Chattisgarh farmers. *Agric. Sci. Digest., 21*, 273-274.

Patnaik, H. (1956). Some useful weeds in and around Cuttack. *Journ. Bomb. Nat. Hist. Soc., 54*, 141-152.

Pragada, P. M., & Venkaaiah, M. (2012). Phytosociological attributes of weed flora in major crops of North coastal Andhrapradesh, India. *Pak. Journ. Weed Sci. Res., 18*(1), 107-126.

Pragada, P. M., & Venkaiah, M. (2010). Phytosociological attributes of weeds in rice fields of North coastal Andhra Pradesh, India. *International Journ. of Curr. Res., 7*, 1-4.

Saxena, H. O., & Brahmam, M. (1994-1996). *Flora of Orissa* (Vols. 1-4). Regional Research Laboratory and Forest Development Corporation, Govt. of Orissa, Bhubaneswar.

Singh, D. J. C., & Rao, K. N. (1973). Weed flora of Andhrapradesh. *Proc. National Academy of Sciences, 19*(2), 223-229.

Subudhi, H. N., & Coudhury, B. P. (2000). Vegetation of rice fields in Cuttack districts (Orissa). *Jour. Econ. Tax. Bot., 24*(1), 157-159.

Subudhi, H. N., Panda, S. P., Naik, R. K., & Choudhury, B. P. (2002). Weed flora of deep water rice fields in Orissa. *Pl. Sci. Res., 24*(1&2), 50-52.

Tripathy, G., & Mishra, G. (1988). Studies on the weed flora associated with wheat crops. *Ind. Bot. Soc., 67*, 227-228.

Appendix

Appendix 1. Enumeration

Nymphaeaceae

Nymphaea nouchali Burm.f.

Nymphaea pubescens Willd.

Nelumbonaceae

Nelumbo nucifera Gaertn.

Papaveraceae

Argemone mexicana L.

Capparaceae

Cleome gynandra L.

Cleome monophylla L.

Cleome rutidosperma Roxb.

Cleome viscosa L.

Violaceae

Hybanthus enneaspermus (L.) Muell.

Polygalaceae

Polygala arvensis Willd.

Polycarpaea corymbosa (L.) Lamk.

Polycarpon prostratum (Forsk.) Asch. & Sch.

Portulacaceae

Portulaca oleracea L.

Portulaca quadrifida L.

Elatinaceae

Bergia ammannioides Roxb.

Bergia capensis L.

Hypericaceae

Hypericum jopinicum Thunb. exMurry

Sterculiaceae

Melochia corchorifolia L.

Zygophyllaceae

Tribulus terrestris L.

Oxalidaceae

Biophytum sensitivum (L.) DC.

Oxalis corniculata L.

Balsaminaceae

Hydrocera triflora (L.) Wt. &Arn.

Fabaceae

Aeschynomene indica L.

Aeschynomene aspera L.

Aeschynomene americana

Alysicarpus monilifer (L.) DC.

Alysicarpus vaginalis (L.) DC.

Crotalaria prostrata Rottl. exWilld.

Crotalaria quinquefolia L.

Desmodium triflorum (L.) DC.

Indigofera linnaei Ali

Indigofera glabra L.

Melilotus indica (L.) All

Sesbania bispinosa (Jacq.) Wt.&Arn.

Smithia conferta Smith

Smithia sensitiva Ait.

Vigna trilobata (L.) Verd.

Vigna sublobata Ravi

Zornia gibbosa Span.

Caesalpiniaceae

Cassia mimosoides L.

Cassia pumila Lam.

Mimosaceae

Mimosa pudica L.

Neptunia oleracea Lour.

Droseraceae

Drosera burmani Vahl

Drosera indica L.

Haloragaceae

Myriophyllum tetrandrum Roxb.

Lythraceae

Ammannia baccifera L.

Ammannia multiflora Roxb.

Rotala indica (Willd.) Koehne

Onagraceae

Ludwigia adscendens (L.) H. Hara

Ludwigia perennis L.

Trapaceae

Trapa natans L. var. *bispinosa* (Roxb.) Makino

Aizoaceae

Glinus lotoides L.

Glinus oppositifolius (L.) DC.

Mollugo pentaphylla L.

Apiaceae

Centella asiatica (L.) Urb.

Rubiaceae

Dentella repens (L.) *Forst. et Forst.*

***Hedyotis brachiata* Wt. &Arn.**

Hedyotis corymbosa (L.) Lamk.

Spermacoce articularis L.f.

Spermacoce pusilla Wall.

Asteraceae

Ageratum conyzoides L.

Blumea membranacea DC.

Eclipta prostrata (L.) L.

Emilia sonchifolia (L.) DC. ex Wt.

Enydra fluctuans Lour.

Gnaphalium polycaulon Pers.

Grangea maderaspatana (L.) Lour.

Mikania cordata (Burm.f.) Robinson

Sphaeranthus indicus L.

Tridax procumbens L.

Vernonia cinerea (L.) Less.

Sphenocleaceae

Sphenoclea zeylanica Gaertn.

Gentianaceae

Canscora decussata (Roxb.) Sch. & Sch.

Canscora diffusa (Vahl) R. Br.

Hoppea dichotoma Willd.

Menyanthaceae

Nymphoides hydrophylla (Lour.) Kuntze

Nymphoides indicum (L.) Kuntze

Hydrophyllaceae

Hydrolea zeylanica (L.) Vahl.

Boraginaceae

Coldenia procumbens L.

Heliotropium indicum L.

Heliotropium strigosum Willd.

Convolvulaceae

Ipomoea aquatic Forssk.

Ipomoea cymosa (Desr.) Roem. & Sch.

Merremia tridentata (L.) Hall. f.

Solanaceae

Solanum surattense Burm. f.

Scrophulariaceae

Bacopa monnieri (L.) Pennell.

Centranthera tranquebarica (Sperigl.) Merr

***Dopatrium junceum* (Roxb.) Buch-Ham. exBenth.**

Limnophila aquatica (Roxb.) Alston

Limnophila heterophylla (Roxb.) Benth.

Limnophila indica (L.) Druce

Lindernia anagallis (Burm.f.) Pennel

Lindernia crustacea (L.) Muell.

Lindernia viscosa (Horn) Bold

Mecardonia procumbens (Mills.) Small

Scoparia dulcis L.

Lentibulariaceae

Utricularia inflexaForssk. var. stellaris (L.f.) P.Taylor

Acanthaceae

Hygrophila auriculata (Schum.) Haine

Hygrophila polysperma (Roxb.) Anders.

Justicia betonica L.

Justicia diffusa Willd.

Lippia geminata Roxb.

Rungia pectinata (L.) Nees

Rungia repens (L.) Nees

Phyla nudiflora (L.) Rich

Nyctaginaceae

Boerhavia diffusa L.

Amaranthaceae

Achyranthes aspera L.

Aerva lanata (L.) Juss ex Sch.

Alternanthera philoxeroides (Mart) Griseb.

Alternanthera sessilis (L.) R. Br. ex DC.

Celosia argentea L.

Gomphrena celosoides L.

Polygonaceae

Polygonum hydropiper L.

Polygonum plebeium R. Br.

Euphorbiaceae

Crozophora rottleri (Geisel) Juss

Croton bonplandianum Baill

Euphorbia hirta L.

Euphorbia rosea Retz.

Phyllanthus urinaria L.

Sebastiania chaemaelea (L.) Muell.

Urticaceae

Pilea microphylla (L.) Liebr

Pouzolzia indica (L.) Benn

Ceratophyllaceae

Ceratophyllum demersum L.

Hydrocharitaceae

Blyxa echinosperma (Clarke) Hook.f.

Hydrilla verticillata (L.f.) Royle

Nechamandra alternifolia (Roxb.) Thw.

Ottelia alismoides (L.) Pers.

Vallisneria natans (Lour.) H. Hara

Orchidaceae

Spiranthes sinensis (Pers.) Ames

Zeuxine strateumatica (L.) Schltr.

Amaryllidaceae

Crinum defixum Ker-Gawl.

Crinum latifolium L.

Curculigo orchioides Gaertn.

Pontederiaceae

Eichhornia crassipes (Mart.) Solm-Laub.

Monochoria hastata (L.) Solm.

Monochoria vaginalis (Burm.f.) Presl. ex Kunth.

Xyridaceae

Xyris indica L.

Commelinaceae

Commelina benghalensis L.

Murdannia nudiflora (L.) Brenan

Amiscophacelus axillaris Rao & Kamathy

Typhaceae

Typha angustata Bory&Chaub.

Typha elephantiana Roxb.

Araceae

Pistia stratiotes L.

Lemnaceae

Spirodela polyrrhiza (L.) Schleid.

Wolffia globosa (Roxb.) Hartog& Vander Plas

Alismataceae

Limnophyton obtusifolium (L.) Miq.

Sagittaria guayanensis HBK

Sagittaria trifolia L.

Najadaceae

Najas faveolata A. Br. ex Magam.

Aponogetonaceae

Potamogeton nodosus Poir.

Potamogeton pectinatus L.

Eriocaulaceae

Eriocaulon quinquangulare L.

Cyperaceae

Bulbostylis barbata (Roxb.) Kunth

Cyperus castaneus Willd.

Cyperus cephalotes Vahl

Cyperus compressus L.

Cyperus difformis L.

Cyperus esculentus L.

Cyperus iria L.

Cyperus platystylis R. Br.

Cyperus rotundus L.

Eleocharis acutangula (Roxb.) Schutt.

Eleocharis dulcis (Burm.f.) Henschel

Fimbristylis dichotoma (L.) Vahl

Fimbristylis ovata (Burm.f.) Kern

Fimbristylis schoenoides (Retz.) Vahl

Fuirena ciliaris (L.) Roxb.

Schoenoplectus articulatus (L.) Palla

Poaceae

Brachiaria ramosa (L.) Stapf.

Brachiaria reptans (L.) GardetHubb

Chloris barbata Sw.

Chrysopogon aciculatus (Retz.) Trin

Cynodon dactylon (L.) Pers.

Dactyloctenium eagyptium (L.) Beauv.

Dicanthium pertusum (L.) Clayton

Digitaria ciliaris (Retz.) Koel

Echinochloa colonum (L.) Link

Echinochloa crusgalli (L.) Beauv.

Echinochloa stagnina (Retz.) Beauv.

Eleusine indica (L.) Gaertn.

Elytrophorus spicatus (Willd.) A. Camus

Eragrostis ciliaris (L.) R. Br.

Eragrostis japonica (Thunb.) Trin.

Eriochloa procera (Retz.) Hubb

Imperata cylindrica (L.) Rauch

Isachne globosa (Thunb.) Kuntze

Leersia hexandra Sw.

Leptochloa chinensis (L.) Nees

Oplismenus compositus (L.) Beauv.

Oryza rufipogon Griff.

Oryza nivara **Sharma etShastry**

Panicum paludosum Roxb.

Panicum repens L.

Paspalidium flavidum (Retz.) A. Camus

Perotis indica (L.) Kuntze

Rottboellia co chinchinensis (Lour.) Clayton

Sacciolepis indica (L.) A. Chase

Setaria intermedia (Roth) Roem& Sch.

Sporobolus indicus (L.) R. Br.

Vetiveria zizanioides (L.) Nash

Azollaceae

Azolla pinnata R. Br.

Parkeriaceae

Ceratopteris thalictroides (L.) Brongn

Marsiliaceae

Marsilia minuta L.

2

Spatial Distribution of Nymphs of *Triozoida limbata* Enderlein, 1918 (Hemiptera: Triozidae) in Guava Orchards

Vera Alves de Sá[1] & Marcos Gino Fernandes[1]

[1] Programa de Pós-Graduação em Entomologia e Conservação da Biodiversidade, Universidade Federal da Grande Dourados (UFGD), Dourados, MS, Brazil

Correspondence: Vera Alves de Sá, Universidade Federal da Grande Dourados (UFGD), Faculdade de Ciências Biológicas e Ambientais, CEP: 79804-970, Dourados, MS, Brazil. E-mail: veraalves_bio@yahoo.com.br

Abstract

Triozoida limbata is considered one of the leading pests of guava crop in Brazil. Its nymphs are responsible for sucking leaf borders, causing curling and drying of the leaves, and leaving them with a necrotic appearance. Knowledge of the spatial distribution of nymphs of *T. limbata* is essential for improving sampling and control techniques. The objective of this study was to perform probabilistic analyses of patterns of spatial distribution of nymphs of *T. limbata* in guava orchards. The study was conducted in four guava orchards in Ivinhema, Mato Grosso do Sul, Brazil. Samplings were performed every 15 days, from April 2012 to March 2014. To obtain the nymph counts, a sampling area was demarcated in each orchard, comprising 50 sampling units. In each unit, a sample was taken randomly from a shoot of 10 cm to 15 cm in length at the median height of the central plant. Dispersion rates were calculated (variance/mean ratio, Morisita index, and Exponent k of Negative Binomial Distribution) and the data obtained in the field were adjusted to the theoretical frequency distributions (Poisson and Negative Binomial). Following the analyses, we concluded that nymphs of *T. limbata* in the studied populations were randomly organized in the four areas that were evaluated, and the sampling data have been adjusted to the Poisson distribution model.

Keywords: damage, horticulture, *Psidium guajava* L., random, spatial arrangement

1. Introduction

During its development, guava is attacked by various pest insects that cause different types of damage (Sá, 2011). These insects represent the main obstacles to cultivate guava because they reduce the yield and quality of fruits (Yana et al., 2010; Ndankeu et al., 2011). The species *Triozoida limbata* Enderlein, 1918 (Hemiptera: Triozidae) is considered as one of the leading crop pests in Brazil (Colombi & Galli, 2009).

Adults of *T. limbata* have transparent wings without stigmata, displaying radial wing nervures, and the middle and first cubital nervures emerge from a single point. They are greenish in color and measure around 2.0 mm to 2.4 mm in length (Taylor et al., 2010). Eggs are laid on branches, shoots, and new leaves. Later in the laboratory, we found that 19 to 92 eggs were laid per female with an egg incubation period of 7 days to 9 days and a nymphal stage between 29 days and 35 days.

Nymphs, which are responsible for the damage of guava plants, are flattened in shape with a pinkish color, and are covered by a whitish, waxy excretion. As they suck the sap at the edges of the leaves, they inject toxins (Munyaneza et al., 2010), making leaves curl and become dry, and causing the appearance of necrosis (Dalberto et al., 2004; Yana et al., 2010; Ndankeu et al., 2011).

The control of *T. limbata* is based on the application of insecticides but there is not much concern about the population density and economic losses (Hassani et al., 2009). Furthermore, the knowledge of spatial distribution of the insect is not taken into account, a factor that is of vital importance for establishing the best sampling criteria and determining the most appropriate moment to apply pest control.

To determine the pattern of spatial arrangement of a given species, it is necessary to have data on the number of individuals. For this purpose, the ecosystem in question needs to enable the performance of samplings (Fernandes et al., 2003). These samplings, according to L. J. Young and J. H. Young (1998), can be used to draw inferences about either the form of distribution of the population sampled or the characteristics of this distribution. To describe the distribution patterns of a population, aggregation indices and frequency distributions are used.

There is a need to understand the behavioral patterns of spatial distribution of the population of *T. limbata*, so that better strategies can be proposed for their management. This research therefore aims to perform probabilistic analyses of patterns of spatial distribution of nymphs of *T. limbata* in guava orchards.

2. Material and Methods

Samplings were performed from April 2012 to March of 2014 in four commercial guava orchards, Pedro Sato cultivar, in the municipality of Ivinhema - MS, Brazil, at the following locations: area 1, Gleba Piravevê, with a total of 550 plants: 22°16′32″S and 53°48′59″W at an altitude of 339 m; area 2, located in Gleba Vitória, with 300 plants: 22° 20′51″S and 53°47′59″W at an altitude of 377 m; area 3, in Gleba Azul, with 2,800 plants: 22°16′22″S and 53°54′07″W at an altitude of 400 m; and area 4, in Gleba Ouro verde, with an orchard comprising 300 plants: 22°17′34″S and 53°56′15″W at a latitude of 377 m. Plants were seven and a half years old at the beginning of the sampling period and were planted with a spacing of 5 m × 7 m between plants; the irrigation used was by micro-aspersion. Insecticides were applied only when the average infestations reached 30 % of the leaves with the presence of nymphs of *T. limbata* (Pazini & Galli, 2011).

The soil in the region is classified as dystrophic Red Latosol, which comprises 70% sand and 18% clay. According to the Köppen classification, the climate is Aw, which is characterized as a rainy tropical climate with dry winters. Guava, Pedro Sato cultivar, comprising plants propagated by seeds, probably Red Ogawa no. 1, in Nova Iguaçu, Rio de Janeiro, Brazil. Its main features are as follows: vigorous plants with relatively high yields; slightly oval fruits, good appearance (reaching 400 g in shavings of the branches); rough outer skin; pinkish, thick, firm pulp, pleasant flavor, and few seeds (Pommer et al., 2006).

The evaluation of nymphs in each area comprised 50 sampling units. In each unit, one shoot from 10 cm and 15 cm in length was sampled, which was randomly taken from the median height of the central plant in each plot where the number of nymphs was counted every 15 days.

For data analysis, the square root transformation of x + 0.5 was used (Zucareli et al., 2009). The mean ($\widehat{m}$) and variance (S^2) in the number of nymphs of *T. limbata* were obtained on each sampling date, taking the relationship between these values as an indicator of spatial distribution (Elliott, 1979). The dispersion indices, described below, were calculated for each of the samplings performed.

Variance/mean ratio (I): values equal to the unit indicate random spatial distribution; values lower than the unit indicate uniform distribution, and values greater than the unit represent aggregate distribution (Rabinovich, 1980). Spatial randomness can be tested by the chi-square test with n-1 degrees of freedom, $\chi 2 = (n - 1)\ S^2/m$ (Elliott, 1979).

Morisita Index (I_δ): this index is relatively independent of the average and number of samples. Thus, when $I_\delta = 1$, the distribution is random; when $I_\delta > 1$, the distribution is of the contagious type, and when $I_\delta < 1$, this indicates a regular distribution (Morisita, 1962).

Exponent k of the negative binomial distribution (k): this is an appropriate dispersion index when the size and number of sampling units are the same in each sample. Often, this is influenced by the size of the sample units. This parameter is an inverse measure of the degree of aggregation, and in this case, negative values indicate a regular or uniform distribution; positive values, close to zero, indicate aggregate arrangement; and values greater than eight indicate a random distribution (Southwood, 1978; Elliot, 1979). On this aspect, Poole (1974) uses another interpretation: when $0 < k < 8$, the index indicates aggregate distribution, and when $0 > k > 8$, this indicates random distribution.

The theoretical frequency distributions used to evaluate the spatial distribution of the species observed in the field were also used. These distributions are presented below, according to L. J. Young and J. H. Young (1998): Poisson Distribution, also known as random distribution, is characterized by the variance that equals the mean ($S^2 = m$); Negative Binomial Distribution presents greater variance than the average, thereby indicating aggregate distribution, in addition to having two parameters as follows: the mean ($\widehat{m}$) and parameter k ($k > 0$).

The chi-square adhesion test was performed to check the adjustments f of the data collected in the field regarding the theoretical frequency distributions. Therefore, we used the chi-squared adhesion test, which compares the total frequencies observed in the sample area with the expected frequencies, according to L. J. Young and J. H. Young (1998). These frequencies are defined by the product of the probabilities of each class and the total number of sampling units used. For this test, it was decided to establish a minimum expected frequency that equals the unit. Statistical analysis was performed using the chi-square test at the levels of 1% and 5%

probability.

3. Results and Discussion

Forty-eight samplings were performed in each field. A total of 30,973 nymphs of *T. limbata* were observed in guava leaves. During the two years of samplings, population peaks occurred between October and December in areas 1 and 4. In areas 2 and 3, the highest populations were recorded between March and May. The large number of individuals sampled in the orchards was probably due to the presence of young leaves of guava during the sampling period. This fact promoted ideal conditions for the multiplication of *T. limbata* (Melo, 2009) due to the scaled pruning method, which favors the production of guava fruits all year round (Hojo et al., 2007). The periods in which the peaks of occurrence of the insect were recorded happened simultaneously with the intense presence of the shoots.

It should be noted that the values of the variances were lower than the averages in thirty-six samplings performed in area 1 (Table 1), forty-six samplings in area 2 (Table 2), thirty-seven samplings in area 3 (Table 3), and forty-four samplings in area 4 (Table 4). The variance/mean ratios were significantly equal to the unit in forty-six samplings in area 1 (Table 1); the same occurred in forty-seven samplings in area 2 (Table 2), forty-six samplings in area 3 (Table 3), and forty-eight samplings in area 4 (Table 4).

The values of the Morisita Index found in these samples, and confirmed by the spatial randomness test, demonstrated that results were significantly equal to the unit in forty-two samplings in area 1 (Table 1), forty-seven samplings in area 2 (Table 2), forty-five samplings in area 3 (Table 3), and forty-eight samplings in area 4 (Table 4).

The values of the K parameter in area 1 were negative in thirty-six of the samplings, and positive and higher than 8 in 11 samplings (Table 1); in area 2, the values were negative in forty-six of the samplings (Table 2); in area 3, thirty-seven samplings had negative values and 10 were higher than 8 (Table 3); and in area 4, values in forty-four of the samplings were negative (Table 4).

Table 1. Statistical Analysis (means and variances) and dispersion index for nymphs of *Triozoida limbata* in guava orchard (area 1), in Ivinhema, Mato Grosso do Sul, Brazil, 2012/2014 (N = 50)

Index	Sampling date					
	10/04/12	25/04/12	10/05/12	25/05/12	09/06/12	24/06/12
$\hat{m}$	2.840	2.680	3.260	5.120	3.540	6.880
S^2	2.137	1.896	4.278	7.781	4.539	5.985
I	0.753 ^{ns}	0.707 ^{ns}	1.312 ^{ns}	1.520 *	1.282 ^{ns}	0.870 ^{ns}
I_δ	0.914 ^{ns}	0.892 ^{ns}	1.094 ^{ns}	1.100 *	1.079 ^{ns}	0.981 ^{ns}
K	-11.475 ^{un}	-9.156 ^{un}	10.440 ^{al}	9.851 ^{al}	12.542 ^{al}	-52.906 ^{un}
X^2	36.873	34.657	64.301	74.469	62.831	42.628
	09/07/12	24/07/12	08/08/12	23/08/12	07/09/12	22/09/12
$\hat{m}$	6.660	7.000	3.960	5.760	5.820	3.800
S^2	6.474	6.449	2.651	8.186	8.681	3.959
I	0.972 ^{ns}	0.921 ^{ns}	0.670 ^{ns}	1.421 ^{ns}	1.492 *	1.042 ^{ns}
I_δ	0.996 ^{ns}	0.989 ^{ns}	0.918 ^{ns}	1.072 *	1.083 *	1.011 ^{ns}
K	-238.314 ^{un}	-88.926 ^{un}	-11.984 ^{un}	13.675 ^{al}	11.838 ^{al}	90.713 ^{al}
X^2	47.631	45.143	32.808	69.639	73.089	51.053
	07/10/12	22/10/12	06/11/12	21/11/12	06/12/12	21/12/12
$\hat{m}$	4.780	3.880	3.220	8.820	2.840	2.500
S^2	6.502	6.516	2.461	9.171	2.260	2.010
I	1.360 ^{ns}	1.679 *	0.764 ^{ns}	1.040 ^{ns}	0.796 ^{ns}	0.804 ^{ns}
I_δ	1.074 *	1.172 *	0.928 ^{ns}	1.004 ^{ns}	0.929 ^{ns}	0.923 ^{ns}

K	13.271 ^{al}	5.711 ^{ag}	-13.657 ^{un}	221.618 ^{al}	-13.896 ^{un}	-12.760 ^{un}
X^2	66.649	82.289	37.447	50.950	38.986	39.400
	05/01/13	20/01/13	04/02/13	19/02/13	06/03/13	21/03/13
$\widehat{m}$	6.800	2.580	1.140	1.900	2.680	2.840
S^2	9.347	2.698	0.123	0.622	1.936	2.545
I	1.375 ^{ns}	1.046 ^{ns}	0.108 ^{ns}	0.328 ^{ns}	0.723 ^{ns}	0.896 ^{ns}
I_δ	1.054 *	1.017 ^{ns}	0.219 ^{ns}	0.649 ^{ns}	0.898 ^{ns}	0.964 ^{ns}
K	18.1551 ^{al}	56.626 ^{al}	-1.278 ^{un}	-2.826 ^{un}	-9.658 ^{un}	-27.369 ^{un}
X^2	67.353	51.233	5.281	16.053	35.403	43.915
	05/04/13	20/04/13	05/05/13	20/05/13	04/06/13	19/06/13
$\widehat{m}$	2.680	2.840	2.600	3.060	2.940	3.140
S^2	1.936	2.178	2.163	2.507	2.180	2.776
I	0.723 ^{ns}	0.767 ^{ns}	0.832 ^{ns}	0.819 ^{ns}	0.741 ^{ns}	0.884 ^{ns}
I_δ	0.898 ^{ns}	0.919 ^{ns}	0.936 ^{ns}	0.942 ^{ns}	0.913 ^{ns}	0.964 ^{ns}
K	-9.658 ^{un}	-12.183 ^{un}	-15.479 ^{un}	-16.918 ^{un}	-11.373 ^{un}	-27.081 ^{un}
X^2	35.403	37.577	40.769	40.137	36.333	43.318
	04/07/13	19/07/13	03/08/13	18/08/13	02/09/13	17/09/13
$\widehat{m}$	3.480	3.340	3.140	2.640	3.680	3.140
S^2	2.581	2.311	2.368	2.031	2.834	2.653
I	0.742 ^{ns}	0.692 ^{ns}	0.754 ^{ns}	0.769 ^{ns}	0.770 ^{ns}	0.845 ^{ns}
I_δ	0.927 ^{ns}	0.909 ^{ns}	0.923 ^{ns}	0.914 ^{ns}	0.938 ^{ns}	0.951 ^{ns}
K	-13.474 ^{un}	-10.837 ^{un}	-12.767 ^{un}	-11.445 ^{un}	-16.013 ^{un}	-20.265 ^{un}
X^2	36.345	33.898	36.949	37.697	37.739	41.408
	02/10/13	17/10/13	01/11/13	16/11/13	01/12/13	16/12/13
$\widehat{m}$	3.900	3.340	2.560	2.880	3.300	2.780
S^2	3.561	3.372	2.047	2.189	2.378	1.971
I	0.913 ^{ns}	1.010 ^{ns}	0.800 ^{ns}	0.760 ^{ns}	0.720 ^{ns}	0.709 ^{ns}
I_δ	0.978 ^{ns}	1.003 ^{ns}	0.923 ^{ns}	0.918 ^{ns}	0.916 ^{ns}	0.897 ^{ns}
K	-44.897 ^{un}	350.400 ^{al}	-12.784 ^{un}	-12.010 ^{un}	-11.806 ^{un}	-9.553 ^{un}
X^2	44.744	49.467	39.188	37.250	35.303	34.741
	31/12/13	15/01/14	30/01/14	14/02/14	01/03/14	16/03/14
$\widehat{m}$	1.020	1.080	1.020	1.000	1.000	1.000
S^2	0.020	0.075	0.020	0.000	0.000	0.000
I	0.020 ^{ns}	0.070 ^{ns}	0.020 ^{ns}	0.000 ^{ns}	0.000 ^{ns}	0.000 ^{ns}
I_δ	0.039 ^{ns}	0.140 ^{ns}	0.039 ^{ns}	0.000 ^{ns}	0.000 ^{ns}	0.000 ^{ns}
K	-1.040 ^{un}	-1.161 ^{un}	-1.040 ^{un}	-1.000 ^{un}	-1.000 ^{un}	-1.000 ^{un}
X^2	0.961	3.407	0.961	0.000	0.000	0.000

* Significant at 5% probability; ^{ns} Non-significant at 5% probability; ^{AG} aggregate; ^{un} uniform; ^{al} Random; $\widehat{m}$ -mean; S^2 - Variance; I - Mean-variance ratio; I_δ - Morisita index; K - Exponent of the negative binominal; X^2 - calculated chi-square.

Table 2. Statistical Analysis (means and variances) and dispersion index for nymphs of *Triozoida limbata* in guava orchard (area 2), in Ivinhema, Mato Grosso do Sul, Brazil, 2012/2014 (N = 50)

Index	Sampling date					
	10/04/12	25/04/12	10/05/12	25/05/12	09/06/12	24/06/12
$\widehat{m}$	4.900	4.500	12.260	3.060	2.460	2.460
S^2	4.459	3.888	11.380	2.874	2.131	1.968
I	0.910 ^{ns}	0.864 ^{ns}	0.928 ^{ns}	0.939 ^{ns}	0.866 ^{ns}	0.800 ^{ns}
I_δ	0.982 ^{ns}	0.970 ^{ns}	0.994 ^{ns}	0.980 ^{ns}	0.946 ^{ns}	0.920 ^{ns}
K	-54.467 ^{un}	-33.075 ^{un}	-170.804 ^{un}	-50.309 ^{un}	-18.395 ^{un}	-12.294 ^{un}
X^2	44.592	0.970	45.483	46.020	42.447	39.195
	09/07/12	24/07/12	08/08/12	23/08/12	07/09/12	22/09/12
$\widehat{m}$	2.560	2.720	2.280	1.220	2.580	1.500
S^2	2.211	2.532	1.838	0.175	2.044	0.418
I	0.864 ^{ns}	0.931 ^{ns}	0.806 ^{ns}	0.144 ^{ns}	0.792 ^{ns}	0.279 ^{ns}
I_δ	0.947 ^{ns}	0.975 ^{ns}	0.916 ^{ns}	0.301 ^{ns}	0.921 ^{ns}	0.523 ^{ns}
K	-18.757 ^{un}	-39.405 ^{un}	-11.771 ^{un}	-1.424 ^{un}	-12.430 ^{un}	-2.080 ^{un}
X^2	42.313	45.618	39.509	7.033	38.829	13.667
	07/10/12	22/10/12	06/11/12	21/11/12	06/12/12	21/12/12
$\widehat{m}$	2.440	2.860	2.500	1.840	3.100	3.120
S^2	1.925	2.123	1.888	0.464	2.704	2.720
I	0.789 ^{ns}	0.742 ^{ns}	0.755 ^{ns}	0.252 ^{ns}	0.872 ^{ns}	0.872 ^{ns}
I_δ	0.915 ^{ns}	0.911 ^{ns}	0.903 ^{ns}	0.597 ^{ns}	0.959 ^{ns}	0.959 ^{ns}
K	-11.558 ^{un}	-11.096 ^{un}	-10.208 ^{un}	-2.460 ^{un}	-24.273 ^{un}	-24.336 ^{un}
X^2	38.656	36.371	37.000	12.348	42.742	42.718
	05/01/13	20/01/13	04/02/13	19/02/13	06/03/13	21/03/13
$\widehat{m}$	3.140	2.760	7.000	2.540	4.360	2.560
S^2	2.735	3.043	12.612	2.172	3.133	1.925
I	0.871 ^{ns}	1.103 ^{ns}	1.802 *	0.855 ^{ns}	0.719 ^{ns}	0.752 ^{ns}
I_δ	0.959 ^{ns}	1.037 ^{ns}	1.113 *	0.944 ^{ns}	0.936 ^{ns}	0.904 ^{ns}
K	-24.351 ^{un}	26.892 ^{al}	8.731 ^{al}	-17.524 ^{un}	-15.494 ^{un}	-10.319 ^{un}
X^2	42.682	54.029	88.286	41.898	35.211	36.844
	05/04/13	20/04/13	05/05/13	20/05/13	04/06/13	19/06/13
$\widehat{m}$	7.940	7.260	6.860	6.480	6.140	5.660
S^2	5.935	4.890	4.286	3.642	6.123	3.902
I	0.747 ^{ns}	0.674 ^{ns}	0.625 ^{ns}	0.562 ^{ns}	0.997 ^{ns}	0.689 ^{ns}
I_δ	0.969 ^{ns}	0.956 ^{ns}	0.946 ^{ns}	0.934 ^{ns}	1.000 ^{ns}	0.946 ^{ns}
K	-31.445 ^{un}	-22.241 ^{un}	-18.284 ^{un}	-14.798 ^{un}	-2199.143 ^{un}	-18.227 ^{un}
X^2	36.627	33.006	30.615	27.543	48.863	33.784
	04/07/13	19/07/13	03/08/13	18/08/13	02/09/13	17/09/13
$\widehat{m}$	5.380	4.780	2.740	2.420	2.580	1.040
S^2	3.179	3.032	2.564	2.371	2.044	0.039
I	0.591 ^{ns}	0.634 ^{ns}	0.936 ^{ns}	0.980 ^{ns}	0.792 ^{ns}	0.038 ^{ns}

I_δ	0.925 ns	0.925 ns	0.977 ns	0.992 ns	0.921 ns	0.075 ns
K	-13.152 un	-13.073 un	-42.578 un	-119.568 un	-12.430 un	-1.081 un
X^2	28.955	31.084	45.847	48.008	38.829	1.846
	02/10/13	17/10/13	01/11/13	16/11/13	01/12/13	16/12/13
$\widehat{m}$	1.020	1.020	1.040	1.020	1.040	1.060
S^2	0.020	0.020	0.039	0.020	0.039	0.058
I	0.020 ns	0.020 ns	0.038 ns	0.020 ns	0.038 ns	0.054 ns
I_δ	0.039 ns	0.039 ns	0.075 ns	0.039 ns	0.075 ns	0.109 ns
K	-1.040 un	-1.040 un	-1.081 un	-1.040 un	-1.081 un	-1.121 un
X^2	0.961	0.961	1.846	0.961	1.846	2.660
	31/12/13	15/01/14	30/01/14	14/02/14	01/03/14	16/03/14
$\widehat{m}$	1.120	1.100	2.440	2.500	2.700	2.380
S^2	0.108	0.092	1.802	1.847	2.010	1.955
I	0.096 ns	0.083 ns	0.739 ns	0.739 ns	0.745 ns	0.821 ns
I_δ	0.195 ns	0.168 ns	0.894 ns	0.897 ns	0.907 ns	0.926 ns
K	-1.239 un	-1.200 un	-9.338 un	-9.570 un	-10.568 un	-13.318 un
X^2	4.714	4.091	36.197	36.200	36.481	40.244

* Significant at 5% probability; ns Non-significant at 5% probability; un uniform; al Random; $\widehat{m}$ - mean; S^2 - Variance; I - Mean-variance ratio; I_δ - Morisita index; K - Exponent of the negative binominal; X^2 - calculated chi-square.

Table 3. Statistical Analysis (means and variances) and dispersion index for nymphs of *Triozoida limbata* in guava orchard (area 3), in Ivinhema, Mato Grosso do Sul, Brazil, 2012/2014 (N = 50)

Index	Sampling date					
	10/04/12	25/04/12	10/05/12	25/05/12	09/06/12	24/06/12
$\widehat{m}$	3.480	8.560	3.500	5.840	5.460	3.700
S^2	2.989	7.598	2.418	6.953	6.947	4.663
I	0.859 ns	0.888 ns	0.691 ns	1.191 ns	1.272 ns	1.260 ns
I_δ	0.960 ns	0.987 ns	0.913 ns	1.032 ns	1.049 ns	1.069 ns
K	-24.684 un	-76.197 un	-11.325 un	30.630 al	20.043 al	14.212 al
X^2	42.092	43.495	33.857	58.342	62.348	61.757
	09/07/12	24/07/12	08/08/12	23/08/12	07/09/12	22/09/12
$\widehat{m}$	4.020	4.020	2.400	3.060	2.900	2.580
S^2	7.612	5.408	2.000	2.017	2.418	1.800
I	1.893 *	1.345 ns	0.833 ns	0.659 ns	0.834 ns	0.698 ns
I_δ	1.219 *	1.085 ns	0.931 ns	0.890 ns	0.943 ns	0.884 ns
K	4.499 al	11.645 al	-14.400 un	-8.975 un	-17.461 un	-8.529 un
X^2	92.781	65.915	40.833	32.294	40.862	34.178
	07/10/12	22/10/12	06/11/12	21/11/12	06/12/12	21/12/12
$\widehat{m}$	2.900	3.440	2.880	1.920	2.700	2.960
S^2	2.255	2.456	2.230	1.993	3.031	3.182
I	0.778 ns	0.714 ns	0.774 ns	1.038 ns	1.122 ns	1.075 ns

I_δ	0.924 ^{ns}	0.918 ^{ns}	0.923 ^{ns}	1.020 ^{ns}	1.045 ^{ns}	1.025 ^{ns}
K	-13.041 ^{un}	-12.020 ^{un}	-12.765 ^{un}	50.176 ^{al}	22.050 ^{al}	39.459 ^{al}
X^2	38.103	34.977	37.944	50.875	55.000	52.676
	05/01/13	20/01/13	04/02/13	19/02/13	06/03/13	21/03/13
$\widehat{m}$	2.440	4.740	4.480	6.600	3.380	3.680
S^2	1.843	6.768	6.051	13.796	2.404	2.426
I	0.755 ^{ns}	1.428 ^{ns}	1.351 ^{ns}	2.090 *	0.711 ^{ns}	0.659 ^{ns}
I_δ	0.901 ^{ns}	1.089 *	1.077 ^{ns}	1.162 *	0.916 ^{ns}	0.909 ^{ns}
K	-9.977 ^{un}	11.080 ^{al}	12.779 ^{al}	6.053 ^{ag}	-11.701 ^{un}	-10.800 ^{un}
X^2	37.016	69.962	66.179	102.424	34.846	32.304
	05/04/13	20/04/13	05/05/13	20/05/13	04/06/13	19/06/13
$\widehat{m}$	2.740	2.780	2.820	3.220	2.760	2.600
S^2	2.074	1.971	2.028	2.665	2.268	1.918
I	0.757 ^{ns}	0.709 ^{ns}	0.719 ^{ns}	0.828 ^{ns}	0.822 ^{ns}	0.738 ^{ns}
I_δ	0.912 ^{ns}	0.897 ^{ns}	0.902 ^{ns}	0.947 ^{ns}	0.936 ^{ns}	0.900 ^{ns}
K	-11.271 ^{un}	-9.553 ^{un}	-10.043 ^{un}	-18.678 ^{un}	-15.475 ^{un}	-9.917 ^{un}
X^2	37.088	34.741	35.241	40.553	40.261	36.154
	04/07/13	19/07/13	03/08/13	18/08/13	02/09/13	17/09/13
$\widehat{m}$	3.100	3.260	3.080	2.960	2.880	2.720
S^2	2.500	2.400	2.238	2.202	2.393	2.287
I	0.806 ^{ns}	0.736 ^{ns}	0.727 ^{ns}	0.744 ^{ns}	0.831 ^{ns}	0.841 ^{ns}
I_δ	0.938 ^{ns}	0.920 ^{ns}	0.912 ^{ns}	0.915 ^{ns}	0.942 ^{ns}	0.942 ^{ns}
K	-16.017 ^{un}	-12.364 ^{un}	-11.271 ^{un}	-11.566 ^{un}	-17.048 ^{un}	-17.100 ^{un}
X^2	39.516	36.080	35.610	36.459	40.722	41.206
	02/10/13	17/10/13	01/11/13	16/11/13	01/12/13	16/12/13
$\widehat{m}$	2.920	2.820	3.080	3.060	2.520	2.340
S^2	2.238	2.110	2.524	2.302	1.969	1.862
I	0.767 ^{ns}	0.748 ^{ns}	0.820 ^{ns}	0.752 ^{ns}	0.781 ^{ns}	0.796 ^{ns}
I_δ	0.921 ^{ns}	0.912 ^{ns}	0.942 ^{ns}	0.920 ^{ns}	0.914 ^{ns}	0.914 ^{ns}
K	-12.509 ^{un}	-11.197 ^{un}	-17.064 ^{un}	-12.360 ^{un}	-11.525 ^{un}	-11.446 ^{un}
X^2	37.562	36.660	40.156	36.869	38.286	38.983
	31/12/13	15/01/14	30/01/14	14/02/14	01/03/14	16/03/14
$\widehat{m}$	2.500	2.560	2.720	2.460	3.340	2.540
S^2	2.092	1.884	2.165	1.927	2.556	2.049
I	0.837 ^{ns}	0.736 ^{ns}	0.796 ^{ns}	0.783 ^{ns}	0.765 ^{ns}	0.807 ^{ns}
I_δ	0.935 ^{ns}	0.898 ^{ns}	0.926 ^{ns}	0.913 ^{ns}	0.931 ^{ns}	0.925 ^{ns}
K	-15.313 ^{un}	-9.696 ^{un}	-13.328 ^{un}	-11.353 ^{un}	-14.220 ^{un}	-13.150 ^{un}
X^2	41.000	36.063	39.000	38.382	37.491	39.535

* Significant at 5% probability; ^{ns} Non-significant at 5% probability; ^{AG} aggregate; ^{un} uniform; ^{al} Random; $\widehat{m}$ - mean; S^2 - Variance; I - Mean-variance ratio; I_δ - Morisita index; K - Exponent of the negative binominal; X^2 - calculated chi-square.

Table 4. Statistical Analysis (means and variances) and dispersion index for nymphs of *Triozoida limbata* in guava orchard (area 4), in Ivinhema, Mato Grosso do Sul, Brazil, 2012/2014 (N = 50)

Index	Sampling date					
	10/04/12	25/04/12	10/05/12	25/05/12	09/06/12	24/06/12
$\hat{m}$	3.020	4.060	3.840	4.260	2.940	2.520
S^2	2.306	3.731	3.280	2.890	2.629	1.847
I	0.763^{ns}	0.919^{ns}	0.854^{ns}	0.678^{ns}	0.894^{ns}	0.733^{ns}
I_δ	0.923^{ns}	0.980^{ns}	0.963^{ns}	0.926^{ns}	0.964^{ns}	0.895^{ns}
K	-12.769^{un}	-50.105^{un}	-26.331^{un}	-13.248^{un}	-27.791^{un}	-9.429^{un}
X^2	37.411	45.030	41.854	33.244	43.816	35.905
	09/07/12	24/07/12	08/08/12	23/08/12	07/09/12	22/09/12
$\hat{m}$	2.580	2.820	2.500	2.720	3.460	2.340
S^2	1.922	2.804	1.847	1.961	4.049	1.943
I	0.745^{ns}	0.994^{ns}	0.739^{ns}	0.721^{ns}	1.170^{ns}	0.830^{ns}
I_δ	0.902^{ns}	0.9980^{ns}	0.897^{ns}	0.899^{ns}	1.049^{ns}	0.928^{ns}
K	-10.117^{un}	-487.085^{un}	-9.570^{un}	-9.7452^{un}	20.312^{al}	-13.802^{un}
X^2	36.504	48.716	36.200	35.324	57.347	40.692
	07/10/12	22/10/12	06/11/12	21/11/12	06/12/12	21/12/12
$\hat{m}$	2.820	2.880	3.580	3.200	4.580	2.740
S^2	2.110	2.189	2.575	3.510	4.820	2.156
I	0.748^{ns}	0.760^{ns}	0.719^{ns}	1.097^{ns}	1.052^{ns}	0.787^{ns}
I_δ	0.912^{ns}	0.918^{ns}	0.923^{ns}	1.030^{ns}	1.011^{ns}	0.923^{ns}
K	-11.197^{un}	-12.010^{un}	-12.754^{un}	33.011^{al}	87.402^{al}	-12.845^{un}
X^2	36.660	37.250	35.246	53.750	51.568	38.547
	05/01/13	20/01/13	04/02/13	19/02/13	06/03/13	21/03/13
$\hat{m}$	2.720	2.880	2.440	3.020	3.000	2.860
S^2	1.920	2.679	2.129	2.796	2.531	2.409
I	0.706^{ns}	0.930^{ns}	0.873^{ns}	0.926^{ns}	0.844^{ns}	0.842^{ns}
I_δ	0.893^{ns}	0.976^{ns}	0.948^{ns}	0.976^{ns}	0.949^{ns}	0.946^{ns}
K	-9.248^{un}	-41.303^{un}	-19.142^{un}	-40.627^{un}	-19.174^{un}	-18.119^{un}
X^2	34.588	45.583	42.754	45.358	41.333	41.266
	05/04/13	20/04/13	05/05/13	20/05/13	04/06/13	19/06/13
$\hat{m}$	3.080	1.840	2.720	2.400	2.800	2.720
S^2	2.238	0.586	1.879	1.755	2.000	1.920
I	0.727^{ns}	0.319^{ns}	0.691^{ns}	0.731^{ns}	0.714^{ns}	0.706^{ns}
I_δ	0.912^{ns}	0.633^{ns}	0.888^{ns}	0.889^{ns}	0.899^{ns}	0.893^{ns}
K	-11.271^{un}	-2.700^{un}	-8.799^{un}	-8.932^{un}	-9.800^{un}	-9.248^{un}
X^2	35.610	15.609	33.853	35.833	35.000	34.588
	04/07/13	19/07/13	03/08/13	18/08/13	02/09/13	17/09/13
$\hat{m}$	2.820	2.520	2.900	2.760	2.640	2.620
S^2	2.151	2.051	2.255	2.309	1.868	1.832
I	0.763^{ns}	0.814^{ns}	0.778^{ns}	0.836^{ns}	0.707^{ns}	0.699^{ns}

I_δ	0.917 ns	0.927 ns	0.924 ns	0.942 ns	0.891 ns	0.887 ns
K	-11.880 un	-13.529 un	-13.041 un	-16.874 un	-9.025 un	-8.714 un
X^2	37.369	39.873	38.103	40.986	34.667	34.267
	02/10/13	17/10/13	01/11/13	16/11/13	01/12/13	16/12/13
$\widehat{m}$	2.760	2.720	2.480	2.660	2.620	3.360
S^2	2.104	2.981	1.847	2.392	1.873	2.602
I	0.762 ns	1.096 ns	0.745 ns	0.899 ns	0.715 ns	0.775 ns
I_δ	0.915 ns	1.035 ns	0.898 ns	0.963 ns	0.893 ns	0.934 ns
K	-11.621 un	28.322 al	-9.709 un	-26.426 un	-9.190 un	-14.903 un
X^2	37.362	53.706	36.484	44.068	35.031	37.952
	31/12/13	15/01/14	30/01/14	14/02/14	01/03/14	16/03/14
$\widehat{m}$	2.880	2.940	2.420	2.620	2.820	2.840
S^2	2.230	2.425	2.330	2.281	2.477	2.219
I	0.774 ns	0.825 ns	0.963 ns	0.871 ns	0.878 ns	0.781 ns
I_δ	0.923 ns	0.941 ns	0.985 ns	0.951 ns	0.957 ns	0.924 ns
K	-12.765 un	-16.780 un	-65.219 un	-20.262 un	-23.195 un	-12.983 un
X^2	37.944	40.415	47.182	42.664	43.043	38.282

ns Non-significant at 5% probability; un uniform; al Random; $\widehat{m}$ - mean; S^2 - Variance; I - Mean-variance ratio; I_δ - Morisita index; K - Exponent of the negative binominal; X^2 - calculated chi-square.

The three indices of spatial distribution used in this research indicated that the spatial arrangement of the nymphs of *T. limbata* was random in both areas studied. This is in contrast with the results found for the guava orchard of the Paluma cultivar, in which the pattern of distribution was aggregate (Marcelino, 2013).

Regarding the test of frequency adjustment of numerical classes of the nymphs of *T. limbata* observed in area 1, the values of the chi-square test were not significant for Poisson distribution in forty samplings, indicating that the distribution is random. For negative binomial distribution, all forty-one samplings presented significant chi-square values, indicating that the distribution was not aggregate. We also found that, in thirty-two samples in area 2, the chi-square values were not significant for Poisson distribution, but in sixteen samplings, values were significant. All thirty-six samplings tested for negative binomial distribution were significant, indicating that the distribution was not contagious (Table 5).

Table 5. Chi-square adhesion test of the expected frequencies of Poisson and Negative Binomial (Bn) distributions, spatial arrangement for nymphs of *Triozoida limbata,* in Ivinhema, Mato Grosso do Sul, Brazil, (areas 1 and 2), 2012/2014

Sampling date	Area 1				Area 2			
	Poisson		Bn		Poisson		Bn	
	X^2	DF (nc-2)	X^2	DF (nc-3)	X^2	DF (nc-2)	X^2	DF (nc-3)
10/04/12	7.210 ns	5	378.308 *	4	6.177 ns	8	2523.623 *	8
25/04/12	7.202 ns	5	3374.213 *	1	6.348 ns	7	2082.656 *	7
10/05/12	15.499 ns	8	240.971 *	6	13.272 ns	14	10883.626 *	19
25/05/12	15.896 ns	9	4093.879 *	11	13.311 ns	7	1655.738 *	6
09/06/12	15.453 ns	8	2443.138 *	8	10.310 ns	5	943.610 *	4
24/06/12	4.742 ns	9	3611.833 *	10	11.452 ns	6	1643.483 *	5
09/07/12	10.062 ns	10	4852.907 *	11	10.598 ns	6	1335.673 *	5
24/07/12	17.774 ns	11	5584.338 *	13	11.954 ns	6	1305.088 *	5
08/08/12	7.323 ns	6	1099.699 *	6	10.242 ns	5	385.508 *	4
23/08/12	16.696 ns	10	3491.862 *	10	39.228 ns	1	-	-
07/09/12	18.826 ns	11	4095.147 *	11	5.942 ns	5	698.281 *	4
22/09/12	6.744 ns	8	2023.486 *	7	22.548 *	2	1.511 ns	1
07/10/12	16.404 ns	9	3501.294 *	10	10.272 ns	5	385.020 *	4
22/10/12	16.155 ns	9	3475.152 *	10	6.860 ns	5	97.068 *	4
06/11/12	4.938 ns	5	207.158 *	2	9.211 ns	4	3072.193 *	3
21/11/12	7.421 ns	11	8878.718 *	17	19.811 *	2	-	-
06/12/12	7.515 ns	5	868.501 *	4	8.033 ns	7	2559.688 *	8
21/12/12	10.330 ns	5	742.606 *	4	29.284 *	6	1262.915 *	5
05/01/13	15.537 ns	11	6234.389 *	14	12.682 ns	7	1649.486 *	6
20/01/13	13.245 ns	7	1622.610 *	6	40.463 *	7	1236.693 *	5
04/02/13	50.755 *	1	-	-	20.119 ns	13	6196.112 *	14
19/02/13	13.049 *	2	21093.672 *	1	26.240 *	6	1345.818 *	5
06/03/13	9.047 ns	4	5367.791 *	3	6.294 ns	7	2184.423 *	7
21/03/13	6.101 ns	6	1270.469 *	5	12.312 ns	7	343.656 *	5
05/04/13	7.519 ns	4	5367.976 *	3	9.612 ns	10	4289.781 *	11
20/04/13	4.632 ns	6	1636.208 *	3	15.404 ns	8	3160.597 *	8
05/05/13	9.130 ns	5	918.402 *	4	14.311 ns	8	3227.736 *	8
20/05/13	11.840 ns	6	1344.263 *	5	13.933 ns	7	1189.850 *	5
04/06/13	11.320 ns	6	1967.140 *	5	13.258 ns	10	4183.093 *	11
19/06/13	10.171 ns	5	1247.967 *	5	14.576ns	9	4263.374 *	11
04/07/13	7.692 ns	3	799.455 *	3	22.240 *	7	4355.904 *	7
19/07/13	3.785 ns	6	3111.196 *	5	4.542 ns	6	1009.070 *	5
03/08/13	12.544 ns	6	1550.115 *	5	6.448 ns	6	1665.146 *	6
18/08/13	10.850 ns	5	374.228 *	4	19.140 *	6	1285.863 *	5
02/09/13	11.970 ns	7	2153.918 *	6	7.322 ns	5	698.035 *	4
17/09/13	7.228 ns	6	1312.430 *	5	71.384 *	1	-	-

02/10/13	13.343 ^{ns}	7	2069.202 *	7	76.463 *	1	-	-
17/10/13	11.581 ^{ns}	6	1247.103 *	5	76.463 *	1	-	-
01/11/13	8.280 ^{ns}	6	1544.826 *	5	71.384 *	1	-	-
16/11/13	11.944 ^{ns}	6	1681.976 *	5	76.463 *	1	-	-
01/12/13	9.245 ^{ns}	5	382.281 *	4	71.384 *	1	-	-
16/12/13	7.720 ^{ns}	4	6461.637 *	3	66.643 *	1	-	-
31/12/13	76.463 *	1	-	-	54.293 *	1	-	-
15/01/14	62.224 *	1	-	-	58.112 *	1	-	-
30/01/14	76.463 *	1	-	-	8.444 ^{ns}	4	7841.148 *	2
14/02/14	72.702 ^{i}	0	-	-	8.705 ^{ns}	4	5771.259 *	3
01/03/14	72.702 ^{i}	0	-	-	6.643 ^{ns}	6	2756.223*	5
16/03/14	72.702 ^{i}	0	-	-	8.898 ^{ns}	5	790.810 *	4

* Significant at 5% probability; ^{ns} Non-significant; ^{i} insufficient of classes; X^2 - chi-square value calculated; DF - degree of freedom; nc - number of classes observed at field.

Considering adjustments of the frequencies for areas 3 and 4, forty-seven samplings were not significant for Poisson distribution. For negative binomial distribution, forty-eight samplings were significant in area 3 and forty-seven samplings in area 4, indicating that the distribution was not aggregate (Table 6).

Table 6. Chi-square adhesion test of the expected frequencies of Poisson and Negative Binomial (Bn) distributions, spatial arrangement for nymphs of *Triozoida limbata,* in Ivinhema, Mato Grosso do Sul, Brazil, (areas 3 and 4), 2012/2014

Sampling date	Area 3				Area 4			
	Poisson		Bn		Poisson		Bn	
	X^2	DF (nc-2)	X^2	DF (nc-3)	X^2	DF (nc-2)	X^2	DF (nc-3)
10/04/12	6.520 ^{ns}	7	1661.443 *	6	11.456 ^{ns}	7	2237.595 *	7
25/04/12	18.141 ^{ns}	11	6355.919 *	14	7.477 ^{ns}	7	2062.243 *	7
10/05/12	12.950 ^{ns}	7	1003.630 *	6	10.309 ^{ns}	8	2076.470 *	7
25/05/12	18.446 ^{ns}	11	4797.295 *	12	5.699 ^{ns}	6	1524.442 *	6
09/06/12	16.183 ^{ns}	10	4783.458 *	12	13.947 ^{ns}	7	2032.698 *	7
24/06/12	13.477 ^{ns}	7	1563.932 *	6	8.467 ^{ns}	4	6892.334 *	3
09/07/12	11.686 ^{ns}	8	5439.167 *	13	8.581 ^{ns}	4	3297.221 *	3
24/07/12	16.279 ^{ns}	9	2431.250 *	8	10.679 ^{ns}	7	1633.014 *	6
08/08/12	9.072 ^{ns}	6	1436.840 *	5	9.304 ^{ns}	4	5771.067 *	3
23/08/12	9.281 ^{ns}	4	25056.732 *	3	7.703 ^{ns}	4	4928.765 *	3
07/09/12	10.729 ^{ns}	5	943.709 *	4	11.452 ^{ns}	7	1579.241 *	6
22/09/12	8.724 ^{ns}	4	36288.768 *	3	11.385 ^{ns}	6	1481.883 *	5
07/10/12	6.430 ^{ns}	5	799.198 *	4	5.239 ^{ns}	5	197.113 *	4
22/10/12	8.140 ^{ns}	5	421.603 *	4	4.415 ^{ns}	6	1683.111 *	5
06/11/12	7.896 ^{ns}	5	763.303 *	4	5.744 ^{ns}	5	644.513 *	4
21/11/12	32.221 *	5	284.814 *	1	10.469 ^{ns}	6	1226.190 *	5
06/12/12	13.650 ^{ns}	7	1598.586 *	6	14.817 ^{ns}	8	2996.997 *	9

21/12/12	13.577 ^{ns}	7	1606.544 *	6	5.339 ^{ns}	6	1525.075 *	5
05/01/13	7.358 ^{ns}	5	4653.618 *	3	8.571 ^{ns}	4	9666.293 *	3
20/01/13	15.756 ^{ns}	10	3493.493 *	10	8.506 ^{ns}	6	1292.287 *	5
04/02/13	7.006 ^{ns}	8	2941.664 *	9	10.767 ^{ns}	5	944.865 *	4
19/02/13	22.249 ^{ns}	13	5427.198 *	13	7.709 ^{ns}	6	1285.986 *	5
06/03/13	6.794 ^{ns}	5	262.095 *	4	4.472 ^{ns}	6	1327.025 *	5
21/03/13	4.227 ^{ns}	6	4711.235 *	5	8.653 ^{ns}	6	1333.294 *	5
05/04/13	8.782 ^{ns}	5	267.049 *	4	4.041 ^{ns}	5	134.385 *	4
20/04/13	8.396 ^{ns}	4	6461.584 *	3	13.379 *	2	-	-
05/05/13	8.696 ^{ns}	4	3767.426 *	3	8.850 ^{ns}	4	22202.383 *	3
20/05/13	8.297 ^{ns}	7	1709.985 *	6	9.060 ^{ns}	4	14984.102 *	3
04/06/13	7.277 ^{ns}	5	921.996 *	4	8.994 ^{ns}	4	4823.654 *	3
19/06/13	8.859 ^{ns}	4	3968.711 *	3	12.276 ^{ns}	6	9666.144 *	3
04/07/13	4.946 ^{ns}	5	1361.909 *	5	4.300 ^{ns}	5	561.593 *	4
19/07/13	3.303 ^{ns}	6	1651.657 *	5	8.121 ^{ns}	6	1474.980 *	5
03/08/13	4.696 ^{ns}	5	134.519 *	4	8.437 ^{ns}	6	1501.986 *	5
18/08/13	8.336 ^{ns}	5	397.481 *	4	10.527 ^{ns}	5	938.324 *	4
02/09/13	10.320 ^{ns}	5	938.888 *	4	7.917 ^{ns}	4	13334.736 *	3
17/09/13	9.790 ^{ns}	5	938.293 *	4	7.543 ^{ns}	4	24357.434 *	3
02/10/13	9.384 ^{ns}	5	717.158 *	4	6.333 ^{ns}	5	462.652 *	4
17/10/13	8.556 ^{ns}	5	196.911 *	4	12.252 ^{ns}	6	1234.195 *	5
01/11/13	9.857 ^{ns}	5	939.603 *	4	8.917 ^{ns}	4	4930.936 *	3
16/11/13	9.478 ^{ns}	5	669.762 *	4	10.119 ^{ns}	6	1284.300 *	5
01/12/13	9.828 ^{ns}	5	396.885 *	4	9.307 ^{ns}	4	9993.327 *	3
16/12/13	10.174 ^{ns}	5	261.762 *	4	5.000 ^{ns}	6	878.942 *	5
31/12/13	10.008 ^{ns}	5	911.887 *	4	8.660 ^{ns}	6	1534.019 *	5
15/01/14	9.238 ^{ns}	4	4991.929 *	3	11.275 ^{ns}	6	1347.442 *	5
30/01/14	6.270 ^{ns}	5	827.104 *	4	11.484 ^{ns}	6	1291.968 *	5
14/02/14	8.633 ^{ns}	6	1984.314 *	5	10.766 ^{ns}	6	1319.584 *	5
01/03/14	5.101 ^{ns}	5	873.838 *	4	11.705 ^{ns}	6	1291.992 *	5
16/03/14	7.980 ^{ns}	5	798.360 *	4	10.343 ^{ns}	6	1508.077 *	5

* Significant at 5% probability; ^{ns} Non-significant; X^2 - chi-square value calculated; DF - degree of freedom; nc - number of classes observed at field.

Data obtained in 86.46% of samplings adjusted the Poisson distribution, indicating a random distribution model for nymphs of *T. limbata*. This contradicts the values of 86.36% of the samplings found by Marcelino (2013), which adjusted the negative binomial distribution. Similar results were also reported for another representative of the family Triozidae, *Bactericera cockerelli*, which is considered a pest of potato, bell pepper, and tomato (Prager et al., 2013, 2014).

The random distribution is considered the least common distribution of contagion in nature, occurring when members of the species are located, generally, in a homogenous environment, where an individual's position is independent of others' positions (Taylor, 1984). In this form of arrangement, the energy expenditure on reproduction is lower, as males can find females without having to extensively search in the area (Shea et al., 1993). In addition, the population gains greater genetic variability because insects that come into the crop can find reproductive partners more easily (Diekötter et al., 2008). Therefore, it would be difficult for the entire population to be affected (Courtney, 1986).

Taking into account that the damage is also distributed in a random manner, the applications of insecticides, either at the wrong time or in an uneven manner, could undermine the efficiency of the integrated pest control as various individuals in the population may not be reached. Under these circumstances, the surviving insects could remain in the crop with enough energy to reproduce and begin a new cycle of attack (Alves, 2012).

Information of the spatial arrangement of the nymphs of *T. limbata* is of vital importance to establish the best sampling criteria and determine the best moment to apply the pest control. Our results will contribute to the development of future sequential sampling plans for T. *limbata* in order to define the exact number of sampling units that should be used.

4. Conclusion

Nymphs of *Triozoida limbata* in the studied populations were randomly organized in the four areas that were evaluated, and the sampling data have been adjusted to the Poisson distribution model.

Acknowledgements

The authors are grateful to *Coordenação de Aperfeiçoamento de Pessoal de Nível Superior* (Coordination for the Improvement of Higher Education Personnel - CAPES) for granting a PhD scholarship to the first author.

References

Alves, T. C. (2012). *Distribuição espacial do percevejo-do-colmo (Tibraca limbativentris Stål) em arroz irrigado.* (Dissertação de Mestrado). Escola de Agronomia e Engenharia de Alimentos, Universidade Federal de Goiás, Goiânia, Brasil.

Colombi, C. A., & Galli, J. C. (2009). Dinâmica populacional e evolução de dano de *Triozoida limbata* (Hemiptera: Psillydae) em goiabeira, em Jaboticabal, SP. *Ciência e Agrotecnologia, 33*, 412-416. http://dx.doi.org/10.1590/S1413-70542009000200008

Courtney, S. P. (1986). Why insects move between host patches: some comments on risck-spreading. *Oikos, 47*(1), 112-114. http://dx.doi.org/10.2307/3565925

Dalberto, F. M. S., Menezes Junior, A. O., Simões, H. C., Benito, N. P., & Pitwak, J. (2004). Flutuação populacional do psilídeo-da-goiabeira, *Triozoida limbata* (Hemiptera: Pslilydae) na região de Londrina Paraná, PR. *Semina: Ciências Agrárias, 25*, 87-92. http://dx.doi.org/10.5433/1679-0359.2004v25n2p87

Diekötter, T., Billeter, R., & Crist, T. O. (2008). Effects of landscape connectivity on the spatial distribution of insect diversity in agricultural mosaic landscapes. *Basic and Applied Ecology, 9*(3), 298-307. http://dx.doi.org/10.1016/j.baae.2007.03.003

Elliott, J. M. (1979). *Some methods for the statistical analysis of sample benthic invertebrates* (1st ed., p. 157). Ambleside: Freshwater Biological Association.

Fernandes, M. G., Busoli, A. C., & Barbosa, J. C. (2003). Distribuição espacial de *Alabama argillacea* (Hübner) (Lepidoptera: Noctuidae) em algodoeiro. *Neotropical Entomology, 32*(1), 107-115. http://dx.doi.org/10.1590/S1519-566X2003000100016

Hassani, M. R., Nouri-Ganbalani, G., Izadi, H., Shojai, M., & Basirat, M. (2009). Economic injury level of the psyllid, *Agonoscena pistaciae*, on pistachio, *Pistacia vera* cv. Ohadi. *Journal of Insect Science, 9*, 1-4. http://dx.doi.org/10.1673/031.009.4001

Hojo, R. H., Chalfun N. N. J., Hojo, T. D., Souza, H. A., Paglis, C. M., & São José, A. R. (2007) Caracterização fenológica da goiabeira 'Pedro Sato' sob diferentes épocas de poda. *Revista Brasileira de Fruticultura, 29*, 20-24. http://dx.doi.org/10.1590/S0100-29452007000100007

Marcelino, M. C. S. (2013). *Distribuição espacial e amostragem sequencial de Triozoida limbata (Hemiptera: Triozidae) em goiabeira* (Tese de Doutorado). Universidade Estadual Paulista, Jaboticabal, Brasil.

Melo, G. (2009). *Dinâmica populacional e inimigos naturais de Triozoida limbata (Hemiptera: Triozidae) e diversidade de famílias de himenópteros parasitoides em pomar convencional e orgânico de goiaba na região de Campinas, SP.* (Dissertação de Mestrado). Instituto Biológico. Campinas, Brasil.

Morisita, M. (1962). Id-index, a measure of dispersion of individuals. *Researches on Population Ecology, 4*(1), 1-7. http://dx.doi.org/10.1007/BF02533903

Munyaneza, J., Fisher, T. W., Sengoda, V. G., Garczynski, S. F., Nissinen, A., & Lemmetty, A. (2010). Association of "*Candidatus* Liberibacter solanacearum" With the Psyllid, *Trioza apicalis* (Hemiptera:

Triozidae) in Europe. *Journal of Economic Entomology, 103*, 1060-1070. http://dx.doi.org/10.1603/EC10027

Ndankeu, Y. P. M., Tamesse, J. L., Burckhardt, D., & Messi, J. (2011). Biodiversity of jumping plant-lice of the Pslyllidae family (Hemiptera: Psylloidea) from the South Region of Cameroon: Faunistics, phenology and host plants. *Journal of Entomology, 8*, 123-138. http://dx.doi.org/10.3923/je.2011.123.138

Pazini, W. C., & Galli, J. C. (2011). Redução de aplicações de inseticidas através da adoção de táticas de manejo integrado do *Triozoida limbata* (Enderlein, 1918) (Hemiptera: Triozidae) em goiabeira. *Revista Brasileira de Fruticultura, 33*(1), 66-72. http://dx.doi.org/10.1590/S0100-29452011000100010

Pommer, C. V., Murakami, K. R. N., & Watlington, F. (2006). Goiaba no mundo. *O Agronômico, 58*, 22-26.

Poole, R. W. (1974). *An introduction to quantitative ecology* (1st ed., p. 525). New York: McGraw Hill.

Prager, S. M., Butler, C. D., & Trumble, J. T. (2013). A sequential binomial sampling plan for potato psyllid (Hemiptera: Triozidae) on bell pepper (*Capsicum annum*). *Pest Management Science, 69*(10), 1131-1135. http://dx.doi.org/10.1002/ps.3475

Prager, S. M., Butler, C. D., & Trumble, J. T. (2014). Binomial Sequential Sampling Plan for *Bactericera cockerelli* (Hemiptera: Triozidae) in *Solanum lycopersicum* (Solanales: Solanacea). *Journal of Economic Entomology, 107*(2), 838- 845. http://dx.doi.org/10.1603/EC13328

Rabinovich, J. E. (1980). *Introducion a la ecologia de problaciones animales* (1st ed., p. 313). México, CECSA.

Sá, V. A. (2011). *Comportamento de acasalamento, níveis de infestação e parasitismo de Triozoida limbata Enderlein, 1918 (Hemiptera: Triozidae) em Psidium guajava L. (Myrtaceae)* (Dissertação de Mestrado), Universidade Federal da Grande Dourados, Dourados, Brasil.

Shea, M. M., Dixon, P. M., & Sharitz, R. R. (1993). Size differences, sex ratio, and spatial distribution of male and female Water Tupelo, *Nyssa aquatica* (Nyssaceae). *American Journal of Botany, 80*(1), 26-30. http://dx.doi.org/10.2307/2445116

Southwood, T. R. E. (1978). *Ecological methods* (2nd ed., p. 525). New York: John Wiley & Sons. http://dx.doi.org/10.1007/978-94-009-1225-0_2

Taylor, G. S., Austin, A. D., Jennings, J. T., Purcell, M. F., & Wheeler, G. S. (2010). Casuarinicola, a new genus of jumping plant lice (Hemiptera: Triozidae) from Casuarina (Casuarinaceae). *Zootaxa, 2601*, 1-27.

Taylor, L. R. (1984). Assessing and interpreting the spatial distributions of insect populations. *Annual Review of Entomology, 29*(1), 321-357. http://dx.doi.org/10.1146/annurev.en.29.010184.001541

Yana, W., Tamesse, J. L., & Burckhardt, D. (2010). Jumping plant-lice of the family Psyllidae Latreille (Hemiptera: Psylloidea) from the Center region of Camerroon: Faunistics, phenology and host plants. *Journal of Entomology, 7*, 1-18. http://dx.doi.org/10.3923/je.2010.1.18

Young, L. J., & Young, J. H. (1998). *Statistical ecology: A population perspective* (1st ed., p. 565). Boston: Kluwer Academic Publishers. http://dx.doi.org/10.1007/978-1-4757-2829-3

Zucareli, V., Bonjovani, M. R., Cavariani, C., & Nakagawa, J. (2009). Tolerância à dessecação e influência do tegumento na germinação de sementes de citrumelo 'swingle' (*Citrus paradisi* MACF X *Poncirus trifoliata* (L) RAF.). *Revista Brasileira de Fruticultura, 31*(1), 291-295. http://dx.doi.org/10.1590/S0100-29452009000100042

3

Impact of Normal and Shear Stresses Due to Wheel Slip on Hydrological Properties of an Agricultural Clay Loam: Experimental and New Computerized Approach

Andrea Battiato[1,3], Abdallah Alaoui[2] & Etienne Diserens[1]

[1] Agroscope, Institute for Sustainability Sciences (ISS), Ettenhausen, Switzerland

[2] Centre for Development and Environment (CDE), University of Bern, Bern, Switzerland

[3] Department of Land, Environment, Agriculture and Forestry, University of Padua, Legnaro, Italy

Correspondence: Etienne Diserens, Agroscope, Institute for Sustainability Sciences ISS, Tänikon 1, 8356 Ettenhausen, Switzerland. E-mail: etienne.diserens@agroscope.admin.ch

Abstract

The main purpose of this study was to evaluate the effect that mechanical stresses acting under the slipping driving wheels of agricultural equipment have on the soil's pore system and water flow process (surface runoff generation during extreme event). The field experiment simulated low slip (1%) and high slip (27%) on a clay loam. The stress on the soil surface and changes in the amounts of water flowing from macropores were simulated using the Tires/tracks And Soil Compaction (TASC) tool and the MACRO model, respectively. Taking a 65 kW tractor on a clay loam as a reference, results showed that an increase in slip of the rear wheels from 1% to 27% caused normal stress to increase from 90.6 kPa to 104.4 kPa at the topsoil level, and the maximum shear contact stress to rise drastically from 6.0 kPa to 61.6 kPa. At 27% slip, topsoil was sheared and displaced over a distance of 0.35 m. Excessive normal and shear stress values with high slip caused severe reductions of the soil's macroporosity, saturated hydraulic conductivity, and water quantities flowing from topsoil macropores. Assuming that, under conditions of intense rainfall on sloping land, a loss in vertical water flow would mean an increase in surface runoff, we calculated that a rainfall intensity of 100 mm h^{-1} and a rainfall duration of 1 h would increase the runoff coefficient to 0.79 at low slip and to 1.00 at high slip, indicating that 100% of rainwater would be transformed into surface runoff at high slip. We expect that these effects have a significant impact on soil erosion and floods in steeper terrain (slope > 15°) and across larger surface areas (> 16 m^2) than those included in our study.

Keywords: MACRO model, macroporosity, surface runoff, saturated hydraulic conductivity, normal stress, shear stress, TASC tool, wheel slip

1. Introduction

The structure of the soil profile down to a depth of 0.6 m can be affected by farming in many ways depending on the characteristics of the machines, the soil type, and the initial soil conditions (e.g. Ansorge & Godwin, 2006; Arvidsson, Trautner, & Keller, 2002; Horn, 2003). Considering topsoil alteration, wheeled agricultural equipment, especially tractors with mounted or trailed implements, exert traction force through a stress-strain interaction between the tires and the topsoil. In this contact surface interaction between soil and tire, soil deforms under normal and shear stresses (eg. Horn, 2002; Horn & Rostek, 2000). The shear stress rises sharply with an increase in traction force and wheel slip, which may lead to the detachment of a weak topsoil layer vulnerable to erosion due to soil failure (Battiato, Diserens, Laloui, & Sartori, 2013).

The effect of soil deformation on soil structure can be assessed with several types of parameters: bulk density and total porosity (Boone, 1988; da Silva, Kay, & Perfect, 1994), macroporosity (Alakukku, 1996), penetration resistance (Pagliai, 1998), air permeability (Ball, 1981; Reszkowska, Krümmelbein, Gan, Peth, & Horn, 2011), saturated hydraulic conductivity (Alakukku, 1996), pre-consolidation pressure (Horn, 1981; Kirby, 1991), dye surface density (Kulli, Gysi, & Flühler, 2003; Alaoui & Helbling, 2006; Alaoui & Goetz, 2008), and infiltration capacity (Alaoui & Helbling, 2006; Blanco-Canqui, Claassen, & Stone, 2010).

The most-used ones are soil cohesion, soil structural strength, bulk density, water potential, and pre-compressive stress. However, these parameters are general indicators that integrate information about the total change in the volume of voids of soil under consideration, but they cannot account for changes in the volume distribution of these voids, their connectivity, or the changes in this connectivity (e.g., Vogeler, Horn, Wetzel, & Kruemmelbein, 2006).

Previous investigations showed that shear stress induces a tortuous pore system with vertical and horizontal soil displacement (Wiermann, 1998), a decrease in air permeability (O'Sullivan, Robertson, & Henshall, 1999), and an excess in surface runoff during strong rainfall (Horn & Rostek, 2000). Models of the driving forces acting at the soil-tire interface related to slip (Steiner, 1979; Osetinsky & Shmulevich, 2004), to the deformation of the tires and the topsoil during tractive performance (Schwieger, 1996), or to the management of the energy requirement of the tractor (Pichlmaier, 2012) described the stresses and deformation distribution of soil and tire but did not provide direct information useful to the practitioner.

Based on field and laboratory studies, computerized simulations emerged as appropriate tools to evaluate the effects of heavy-load machines on soil compaction. They included mapping on a national scale with the Soil Compaction Model (SOCOMO) (van den Akker, 1997, 2004) or on a plot scale (Diserens, Chanet, & Marionneau, 2010) and can be designed as practical applications such as Terranimo® (Stettler et al., 2010) and Tyres/tracks And Soil Compaction (TASC) (Diserens & Spiess, 2004). A practical forecasting module integrated into TASC V3.0 (Battiato, 2014; Diserens & Battiato, 2013) calculates the slip-rate limit beyond which the topsoil failure occurs with the corresponding traction force.

Although the above approaches provide useful information on the mechanical disturbance of soil structure in order to better manage agricultural land, they do not elucidate the risk of erosion or flooding during heavy rainfall in hillslope soils.

Today, several hydrological methods are used to estimate surface runoff on different landscape scales. These methods are based on soil topography, geology, vegetation cover, and soil type (Peschke, Etzenberg, Müller, Töpfer, & Zimmermann, 1999; Scherrer & Naef, 2003; Markart et al., 2004). Additional tools that assess the flow processes in topsoil by using input soil parameters in response to a given rainfall or irrigation intensity (e.g., Jarvis, 1994; Simunek, Sejna, & van Genuchten, 2007) do not consider external perturbations caused by land use. Surprisingly, no or insufficient attempts have been made in field experiments or computer simulations to identify cause-and-effect relationships between the magnitude of shear stress and normal stress applied by agricultural machines and the changes in the soil water dynamics. Knowledge of these relationships would be of great relevance to the practitioner for a better soil and water management on steep grounds used for agriculture, especially in conservation systems without plowing.

In addition to mechanical components, climate change will modify the hydrological regime in many regions and consequently the water dynamics in unsaturated soil. Based on regional climate models (Christensen & Christensen, 2003; Graham, Hagemann, Jaun, & Beniston, 2007), future summers are likely to show more frequent extreme events that result in catastrophic flooding, despite a general trend toward drier summer conditions. Thus, increasing rainfall intensity and soil compaction magnitude should drastically increase surface runoff that would promote flood events and erosion during extreme rainfall events.

The aim of this study was to evaluate the impact of mechanical stresses acting at the soil-tire interface during wheel slip on soil structure – parameterized as total porosity, pore-size distribution, saturated hydraulic conductivity, and Brilliant Blue dye-tracer distribution – and, consequently, on water flow process (water flowing from macropores and surface runoff). To achieve this goal, we measured and modelled both, mechanical and hydraulic properties.

2. Material and Methods

2.1 Location and Soil Description

The two study sites were located in Tänikon (47°29′0″N, 8°54′44″E) in the canton of Thurgovie, in Eastern Switzerland, on the border of the Prealps at 536 m.a.s.l. The annual mean rainfall values over the last decade amounted to 1140 mm with peaks in May and July/August. The study area was situated on a plain where land cover consisted of wheat stubble. The soil was described as albic luvisol (USA classification) (Scheffer & Schachtschabel, 1998). At both sites, soil texture consisted of clay loam to a depth of 0.45 m. Soil organic carbon content (determined by weight loss on ignition) varied from 2% (topsoil) to 0.2% (subsoil). A pH of 7.0-7.2 was measured at the soil surface, increasing slightly to 7.4 in the subsoil (Table 1). The treatments were carried out on August 26th, 2008, under dry topsoil conditions above 0.15 m and relatively wet conditions below this depth.

Table 1. Physical and hydraulic parameters of soil under consideration

Depth interval (m)	Particle-size distribution (%)			Texture	Organic carbon OC (%)	pH	Water content* (%)	Water potential* (hPa)
	Clay (< 2 μm)	Silt (2-50 μm)	Sand (> 50 μm)					
Site 1								
0.00–0.15	37.0	39.2	23.8	clay loam	2.0	7.0	29.05	28.75
0.31–0.45	36.7	37.2	26.1	clay loam	1.0	7.4	36.30	5.31
0.46–0.50	44.6	23.3	32.1	clay	0.2	7.4	27.05	0.88
Site 2								
0.00–0.15	36.3	41.0	22.7	clay loam	2.0	7.2	-	-
0.31–0.45	37.3	40.2	22.5	clay loam	1.0	7.4	-	-
0.46–0.50	35.2	30.1	34.7	clay loam	0.2	7.4	-	-

Note. * Recorded data only at site 1 - treatment with 1% slip; Site 2: treatment with 27% slip; Textural classification according to the soil taxonomy of the U.S. Department of Agriculture (USDA).

2.2 Measurements

2.2.1 Mechanical Properties

1) Traction Tests

The aim of the field traction tests was to investigate the effects of tractor-tire slip on the soil. Two corridors, each 4 m wide and 70 m long, were marked in the field (Figure 1). In the first corridor (site 1), the tractor moved with minimal slip in a self-propelled, steady-state condition, whereas in the second corridor (site 2), the tractor moved with slip in a steady-state condition with high drawbar pull. A drawbar pull of 21.8 kN was controlled by a braking tractor. Whereas almost no wheel slip (1%) was measured at site 1, a wheel slip of 27% was measured at site 2. The pulling tractor and the braking tractor did not move in alignment. This allowed the two tractors to have independent tracks, so that the soil at site 2 was trafficked by the pulling tractor only (Figure 1). An area adjacent to the two sites served as an untreated control site (without traffic).

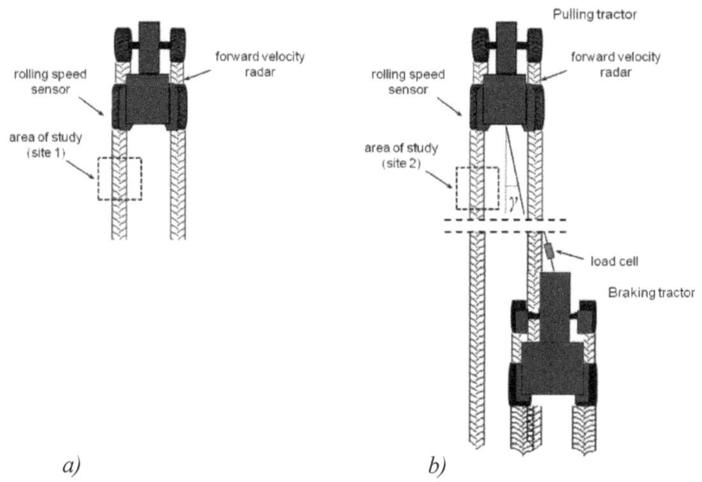

Figure 1. Experimental design of the treatments, *a)* traffic with low slip 1% (site 1), *b)* traffic with high slip 27% (site 2)

A mechanical front-wheel drive Hürlimann H488 DT tractor with 65 kW engine power and 40.0 kN weight was employed as a pulling tractor. The braking tractor was a Massey Ferguson 8470 with 250 kW engine power and 90.6 kN weight (Table 2).

Table 2. Technical data of the tractors used in the traction tests

Braking tractor	Massey Ferguson 8470
Power (kW)	250
Weight (kN)	90.6
Pulling tractor	Hürlimann H488 DT
Power (kW)	65
Weight (kN)	40
Wheelbase (m)	2.34
Tire (front; rear)	380/85R24; 420/85R34
Tire width (front; rear) (m)	0.38; 0.44
Tire unloaded radius (front; rear) (m)	0.63; 0.79
Tire rolling radius (front; rear) (m)	0.59; 0.78
Tire inflation pressure (front; rear) (kPa)	160; 160
Tire stiffness (front; rear) (kN/m)	325; 432
Stationary wheel load (front; rear) (kN)	9.1; 10.9
Drawbar height (m)	0.83
Theoretical speed ratio	0.99

The traction force was measured by a 200 kN load cell (HBM U2B, Darmstadt, Germany) in section with the steel cable that connected the two tractors. The actual forward speed was measured by a radar velocity sensor (Dickey-john RVS2, Auburn, Illinois, U.S.A.), and the wheel rolling speed was recorded with a wireless wheel-speed sensor at two pulses per wheel revolution, set on a rear wheel of the pulling tractor. All of the above parameters were recorded by a data logger. The load acting on the wheels of the pulling tractor in the stationary condition was measured with a flat-bed wheel-load scale (Haenni WL 103, Jegenstorf, Switzerland). The pulling tractor was equipped with 380/85R24 front tires and 420/85R34 rear tires. Tire inflation pressure was measured with a Motometer tire-pressure gauge (Mühlacker-Lomersheim, Germany) and adjusted to 160 kPa on front and rear wheels. The rolling radius and the theoretical speed ratio of the tractor wheels, defined as the distance traveled per revolution of the wheel divided by 2, and as the ratio of the theoretical speed of the front wheel to that of the rear wheel, respectively, were determined during preliminary tests at zero drawbar pull on a smooth road. Tire stiffness was calculated according to Lines and Murphy (1991). The technical specifications of the pulling and braking tractors are reported in Table 2. The pulling tractor moved in a rectilinear direction with locked differential. The traction force in the longitudinal direction was obtained from the measured force by taking into account the angle γ (approx. 3°) of the steel cable that connected the two tractors (Figure 1). The actual forward speed of the tractor was 1.6 m s^{-1} at site 1 and 0.9 m s^{-1} at site 2.

The shear magnitude related to the slip was evaluated by marking the soil surface with two differently colored dyes, blue and red, before the passage of the tractor. The food color Brilliant Blue FCF (E-133) was applied at 120 L m^{-2} and 3.7 g/10^{-3} m^3; the food color Allura Red (E-129) was applied at 120 L m^{-2} and 7.4 g/10^{-3} m^3. After the treatment, soil-layer displacement was measured.

2) Topsoil Mechanical Parameters

Topsoil mechanical parameters for simulating soil-tire interaction were obtained by vertical plate penetration tests and horizontal plate shear deformation tests according to Bekker (1960), with a tractor-mounted hydraulic press (bevameter) described by Diserens and Steinmann (2003) (Table 3).

The vertical plate penetration tests were carried out with two circular plates of diameter 0.2 m and 0.3 m at a penetration rate of approx. 0.02 m s^{-1}. The cohesive and frictional moduli of deformation, Kc and $K\varphi$, respectively, and the exponent of deformation n were determined according to Wong (1980).

The horizontal plate shear deformation tests were performed by using an annular plate with an outer diameter of 0.3 m and an inner diameter of 0.2 m, at different vertical pressures. Values of cohesion c, angle of shear resistance φ, and shear deformation modulus k were determined according to the procedure described by Wong

(1980).

Penetration resistance was determined by using a penetrometer with a screwdriver head (tip width: 6×10^{-3} m) developed by Agroscope Tänikon (Bastgen & Diserens, 2009). All the measured mechanical soil properties are reported in Table 3. Based on measured tractor and soil parameters, the shear stress τ and normal stress σ occurring at the soil-tire interface were calculated according to Osetinsky and Shmulevich (2004).

Table 3. Mechanical parameters of soil under consideration

Depth interval (m)	0-0.20
Cohesive modulus of deformation K_c (kN m$^{-(n+1)}$)	4555
Frictional modulus of deformation K_φ (kN m$^{-(n+2)}$)	-3037
Modulus of deformation K (kN m$^{-(n+2)}$)	8073
Exponent of deformation n	0.90
Cohesion c (kPa)	5.0
Angle of shear resistancc φ (°)	30.0
Shear deformation modulus k (m)	0.01
Mean penetration resistance $PR_{10\ cm}$ (daN) (n = 24)	7.51

2.2.2 Hydrological Properties

1) Total Porosity and Pore-Size Distribution

Total porosity (TP) and pore-size distribution (PSD) were determined for samples of undisturbed soil (five per depth) with a diameter of 55 mm and a height of 42 mm. The TP was measured at depths of 0–0.15, 0.16–0.30, 0.31–0.45, and 0.46–0.60 m, whereas PSD was measured at depths of 0.05, 0.15, and 0.35 m. The samples were taken at intervals of about 0.15 m in a horizontal direction. Three pore classes were defined: macropores, i.e., pores larger than 50 μm; mesopores, measuring between 0.2 and 50 μm; and micropores, i.e., those smaller than 0.2 μm.

The PSD was determined by a vacuum pressure membrane apparatus described by Dirksen (1999) with a hanging water column for a water suction $h < 31.6$ kPa and with a gas adsorption porosimeter using N_2 for 31.6 kPa $< h <$ 1585 kPa.

2) Saturated Hydraulic Conductivity

Saturated hydraulic conductivity (K_{sat}) was measured in undisturbed soil samples (five per depth) taken along the track trafficked with 1% slip (site 1), along the track trafficked with 27% slip (site 2) (Figure 1), and from an untrafficked area of the field (control site). Water conductivity measurements were carried out according to the constant head method (Klute & Dirksen, 1986).

Samples from the topsoil surface below the loose layer (0.03–0.07 m) were taken after removal of the soil cut by the tire lugs. Additional samples were collected deeper in the topsoil (0.13–0.17 m) and in the subsoil (0.33–0.37 m).

3) Irrigation Experiments

From a pit, 1 m deep, three water content probes were installed horizontally under both sites (site 1 and site 2). The two sites were irrigated with a rainfall simulator covering a 1 m^2 area with an intensity of 30 mm h^{-1} for 1 h. Water content was measured at the three depths of 0.15, 0.35, and 0.50 m with a time domain reflectometer (TDR) (CR10X & TDR100, Campbell Scientific Inc. Logan, Utah, USA) with 0.20 m wave guides (two parallel rods of 6 mm diameter). Calibration was performed according to Roth, Schulin, Flühler, and Attinger (1990). The TDR measurements were recorded every 60 s.

The results obtained were analyzed according to the decrease in water content (θ) after cessation of infiltration during the drainage phase ($\Delta\theta_d$, defined as the difference between maximum water content reached at the infiltration phase and minimum water content measured within 2 h after the cessation of the irrigation) (Alaoui & Helbling, 2006). Low $\Delta\theta_\delta$ values indicated the draining of micropores, whereas high values indicated the draining of larger pores, such as macropores or cracks, and pointed out preferential flow (Alaoui, Germann, Jarvis, & Acutis, 2003).

4) Dye Infiltration Experiment

A dye infiltration experiment was carried out to visualize the heterogeneity of dye-tracer distribution pathways. To this end, 240×10^{-3} m^3 of dye-tracer solution were prepared by diluting 880 g of Brilliant Blue FCF powder - also known as food dye E-133 (Flury & Flühler, 1995) - in tap water (final concentration = 3.7 g/10^{-3} m^3). The prepared solution was applied at site 1 and site 2 for 3 h at a constant rate of 24 mm h^{-1} by using a rainfall simulator covering a 1 m^2 area. Being neutral or anionic (depending on the pH), Brilliant Blue (BB) is not heavily adsorbed by negatively charged soil constituents.

One day after infiltration, a soil pit was excavated and six vertical profiles (0, 0.20, 0.40, 0.60, 0.80, and 1.00 m from the edge of the area treated with the rainfall simulator) were prepared. A rubber string grid (1.0 × 0.7 m) was attached in front of each profile. The profiles were photographed with a digital camera (HP Photosmart 945; resolution: 5 megapixels). The resultant digital images had a resolution of approx. 2000 × 2000 pixels. The final coverage of the stained areas was determined from profile images according to the procedure described by Alaoui and Goetz (2008). Afterwards, the distribution of the dye coverage with depth was calculated for each image by counting the pixels indicating stained soil for each horizontal pixel row of the image.

To allow quantitative estimates of different BB concentrations, we performed a calibration linking specific colors with corresponding BB concentration ranges. Ten standard solutions were prepared (BB concentration: 0.1, 0.5, 1, 2, 4, 6, 20, 40, 80 and 150 g/10^{-3} m^3). These solutions were photographed with the same camera under the same light conditions as in the field. The surface density of BB was then estimated to a depth of 0.80 m for both sites (site 1 and site 2).

5) Surface Runoff Estimation

To estimate surface runoff resulting from shear stress, we considered previous measurements of surface runoff on a 16 m^2 surface area with a 15° slope, carried out on similar soil texture (clay loam) and with a similar tractor (Hürlimann Prestige 88) and treatment design (Alaoui, Spiess, Beyeler, & Weingartner, 2012). In the previous study, the surface runoff was measured at different irrigation intensities of 24, 36, and 48 mm h^{-1}. In this study, we calculated surface runoff at an irrigation intensity of 100 mm h^{-1} according to a linear regression obtained from the measurements of the previous study (24, 36, and 48 mm h^{-1}) with R^2 equal to 0.91. We hypothesized in this study that under similar conditions such as a field slope of 15° and shear stress (treatment with 27% slip), the observed-reduction in water flowing from macropores from site 1 to site 2, calculated at 0.15 m topsoil depth (see section 3.4), is expected to be transferred to surface runoff. The main aim of these calculations was to evaluate the impact of shear stress at different rainfall intensities in hillslope areas with a 15° field slope. The theoretical rainfall intensity of 100 mm h^{-1} represented heavy storms corresponding to hourly annual maximum intensities with return periods between 10 and 100 yrs. for the northern part of Switzerland.

2.3 Models

2.3.1 Mechanical Model

The employed model simulated soil-tire contact stress and traction performance of the tractor. It was based on the following assumptions: the soil behaves like a plastic non-linear medium; the wheel spins in steady-state motion at a low velocity; the tire deforms in linear elasticity; the soil-tire contact surface in the longitudinal direction is parabolic in shape, with the apex at the rear point of contact; and the wheel-soil interaction is two dimensional (plane strain problem).

The stress state (shear stress and normal stress) distribution at the soil-tire contact surface was simulated according to the soil-tire interaction model presented by Shmulevich and Osetinsky (2003) and Osetinsky and Shmulevich (2004). Load transfer from the front axle to the rear axle by pulling, and the lowest threshold slip beyond which topsoil shear would occur based on the soil shear stress–maximal soil strength concept of Mohr-Coulomb (Battiato, Diserens, Laloui, & Sartori, 2013) were taken into account in the simulation. The model was adapted for a mechanical front-wheel drive vehicle according to Battiato and Diserens (2013) and integrated in the tool TASC V3.0 (second module) (Battiato, 2014). The soil and the tractor specifications included in the simulation are reported in Table 4.

Table 4. Soil and tractor parameters for the simulation of slip value beyond which soil shear occurs with corresponding traction force, shear stress, and normal stress (TASCV3.0 – module 2)

Tractor parameters	
Traction	4 driven wheels
Wheelbase (m)	2.34
Tire description (front; rear)	380/85R24; 420/85R34
Tire width (front; rear) (m)	0.39; 0.44
Tire unloaded radius (front; rear) (m)	0.63; 079
Tire rolling radius (front; rear) (m)	0.61; 0.76
Tire inflation pressure (front; rear) (kPa)	160; 160
Tire stiffness (front; rear) (kN m^{-1})	330; 434
Stationary wheel load (front; rear) (kN)	9.1; 10.9
Drawbar height (m)	0.83
Theoretical speed ratio	0.99
Soil parameters	
Soil texture	clay loam and loam
Topsoil hardness (penetration resistance to screwdriver)	semi-firm
Modulus of deformation K (kN $m^{-(n+2)}$)	5500
Exponent of deformation n	0.85
Cohesion c (kPa)	14.70
Angle of shear resistance φ (°)	25
Shear deformation modulus k (m)	0.0093

2.3.2 Hydrological Model

The MACRO model (Jarvis, 1994) is a dual-porosity model that simulates water and solute transport in macroporous soil. The model divides the total porosity into macropores and micropores. Water flow in micropores is calculated with the Richards' equation (Richards, 1931), whereas macropore flow is simulated as a power-law function of the saturation level in macropores. The two domains are separated by a boundary water content (θ_b), a boundary saturated hydraulic conductivity (K_b), and a boundary tension (Ψ_b). An effective diffusion-path length d controls mass exchange between the domains (Table 5).

In this study, the MACRO model was used to quantify the water flowing from macropores at both sites (site 1 with 1% slip and site 2 with 27% slip) in order to evaluate the macropore damage due to the applied shear and compressive stress.

Table 5. MACRO model input parameters: soil hydraulic properties at site 1 (1% slip) and site 2 (27% slip)

Depth (m)	$\theta_s^{\dagger}$ (m^3 m^{-3})	$\theta_b^{\ddagger}$ (m^3 m^{-3})	$\theta_r^{\ddagger}$ (m^3 m^{-3})	$\Psi_b^{\ddagger}$ (cm)	$K_{sat}^{\dagger}$ (mm h^{-1})	$K_b^{\ddagger}$ (mm h^{-1})	$d^{\ddagger}$ (mm)	$n^{*\ddagger}$	$n^{\ddagger}$	$\alpha^{\ddagger}$ (cm^{-1})
Site 1										
0–0.15	0.54	0.45	0	1	70	0.1	20	4	1.108	0.027 102
0.16–0.30	0.54	0.45	0	1	70	0.1	20	4	1.101	0.031 269
0.31–0.45	0.50	0.43	0	1	50	0.1	20	4	1.096	0.027 632
0.46–0.60	0.50	0.35	0	1	40	0.1	20	4	1.079	0.035 064
Site 2										
0–0.15	0.45	0.20	0	1	40	0.1	6	4	1.050	0.000 025
0.16–0.30	0.45	0.35	0	1	40	0.1	6	2	1.050	0.000 062
0.31–0.45	0.45	0.33	0	1	40	0.1	5	2	1.140	0.022 394
0.46–0.60	0.45	0.35	0	1	30	0.1	5	2	1.148	0.028 195

Note. † Measured parameters; ‡ Parameters derived from calibration; θ_s: saturated water content; θ_b: boundary water content; θ_r: residual water content; Ψ_b: boundary tension; K_{sat}: saturated water conductivity; K_b: boundary hydraulic conductivity; *d*: effective diffusion-path length; n^*: reflects pore-size distribution index and tortuosity in the macropore system; *n* and α are van Genuchten's parameters (Simunek et al., 2007).

3. Results and Discussion

Results of mechanical and hydraulic measurements are presented in paragraphs 3.1 and 3.2, respectively, whereas modeling results are presented and discussed in paragraph 3.3 for soil stresses and in 3.4 for water dynamics.

3.1 Mechanical Aspects

Measured values from tables 1 to 3 were collected for the determination of the stress distribution at the interface soil-tire. At 27% slip (site 2), the soil was sheared significantly and displaced over a distance of 0.35 m (Figure 2) under the wheels of a 65 kW tractor (Table 2). The load acting on the front wheel decreased whereas the load acting on the rear wheel increased with increasing drawbar pull and slip due to the load transfer from the front to the rear axle. This shift caused changes in the geometry of the contact surface, which became smaller under the front wheel and larger under the rear wheel.

Figure 2. Topsoil deformation owing to tractor traffic, *a)* longitudinal displacement by 1% slip (site 1), *b)* longitudinal displacement by 27% slip (site 2)

Normal stress increased with increasing slip from 90.6 kPa (site 1) to 104.4 kPa (site 2) (Figure 3) at topsoil level, exceeding the indicative value of 85 kPa for clay loam soil (Table 1) (Diserens, Chanet, & Marionneau, 2010). This observation was valid for both axles at 1% slip (site 1), and for only the rear axle at 27% slip (site 2). As the slip increased, the maximum shear contact stress rose noticeably for front (from 19.7 kPa to 42.6 kPa) and rear wheels (from 6.0 kPa to 61.6 kPa) and approached maximal soil strength over a large part of the contact surface (Figure 3).

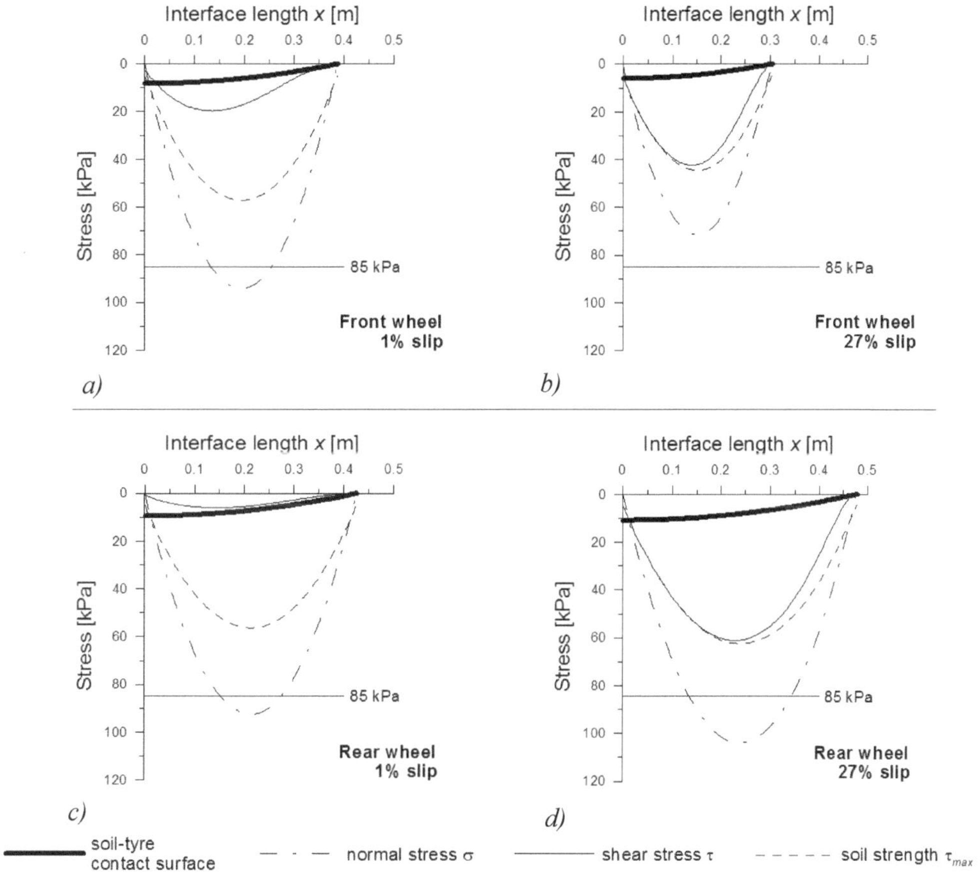

Figure 3. Simulation of the geometry of the soil-tire contact surface, contact stresses included indicative value for normal stress: *a)* front wheel at 1% slip; *b)* front wheel at 27% slip; *c)* rear wheel at 1% slip; *d)* rear wheel at 27% slip

3.2 Hydrological Aspects

According to the statistical analysis, TP at the soil surface (0–0.15 m) was significantly lower at site 2 (TP = 43%) than at site 1 (TP = 54%) (Figure 4). The TP values measured at both trafficked sites were significantly lower than the value measured at the control site at 0.15 m depth (TP = 60%). This difference tended to persist, albeit to a lesser extent, below 0.15 m. These changes in soil porosity likely resulted from the effects of soil shear acting down to a depth of 0.08 m and from a normal stress value exceeding 100 kPa at 27% slip, and reaching already a high value of 93 kPa at 1% slip (Figure 3).

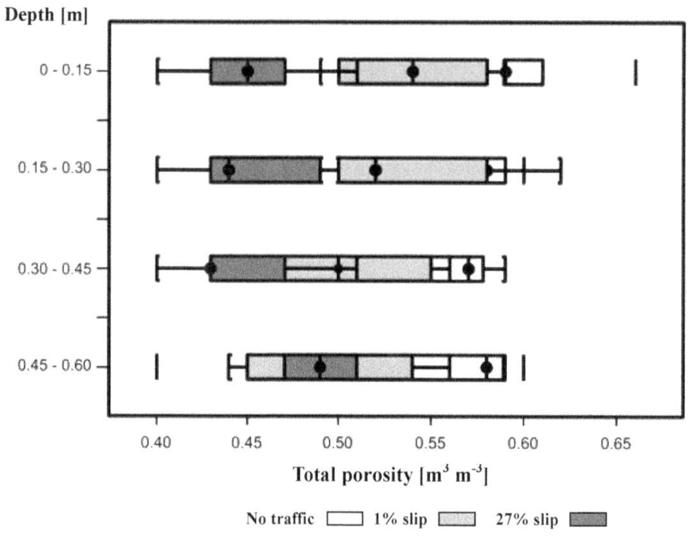

Figure 4. Total porosity measured by no traffic (control site), by 1% slip (site 1), and by 27% slip (site 2), at different depths

At the soil surface (0–0.05 m depth), the macroporosity decreased from 5% at the control site to 2% and 0% by 1 % slip and by 27% slip respectively. In all cases, especially at site 2, the macroporosity values were lower than the indicative value of 7% (pF 1.8) for agricultural soil (SSSS, 2004) (Figure 5). In comparison, at 0.35 m depth (subsoil), a causal effect of traffic-related stresses on soil porosity could not be shown. Pore-size distribution was almost identical between the control site and site 1 (Figure 5). However, the decrease of total porosity and the almost complete destruction of macropores at site 2 in the topsoil might be explained by the effect of the high slip level coupled with increasing load at the rear axle exerted by the 65 kW tractor. The results demonstrated that excessive normal and shear stresses values led to severe reduction of the macroporosity in the topsoil, near the surface.

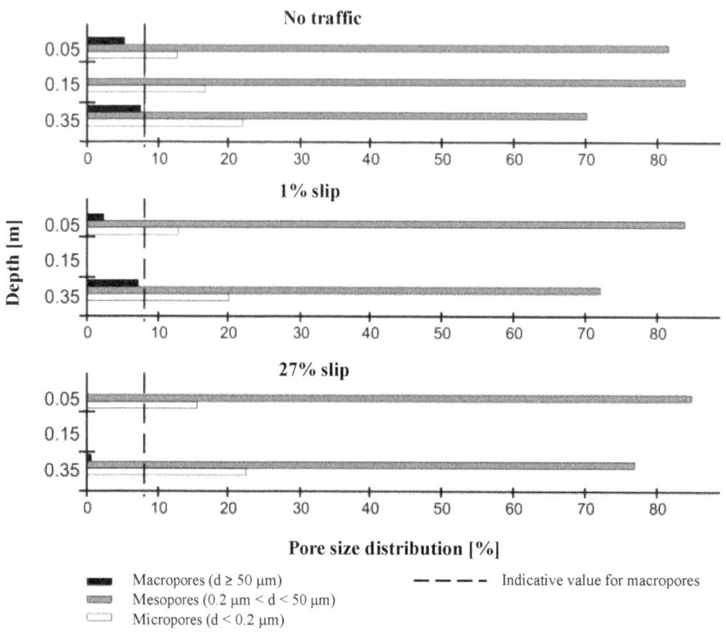

Figure 5. Pore-size distribution by no traffic (control site), by 1% slip (site 1), and by 27% slip (site 2), and threshold for macropores at different depths

Otherwise, strong correlations between normal stress and penetration resistance on one part and between penetration resistance and macroporosity on the other part were also be observed by Servadio, Marsili, Vignozzi, Pellegrini, and Pagliai (2005).

A significant decrease in K_{sat} (mean value) was observed from the control site (8.3×10^{-6} m s^{-1}) to site 1 (2.8×10^{-6} m s^{-1}) and to site 2 (0.14×10^{-6} m s^{-1}) at the soil surface between 0 and 0.10 m depth (Figure 6). At site 2, K_{sat} was lower than the indicative value of 1.0×10^{-6} m s^{-1} for agricultural soil (SSSS, 2004). Below 0.10 m depth, no significant difference in K_{sat} was observed among all sites. At 1% slip (site 1), the highest stress state was reached under the front tire ($\sigma = 94.2$ kPa and $\tau = 19.7$ kPa), which caused a decrease of about 66% in K_{sat} (compared with the control). At 27% slip (site 2), the highest stress state was reached under the rear tire ($\sigma = 104.4$ kPa and $\tau = 61.6$ kPa), causing a reduction of about 98% in K_{sat} (compared with the control). The decrease in K_{sat} from site 1 to 2 was most likely due primarily to the drastic increase in shear stress (210%) with high friction force of the lugs on the firm soil, below the loose soil layer, and secondarily to the slight increase in normal contact stress (11%).

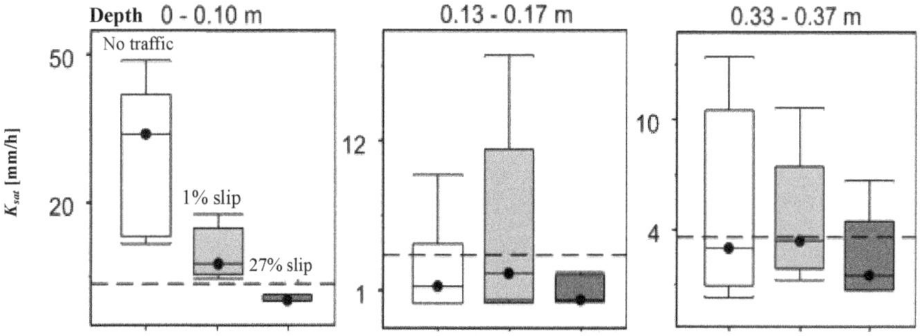

Figure 6. Saturated hydraulic conductivity K_{sat} (mean values, quartiles and extreme values) measured by no traffic (control site), by 1% slip (site 1), and by 27% slip (site 2), at different depths with indicative value; note the different scales on each Y-axis

Between 0.15 m and 0.35 m, the presence of macropores at site 1 indicated by the K_{sat} analysis (Figure 6) was confirmed with the dye-tracer analysis (Figure 7). The weak dye-tracer density at 0.30 m depth at site 1, trafficked with minimal slip (Figure 7), illustrated that few but intact and efficient macropores were present to transport the major portion of water and solute downward (Alaoui & Goetz, 2008; Holden & Gell, 2009), supporting the theory that earthworm burrows act as dominant flow pathways for preferential flow. At the trafficked site with 27% slip (Figure 7), the zone of efficient macropores (0.30 m depth) disappeared, most likely replaced by another type of pores as indicated by the homogeneous distribution of dye-tracer density throughout the soil profile up to 0.50–0.60 m depth.

The evolution of the water content during the drainage phase (after cessation of the irrigation) revealed a significant difference between sites 1 and 2 (Figure 8). In fact, a drastic decrease in water content was observed at site 1 ($\Delta\theta_d = 8\%$), corresponding to a rapid draining of macropores, whereas no significant decrease in water content was observed at site 2 ($\Delta\theta_d = 0.3\%$), attesting to the absence of macropores. These observations related to the function of macropores can be expressed quantitatively in terms of water flowing from macropores simulated with the MACRO model (section 3.4).

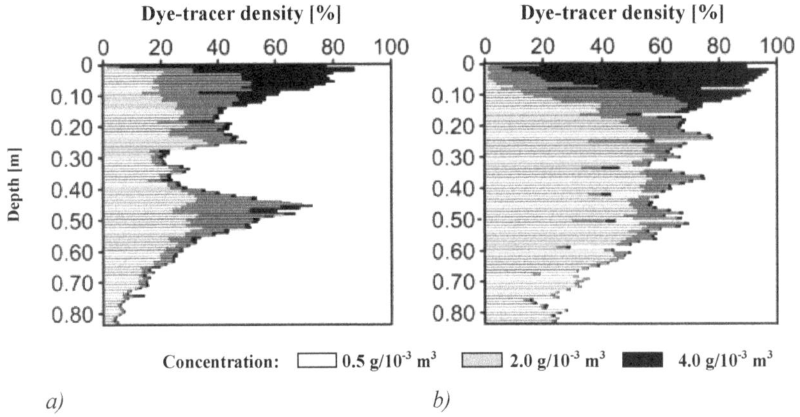

Figure 7. Dye-tracer density versus depth measured *a)* by 1% slip (site 1), *b)* by 27% slip (site 2), the sections presented here were directly under the track of the tractor

These findings show that high mechanical stresses due to slip of a 65 kW tractor drastically disturbed topsoil structure as illustrated by a decrease in total porosity (Figure 4), a reduction in macroporosity shown by pore-size distribution (Figure 5), and a decrease in K_{sat} (Figure 6), and consequently could affect water dynamics throughout the entire soil profile (Figures 7 and 8) (section 3.4).

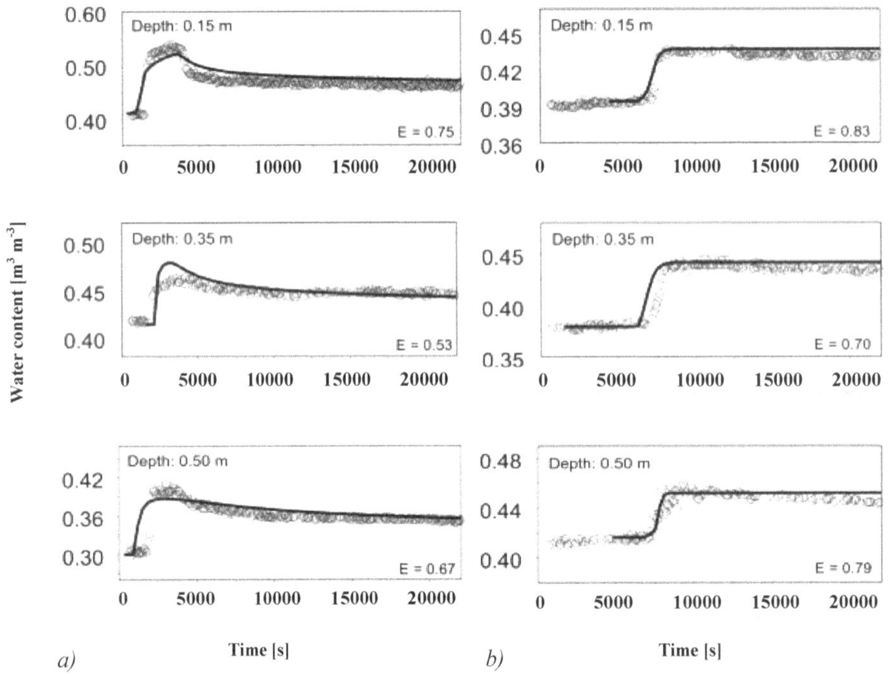

Figure 8. Measured (empty circles) versus simulated (solid lines) water content *a)* by 1% slip (site 1), and *b)* by 27% slip (site 2) at different depths. E = model efficiency

3.3 Modeling Soil Stresses

The tractor specifications and the measured mechanical soil properties are given in Tables 2 and 3, respectively. Based on measured and simulated values (Table 4) with the tool TASC V3.0 (Battiato, 2014; Diserens & Battiato, 2013), the normal and shear stress values were calculated separately for the front and rear wheels according to Osetinsky and Shmulevich (2004). The evolution of the stress with slip is given in Figure 9.

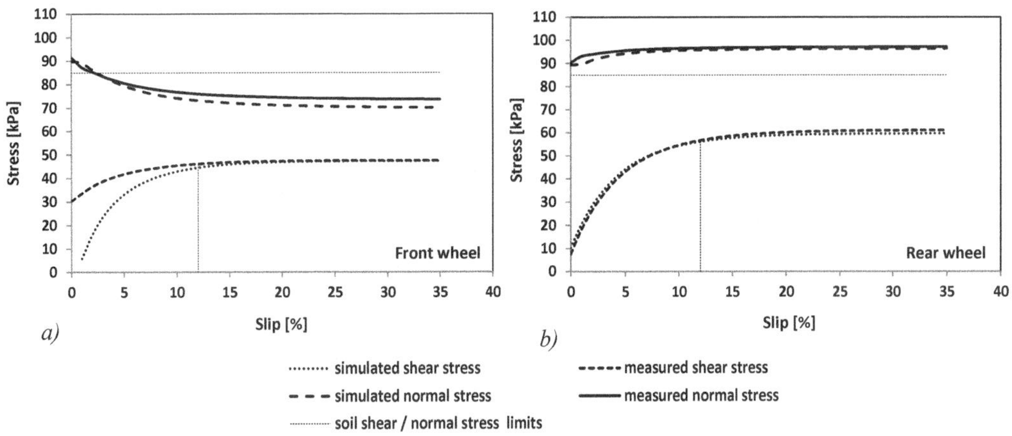

Figure 9. Measured and simulated normal and shear stresses of *a)* front and *b)* rear wheel

Regarding shear stress, a significant discrepancy between the simulation and the measurements at minimal slip level was observed for the front wheel and could be due to the higher value of rolling radius used in the simulation (0.61 m) compared with the measured one (0.59 m) (Tables 2 and 4). For a given shear stress, the simulated slip level on a semi-firm wheat stubble ground was therefore higher than the measured one. Beyond 10 % slip, the importance of the rolling radius decreased significantly. For both wheels, the simulation as well as the measured values indicated a soil-failure limit at 12% slip (Figure 9).

The normal stress was determined through soil hardness, width and diameter of the tires, width and length of the contact surface, tire inflation pressure, wheel load, and amount of load transfer. In the simulation, the topsoil was less firm with a modulus of deformation of 5500 (Table 4) compared with the measured value of 8073 (Table 3). For the front wheel, with a width of 0.39 m used in the simulation (Table 4) instead of 0.38 m obtained from the measurements (Table 2), the length and the width of the contact surface were greater, the traction force and load transfer were higher, and the normal stress acting at soil-tire interface of the front wheel was consequently lower in the simulation than in the calculation based on measurements. Only at low slip (< 5%), the load transfer caused by a larger rolling radius (0.61 m) was lower than that caused by a smaller rolling radius (0.59 m), and therefore simulated wheel load and normal stress were slightly higher than the respective estimates obtained from measurements (Figure 9). For the rear wheel, no significant divergence was observed between measured and simulated values (Figure 9).

3.4 Modeling Water Dynamics

Model calibration with water content (θ) measured at depths of 0.15, 0.35, and 0.50 m yielded satisfactory results. The model reproduced the measured values closely as shown by the model efficiency (E) that was calculated according to Nash and Sutcliffe (1970) and ranged between 0.53 and 0.83 (Figure 8). Once calibrated for θ at different depths (Figure 8 and Table 5), the MACRO model was run to calculate the volume of water flowing from macropores by 1% slip and by 27% slip (Figure 10). A significant difference was observed between sites 1 and 2 at all depths. In the soil surface (0.15 m), the water flowing from macropores was 7.0×10^{-3} m^3 at site 1 and 5.0×10^{-3} m^3 at site 2, corresponding to a difference of about 30%. At 0.35 m depth, the difference was 52% (5.0×10^{-3} m^3 at site 1 and 2.4×10^{-3} m^3 at site 2), and at 0.50 m depth, the difference was 76% (3.0×10^{-3} m^3 at site 1 and 0.7×10^{-3} m^3 at site 2). The difference between the water flowing from macropores at sites 1 and 2 was relatively consistent over time at 0.15 and 0.35 m depth (Figure 10), likely because of the similar soil texture at the two sites (Table 1). The steadily increasing difference in water flowing between sites 1 and 2 at 0.50 m depth might have resulted from the small difference in soil texture (Table 1).

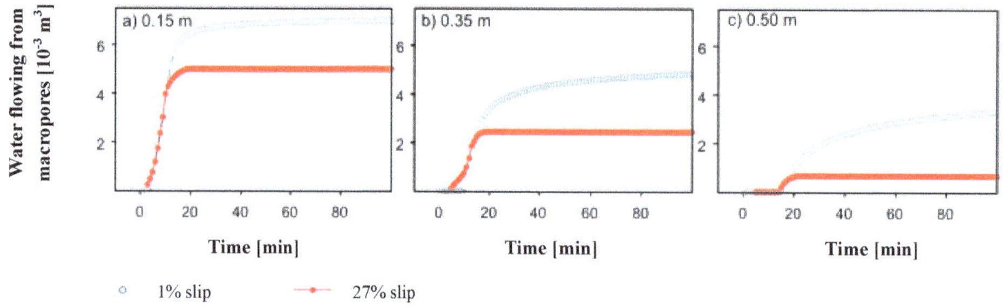

Figure 10. Simulated water flowing from macropores by 1% slip (site 1), and by 27% slip (site 2), estimated at *a)* 0.15, *b)* 0.35, and *c)* 0.50 m depth

As mentioned above, we assumed that on a sloping ground (15°), the reduction in the water flowing from macropores at site 2 is expected to be transferred to surface runoff. Results clearly showed that surface runoff increased with increasing rainfall intensity from 24 to 100 mm h^{-1} and increased more on soil trafficked by 27% slip than on soil trafficked by 1% slip (Table 6). In fact, at a rainfall intensity of 100 mm h^{-1} on sheared soil (site 2), the runoff coefficient was 1.00 indicating a predisposition zone with the highest risk (100%) of surface runoff excess (Markart et al., 2004; Alaoui, Spiess, Beyeler, & Weingartner, 2012). This estimate implies that the entire amount of rainfall (100%) would be routed to surface runoff (= 100 000 m^3 in the case of a grassland with a surface area of 1 km^2 and rainfall intensity of 100 mm h^{-1}) (Table 6). We expect that in soil conservation systems without plowing, the herein identified changes could have significant impacts on soil erosion and flooding for steeper grounds (> 15°) and larger surface areas than those considered in our study.

Table 6. Runoff coefficient (RC) and surface runoff (SR) as affected by soil deformation under different rainfall intensities

Treatment	Intensity (mm h^{-1})							
	24		36		48		100	
	RC (-)	SR (m^3)	RC (-)	SR (m^3)	RC (-)	SR (m^3)	RC (-)	SR (m^3)
1% slip	0.07	1680	0.24	8640	0.29	13 920	0.79	79 000
27% slip	0.09	2184	0.31	11 232	0.38	18 096	1.00	10^5

Note. RC and SR for 24, 36, and 48 mm h^{-1} were measured in situ for minimal slip (Alaoui et al., 2012) and extrapolated (§ 2.2.2) by 27% slip, whereas RC and SR for 100 mm h^{-1} were obtained by linear regression (R^2: 0.91) for a grassland-specific surface area of 1 km^2.

Figure 11 shows how an increase of 6.6 kPa in normal stress under the rear wheel resulted in a loss of macropore water flowing of 2.1×10^{-3} m^3. For comparison, to produce the same loss in water flowing from macropores, an increase of 38.7 kPa in shear stress would be required. It is noteworthy that both stresses (shear and normal) were combined in the deformation process in which normal stress dominated. These results agree with those from laboratory experiments carried out with a simple shear apparatus (Battiato, 2014). The soil-failure limit indicated in figure 11 corresponded to 12% slip and was in line with that identified by Battiato (2014). It implied that major loss of water flowing from macropores occurred between 1% and 12% slip over the modeled range of stresses.

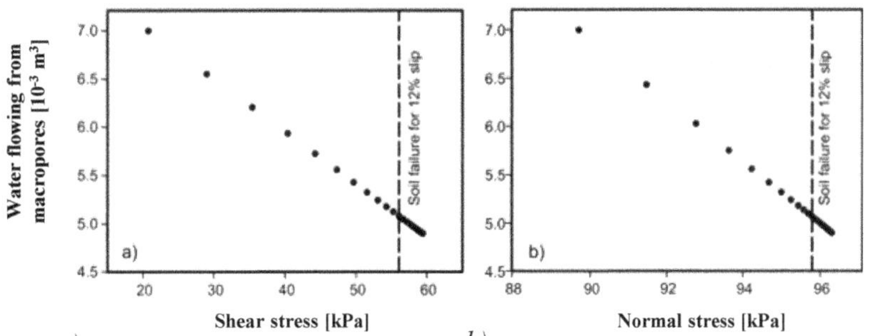

a) *b)*

Figure 11. Simulated water flowing from macropores *a)* vs. simulated shear stress and *b)* vs. simulated normal stress

Because plastic volumetric strain is always linked to proportional alteration in a three-phase system (air, water, soil matrix), hydraulic soil properties have become an essential feature in geotechnical engineering with major interest in soil liquefaction processes. A number of hydro-mechanical elastoplastic constitutive models for saturated soils have been proposed. However, not only saturated soils but also unsaturated soils are affected by liquefaction (Khalili, Habte, & Zargarbashi, 2008; Bian & Shahrour, 2009; Arairo, Prunier, Djeran-Maigre, & Millard, 2014). Furthermore, most of the investigations have been carried out under laboratory conditions and have yet to be conducted in the field. In addition, each soil deformation has to be coupled to hydraulic properties, because deformation and failure in soils depend on the advection and diffusion processes and their history and vice versa (Horn, Richards, Baumgartl, & Wiermann, 1998).

Despite the tremendous effort in this area of research, no attempts have been made to link the changes in the mechanical and hydraulic properties caused by soil deformation with surface runoff generation in hillslope soils, likely because it is difficult to bridge the gap between lab (measurements of mechanical stresses) and field investigations (measurements of surface runoff).

Although we assume a linear relationship between shear stress and water flowing from macropores (between two single values, namely 1% and 27% slip), it will be necessary to conduct detailed measurements at different slip levels and to determine the corresponding surface runoff under various natural conditions. This will help to set a threshold value of the stress beyond which surface runoff will occur for various soil and external conditions in order to better protect soil from erosion and floods on the landscape scale.

4. Conclusions

A combination of mechanical and hydraulic approaches and of measurements and modeling enabled us to observe close causal effects between mechanical stress at the tire-soil interface and surface runoff generation. Our results showed that excessive normal and shear stress severely affect the soil structure by drastically reducing macroporosity and saturated hydraulic conductivity in the topsoil. Modeling results also showed that the amount of water flowing from topsoil macropores decreases markedly in soil affected by a high level of wheel slip. Considering a 15° slope, and assuming that a loss in vertical water flow would be transformed into surface runoff under the combined effects of field slope and increased rainfall intensity, we calculated that the runoff coefficient for a rainfall intensity of 100 mm h^{-1} and duration of 1 h would increase to 0.79 at site 1 with minimum wheel slip (1%) and to 1.00 at site 2 with high wheel slip (27%). Further field tests are needed to establish consistent computerized links between mechanical stress and surface runoff at different slip levels. We expect that this will make it possible to identify the threshold stress value beyond which surface runoff and erosion are generated: a key piece of information when it comes to supporting farmers in the use of machinery.

Acknowledgements

All the authors wish to acknowledge the Swiss Federal Office for the Environment FOEN and the tire manufacturer Michelin for their respective financial contributions.

References

Alakukku, L. (1996). Persistence of soil compaction due to high axle load traffic: I. Short-term effects on the properties of clay and organic soils. *Soil Till. Res., 37*, 211-222.

Alaoui, A., Spiess, P., Beyeler, M., & Weingartner, R. (2012). Up-scaling surface runoff from plot to catchment scale. *Hydrol. Res., 43*(4), 531-546. http://dx.doi.org/doi:10.2166/nh.2012.057

Alaoui, A., & Goetz, B. (2008). Dye tracer and infiltration experiments to investigate macropore flow. *Geoderma, 144*, 279-286. http://dx.doi.org/10.1016/j.geoderma.2007.11.020

Alaoui, A., & Helbling, A. (2006). Evaluation of soil compaction using hydrodynamic water content variation: comparison between compacted and non-compacted soil. *Geoderma, 134*, 97-108. http://dx.doi.org/10.1016/j.geoderma.2005.08.016

Alaoui, A., Germann, P., Jarvis N., & Acutis, M. (2003). Dual-porosity and kinematic wave approaches to assess the degree of preferential flow in an unsaturated soil. *Hydrol. Sci. J., 48*(3), 455-472.

Ansorge, D., & Godwin, R. (2006). High Axle Load – Track – Tire Comparison. In R. Horn, H. Fleige, S. Peth & X. Peng (Eds.), *Soil Management for Sustainability* (pp. 9-14). Advances in Geoecology 38. Reiskirchen, Germany, Catena Verlag, GeoEcology.

Arairo, W., Prunier, F., Djeran-Maigre, I., & Millard, A. (2014). On the use of effective stress in three-dimensional hydro-mechanical coupled model. *Comput. Geotech., 58*, 56-68. http://dx.doi.org/10.1016/j.compgeo.2014.01.014

Arvidsson, J., Trautner, A., & Keller, T. (2002). Influence of Tyre Inflation Pressure on Stress and Displacement in the Subsoil. In M. Pagliai & R. Jones (Eds.), *Sustainable Land Management – Environmental Protection. A Soil Physical Approach* (pp. 331-338). Advances in Geoecology 35. Reiskirchen, Germany, Catena Verlag, GeoEcology.

Ball, B. C. (1981). Pore characteristics of soils from two cultivation experiments as shown by gas diffusivities and permeabilities and air-filled porosities. *J. Soil Sci., 32*, 465-481.

Bastgen, H. M., & Diserens, E. (2009). q-value for calculation of pressure propagation in arable soils taking topsoil stability into account. *Soil Till. Res., 102*, 138-143. http://dx.doi.org/10.1016/j.still.2008.08.005

Battiato, A. (2014). *Soil-Tyre Interaction Analysis for Agricultural Tractors: Modelling of Traction Performance and Soil Damage* (PhD Thesis, p. 165). Università degli Studi di Padova. Dipartimento di Territorio e Sistemi Agro Forestali (T.E.S.A.F.). Scuola di Dottorato di Ricerca in Territorio Ambiente Risorse e Salute (T.A.R.S.) Indirizzo: Tecnologie Meccaniche dei Processi Agricoli e Forestali, Padova, Italy. Ciclo XXVI.

Battiato, A., & Diserens, E. (2013). Influence of tyre inflation pressure and wheel load on the traction performance of a 65 kW MFWD tractor on a cohesive soil. *Journal of Agricultural Science, 5*(8), 197-215. http://dx.doi.org/10.5539/jas.v5n8p197

Battiato, A., Diserens, E., Laloui, L., & Sartori, L. (2013). A mechanistic approach to topsoil damage due to slip of tractor tyres. *J. Agric. Sci. Appl., 2*(3), 160-168.

Bekker, M. G. (1960). *Off-the-Road Locomotion*. Ann Arbor, Michigan, University of Michigan Press.

Bian, H., & Shahrour, I. (2009). Numerical model for unsaturated sandy soils under cyclic loading: application to liquefaction. *Soil Dyn. Earthq. Eng., 29*, 237-244.

Blanco-Canqui, H., Claassen, M. M., & Stone, L. R. (2010). Controlled traffic impacts on physical and hydraulic properties in an intensively cropped no-till soil. *Soil Sci. Soc. Am. J., 74*, 2142-2150. http://dx.doi.org/doi:10.2136/ssaj2010.0061

Boone, F. R. (1988). Weather and other environmental factors influencing crop responses to tillage and traffic. *Soil Till. Res., 11*, 283-324. http://dx.doi.org/10.1016/0167-1987(88)90004-9

Christensen, J. H., & Christensen, O. B. (2003). Severe summertime flooding in Europe. *Nature, 421*, 805-806.

da Silva, A. P., Kay, B. D., & Perfect, E. (1994). Characterization of the least limiting water range of soils. *Soil Sci. Soc. Am. J., 58*, 1775-1781. http://dx.doi.org/10.2136/sssaj1994.03615995005800060028x

Dirksen, C. (1999). *Soil Physics Measurements*. Reiskirchen, Germany, Catena Verlag GeoEcology.

Diserens, E., & Battiato, A. (2013). TASC V3.0 - Prognose Bodengefährdung und Treibstoffverbrauch. *ART-Berichte Nr. 766* (p. 8).

Diserens, E., Chanet, M., & Marionneau, A. (2010). *Machine Weight and Soil Compaction: TASC V2.0.xls - a Practical Tool for Decision-Making in Farming* (p. 10). AgEng 2010, Clermont-Ferrand, France, Sept. 7-9, 2010, REF 239.

Diserens, E., & Spiess, E. (2004). Wechselwirkung zwischen Fahrwerk und Ackerboden. TASC: Eine

PC-Anwendung zum Beurteilen und Optimieren der Bodenbeanspruchung. *FAT Berichte Nr. 613* (p. 16).

Diserens, E., & Steinmann, G. (2003). *In-situ determination of the fracture point of an agricultural soil using the plate penetration test, comparison with the oedometer method and validation* (pp. 93-106). Proceedings of the International Conference on Geo-Environmental Engineering, Singapore.

Flury, M., & Flühler, H. (1995). Tracer characteristics of Brilliant Blue FCF. *Soil Sci. Soc. Am. J., 59*, 22-27. http://dx.doi.org/10.2136/22aj1995.03615995005900010003x

Graham, L. P., Hagemann, S., Jaun, S. O., & Beniston, M. (2007). On interpreting hydrological change from regional climate models. *Climatic Change, 81*, 97-122. http://dx.doi.org/doi:10.1007/s10584-006-9217-0

Holden, J., & Gell, K. F. (2009). Morphological characterization of solute flow in a brown earth grassland soil with crane fly larvae burrows (leatherjackets). *Geoderma, 152*, 181-186. http://dx.doi.org/10.1016/j.geoderma.2009.06.006

Horn, R. (2003). Stress–strain effects in structured unsaturated soils on coupled mechanical and hydraulic processes. *Geoderma, 116*(1-2), 77-88. http://dx.doi.org/10.1016/S0016-7061(03)00095-8

Horn, R. (2002). Soil Mechanical Properties and Processes in Structured Unsaturated Soils under Various Landuse and Management Systems. In M. Pagliai, & R. Jones (Eds.), *Sustainable Land Management Environmental Protection* (pp. 305-318). Advances in Geoecology: 35, Catena Verlag GmbH, Reiskirchen, Germany.

Horn, R. (1981). Eine Methode zur Ermittlung der Druckbelastung von Böden anhand von Drucksetzungsversuchen. *Zeitschrift für Kulturtechnik und Flurbereinigung, 22*(1), 20-26.

Horn, R., & Rostek, J. (2000). Subsoil Compaction Processes – State of Knowledge. In R. Horn, J. J. H. van den Akker & J. Arvidsson (Eds.), *Subsoil Compaction – Distribution, Processes and Consequences* (pp. 44-55). Advances in Geoecology 32. Reiskirchen, Germany, Catena Verlag GeoEcology.

Horn, R., Richards, B. G., Gräsle, W., Baumgartl, T., & Wiermann, C. (1998). Theoretical principles for modelling soil strength and wheeling effects – A review. *Journal of Plant Nutrition and Soil Science, 161*(4), 333-346. http://dx.doi.org/10.1002/jpln.1998.3581610402

Jarvis, N. J. (1994). *The MACRO Model Version 3.1 – Technical description and sample simulations* (Reports and dissertations No. 19, p. 51). Department of Soil Science, Swedish University of Agricultural Sciences, Uppsala, Sweden.

Khalili, N., Habte, M. A., & Zargarbashi, S. (2008). A fully coupled flow deformation model for cyclic analysis of unsaturated soils including hydraulic and mechanical hystereses. *Comput. Geotech., 35*, 872-889. http://dx.doi.org/10.1016/j.compgeo.2008.08.003

Kirby, J. M. (1991). Strength and deformation of agricultural soil: measurement and practical significance. *Soil Use Management, 7*, 223-229. http://dx.doi.org/10.1111/j.1475-2743.1991.tb00878.x.

Klute, A., & Dirksen, C. (1986). Hydraulic Conductivity of Saturated Soil - Constant Head Method. In A. Klute (Ed.), *Methods of Soil Analysis* (pp. 694-700). *Part 1. Physical and Mineralogical Methods - Agronomy Monograph No. 9* (2nd ed.). Madison, Wisconsin, American Society of Agronomy - Soil Science Society of America.

Kulli, B., Gysi, M., & Flühler, H. (2003). Visualizing soil compaction based on flow pattern analysis. *Soil Till. Res., 70*, 29-40.

Lines, J. A., & Murphy, K. (1991). The stiffness of agricultural tractor tyres. *J. Terramechanics, 28*(1), 49-64. http://dx.doi.org/10.1016/0022-4898(91)90006-R

Markart, G., Kohl, B., Sotier, B., Schauer, T., Bunza, G., & Stern, R. (2004). *Provisorische Geländeanleitung zur Abschätzung des Oberflächenabflussbeiwertes auf alpinen Boden-/Vegetationseinheiten bei konvektiven Starkregen* [A simple Code of Practice for Assessment of Surface Runoff Coefficients for Alpine Soil/Vegetation Units in Torrential Rain] (Version 1.0). BFW Dokumentationen. Bundesamt und Forschungszentrum für Wald, Wien, Austria.

Nash, J. E., & Sutcliffe, J. V. (1970). River flow forecasting through conceptual models, part I - A discussion of principles. *J. Hydrol., 10*, 282-290. http://dx.doi.org/10.1016/0022-1694(70)90255-6

O'Sullivan, M. F., Robertson, E. A. G., & Henshall, J. K. (1999). Shear effects on gas transport in soil. *Soil Till. Res., 50*, 73-83.

Osetinsky, A., & Shmulevich, I. (2004). Traction performance simulation of a pushed/pulled driven wheel. *Trans. ASAE, 47*(4), 981-994.

Pagliai, M. (1998). Soil porosity aspects. *Int. Agrophys., 4*, 215-232.

Peschke, G., Etzenberg, C., Müller, G., Töpfer, J., & Zimmermann, S. (1999). *Die Abflussbildung in ihrer Abhängigkeit von der wirksamen Kombination flächenvariabler Einflussfaktoren und Gebietszustand* [Runoff generation depending from the effective combination of spatial variable factors and antecedent conditions]. Abschlussbericht zur Forschung, DFG-Schwerpunkt Programm "Regionalisierung in der Hydrologie." Internationales Hochschulinstitut Zittau, Zittau, Germany.

Pichlmaier, B. R. (2012). *Traktormanagement für Traktoren* (PhD Thesis, p. 209). Dissertation der Technischen Universität München, Institut für Maschinen- und Fahrzeugtechnik. Lehrstuhl für Fahrzeugtechnik, München, Germany.

Reszkowska, A., Krümmelbein, J., Gan, L., Peth, S., & Horn, R. (2011). Influence of grazing on soil water and gas fluxes of two Inner Mongolian steppe ecosystems. *Soil Till. Res., 111*, 180-189. http://dx.doi.org/10.1016/j.still.2010.10.003

Richards, L. A. (1931). Capillary conduction of liquids in porous mediums. *Physics, 1*, 318-333.

Roth, K., Schulin, R., Flühler, H., & Attinger, W. (1990). Calibration of time domain reflectometry for water content measurement using a composite dielectric approach. *Water Resour. Res., 26*, 2267-2274. http://dx.doi.org/10.1029/WR026i010p02267

Scheffer, F., & Schachtschabel, P. (1998). *Lehrbuch der Bodenkunde*. Ferdinand Enke Verlag Stuttgart. 14 Aufl., 494 S.

Scherrer, S., & Naef, F. (2003). A decision scheme to identify dominant flow processes at the plot-scale for the evaluation of contributing areas at the catchments-scale. *Hydrol. Proc., 17*, 391-401.

Schwieger, H. (1996). *Untersuchung neuartiger Laufwerke und lasergestütze Erfassung der Reifen-/ Bodenverformung* (PhD Thesis, p. 166). Forschungsbericht Agrartechnik des Arbeitskreises Forschung und Lehre der Max-Eyth-Gesellschaft Agrartechnik im VDI (VDI-MEG), Kiel, Germany.

Servadio, P., Marsili, A., Vignozzi, N., Pellegrini, S., & Pagliai, M. (2005). Effects on some soil qualities in central Italy following the passage of four wheel drive tractor fitted with single and dual tires. *Soil Till. Res., 84*, 87-100. http://dx.doi.org/10.1016/j.still.2004.09.018

Simunek, J., Sejna, M., & van Genuchten, M. T. (2007). *The HYDRUS Software Package for Simulating the Two- and Three-Dimensional Movement of Water, Heat, and Multiple Solutes in Variably-Saturated Media*. PC-Progress, Prague, Czech Republic.

Shmulevich, I., & Osetinsky, A. (2003). Traction performance of a pushed/pulled drive wheel. *J. Terramech., 40*, 35-50. http://dx.doi.org/10.1016/j.jterra.2003.09.001

SSSS. (2004). Definition und Erfassung von Bodenschadverdichtungen. *Positionspapier der BGS-Plattform Bodenschutz. Swiss Society of Soil Science (SSSS)* (Document 13, p. 56).

Steiner, M. (1979). *Analyse, Synthese und Berechnungsmethoden der Triebkaft-Schlupf-Kurve von Luftreifen auf nachgiebigem Boden* (PhD Thesis, p. 190). Forschungsbericht Agrartechnik des Arbeitskreises Forschung und Lehre der Max-Eyth-Gesellschaft (MEG) 33. München, Germany.

Stettler, M., Keller, T., Schjønning, P., Lamandé, M., Lassen, P., Pedersen, J., & Weisskopf, P. (2010). *Terranimo® - a web-based tool for assessment of the risk of soil compaction due to agricultural field traffic* (p. 7). AgEng 2010, Clermont-Ferrand, France, Sept. 7-9, 2010, REF 384.

van den Akker, J. J. H. (2004). SOCOMO: a soil compaction model to calculate soil stresses and the subsoil carrying capacity. *Soil Till. Res., 79*, 113-127. http://dx.doi.org/10.1016/j.still.2004.03.021

van den Akker, J. J. H. (1997). *Construction of a wheel–load bearing capacity map of the Netherlands*. Proceedings of the 14th ISTRO Conference July 27 - Aug. 1, 1997, Pulawy, Poland. Polish Society of Agrotechnical Sciences. Bibliotheca Fragmenta Agronomica 2A/97, 15-18.

Vogeler, I., Horn, R., Wetzel, H., & Kruemmelbein, J. (2006). Tillage effects on soil strength and solute transport. *Soil Till. Res., 88*, 193-204. http://dx.doi.org/10.1016/j.still.2005.05.009

Wiermann, C. (1998). *Auswirkungen differenzierter Bodenbearbeitung auf die Bodenstabilität und das Regenerationsvermögen lössbürtiger Ackerstandorte*. Schriftenreihe Institut für Pflanzenernährung und

Bodenkunde. CAU Kiel, Germany.

Wong, J. Y. (1980). Data processing methodology in the characterization of the mechanical properties of terrain. *J. Terramechanics, 17*(1), 13-41. http://dx.doi.org/10.1016/0022-4898(80)90014-2

4

Corn Silage (*Zea mays* L.) Response to Zinc Foliar Spray Concentration When Grown on Sandy Soil

Saad Drissi[1], Abdelhadi Aït Houssa[2], Ahmed Bamouh[1] & Mohamed Benbella[2]

[1] Plant Production, Protection and Biotechnology Department, Hassan II Agronomy and Veterinary Institute, Rabat, Morocco

[2] Department of Agronomy and Plant Breeding, National Agricultural School of Meknes, Meknes, Morocco

Correspondence: Saad Drissi, Plant Production, Protection and Biotechnology Department, Hassan II Agronomy and Veterinary Institute, Rabat, Morocco. E-mail: drissi_agro@yahoo.com

Abstract

The objective of this study is to identify an adequate zinc (Zn) foliar spray concentration which corrects Zn deficiency without disrupting other plant mineral contents, enhances plant growth, and thereby corn silage yield when grown on sandy soil. A field experiment was conducted using five Zn foliar spray concentrations (w/v): 0.03%, 0.07%, 0.10%, 0.14% and 0.18%. Zn sulfate ($ZnSO_4$. $7H_2O$) was used as a source of Zn. Zn foliar application was realized at two growth stages (6-7 and 9-10 leaf stages). A treatment without Zn foliar spray was maintained as control. The results showed a quadratic response of corn silage towards Zn foliar spray concentrations, in which 0.09% was the optimum value for overcoming Zn deficiency. Such level increased shoot Zn concentration at harvest from 15 mg kg^{-1} to 21.8 mg kg^{-1}, didn't decrease plant mineral content below critical levels, enhanced plant growth and raised silage yield by 49.4% compared to control. On the other hand, Zn foliar spray concentration up to 0.10% induced visible leaf damage, growth inhibition and a decrease of 26% in silage yield compared to Zn foliar applications at 0.09%.

Keywords: corn silage, Zn deficiency, Zn foliar spray, growth, mineral content, yield

1. Introduction

Most plant micro-nutrition research has reported the severe effects of Zn deficiency on crop productivity especially for cereals (Cakmak, 2008). Corn (*Zea mays* L.) is one of the very sensitive cereal species to Zn deficiency stress (Lindsay & Norvel, 1977; Gupta, Kening, & Siyuan, 2008). Many authors have reported importance of Zn in completing some vital physiological functions such as energy production and protein synthesis (Hansch & Mendel, 2009; Gupta et al., 2008; Mousavi, Galavi, & Rezaei, 2013; Qiao et al., 2014).

Under Zn deficiency, corn has low Zn concentration, usually below the critical level of 22 mg kg^{-1} (Singh, Natesan, Singh, & Usha, 2005), and shows white linear bands between the midrib and the margin of leaves (Singh et al., 2005), as well as an obvious shortening of plant height and leaf area (De Vasconcelos, Clístenes, & Fernando, 2011). All this result in significant decline in silage and grain yields (Van Biljon, Wright, Fouche, & Botha, 2010; Potarzycki & Grzebisz, 2009). In order to prevent Zn deficiency stress, both soil Zn supply and Zn foliar spray have been advisable. In the context of many dairy farmlands from Loukkos perimeter (North West of Morocco), corn silage grown on a sandy soil, naturally poor in Zn, showed Zn deficiency symptoms that resulted in a silage yield decline compared to adequate fertilized plots receiving either Zn sulfate or animal manures. Zn foliar spray is still not a common fertilization method in this region, except some commercial mixtures of macro and micronutrients that provided a partial recovery from Zn deficiency stress. Many authors have reported the importance of Zn foliar applications at adequate levels to improve yields of many grown crops such as corn (Potarzycki & Grzebisz, 2009), wheat grain (Haslett, Reid, & Rengel, 2001), tomato fruit (Kaya & Higgs, 2002) and safflower (Rajabi et al., 2013). The foliage uptake depends on many factors such as plant species, timing of application, and concentration (Wojcik, 2004). Knoche, Petracek, Bukovac, and Shafer (1994) reported linear positive relationship between nutrient foliar spray concentration and the rate of its uptake by tomato's leaves. Potarzycki and Grzebisz (2009) found positive quadratic relationship between Zn foliar spray concentration and corn grain yield. They have denoted optimal yielding at concentration ranging between 0.25% and 0.37% of Zn applied at 5 leaf stage. Also, Takkar and Wallker (1993) reported that Zn foliar spray concentration at 0.5-1% of

$ZnSO_4{\cdot}7H_2O$ resulted in optimal biomass yields for many crops.

On the other hand, Zn foliar applications at high concentration induced leaf injury (Qiao et al., 2014), inhibited plant growth and declined yield (Kaya & Higgs, 2002). Moreover, it may disrupt the uptake of other macro and micronutrients. In this respect, Kaya and Higgs (2002) found that the increase of Zn foliar spray concentration on tomato increased leaves Zn concentration, but reduced potassium (K), magnesium (Mg) and iron (Fe) contents.

Despite numerous advantages of foliar spray such as being environmentally friendly, inexpensive and fast in correcting deficiency, we found little literature about managing Zn foliar spray for adequate mineral content and corn silage growth. Therefore, the present study was undertaken with the objective of identifying an adequate concentration of Zn to spray on corn silage grown on a sandy soil in order to correct Zn deficiency without disrupting mineral composition and plant growth.

2. Method

2.1 Site Description

A field experiment was conducted on an agricultural farm (Bassita II farm) located in Loukkos perimeter (34°96′N lat., 6°21′W long., 60 m above the sea level, North West of Morocco) where the climate is maritime. We recorded, during the growing season (August 09 to November 20, 2012), 16.6 °C and 24.2 °C as average of maximum and minimum temperatures, respectively. A rainfall amount of 283 mm was recorded especially between September and November. A comparison between the studied growing season weather and the climatic data of the last 40 years (1971-2010)* of the experimental site is reported in Table1.

Table 1. Temperature, evapotranspiration and rainfall during the growing season (August to November) for 2012 and long-term means (1971- 2010) of the experimental site

	Min temperature (°C)		Max temperature (°C)		Evapotranspiration (mm day^{-1})		Rainfall (mm)	
	2012	1971- 2010	2012	1971-2010	2012	1971-2010	2012	1971-2010
August	19.3	18.9	25.7	27.5	4.0	4.0	4	1.8
September	17.5	17.6	24.8	26.4	3.6	3.8	47	17.8
October	16.3	14.9	25.6	23.7	3.5	3.2	88	67.2
November	11.4	11.3	20.4	20.1	2.5	2.5	171	114.1

*Source: Taken from National Directorate of Meteorology (DMN), Morocco.

2.2 Experimental Soil

For basic soil characteristics analysis, we collected a soil sample 0-20 cm. The soil was sandy (88.8% of sand, 7.5% of clay and 5.3% of silt), no calcareous, with a very low amount of DTPA extractable Zn (0.13 mg kg^{-1}), which was below the critical level of 0.8 mg kg^{-1} required for corn (Landsay & Norvel, 1977). The other soil chemical characteristics are listed in Table 2.

Table 2. Soil characteristics

Soil property	
pH H_2O (1/5)	6.1
Cation exchange capacity (meq $100g^{-1}$) (Cobaltihexamine Chloride method)	4.4
Organic matter (%) (Walkley & Black method)	0.4
Extractable P_2O_5 (mg kg^{-1}) (Olsen method)	49
Extractable K_2O (mg kg^{-1})a	81
Extractable MgO (mg kg^{-1})a	101
Extractable CaO (mg Kg^{-1})b	868
Extractable Zn (mg kg^{-1})c	0.13
Extractable Cu (mg kg^{-1})c	0.06
Extractable Mn (mg kg^{-1})c	17.4
Extractable Fe (mg kg^{-1})c	17.45

Extractants: a = Ammonium Acetate.; b = Sodium Acetate; c = Diethylene Triamine Penta-Acetic acid (DTPA).

2.3 Crop Management and Experimental Design

The land was prepared for planting by cultivator tillage. Corn hybrid Panama was planted on 09 August 2012. The sowing was done on twin lines with a spacing of 45 cm, 90 cm between twin lines spacing and the inter-row seeding distance was 12 cm to approximate 120,000 plants ha^{-1}.

A completely randomized block design with five replications was used. Plots included 3 twin lines of plants, 4 m width and 8 m length. There were 4 m between adjacent plots in the same block and 1.4 m between adjacent blocks.

Five Zn foliar spray concentrations (w/v) were tested: 0.03%, 0.07%, 0.10%, 0.14% and 0.18%. Zn sulfate ($ZnSO_4 \cdot 7H_2O$, 22.5% of Zn) was used as source of Zn. The foliar spray, about 625 l ha^{-1}, was done using a hand-held sprayer in an early and windless morning. The spray was done twice during the growth season at: i) 5-6 leaf stage and ii) 9-10 leaf stage. A treatment without Zn foliar spray was maintained as control. In order to ensure an adequate nutrition with other nutrient elements, all treatments received adequate amounts of macro and micro elements during the growing season: 480 Kg ha^{-1} of N as ammonium nitrate and diammonium phosphate (DAP), 345 Kg ha^{-1} of P_2O_5 as DAP, 331 Kg ha^{-1} of K_2O as potassium sulfate, 6.25 Kg ha^{-1} of Cu as copper sulfate, 4.61 Kg ha^{-1} of B as boron sulfate and 3.35 Kg ha^{-1} of Mn as manganese sulfate.

Each line of plants was equipped with a drip line irrigation system using 1.2 l h^{-1} emitters and 0.4 m as emitters spacing. Watering was done whenever required during the growing season. Weeds were controlled using a mixture of pre-emergence herbicides (Pendimethaline, Mesotrione, Terbuthylazine and S-metolachlor) and fungal disease (Helmintosporium) was controlled with Epoxiconazole.

2.4 Measurements

Taking into account that Zn deficiency symptoms appeared on corn as white linear areas between the midrib and margin of leaves (Singh et al., 2005; Alloway, 2008), visual Zn deficiency symptoms evolution was determined each 10 days, on 10 randomly chosen plants for each treatment's replication. The symptoms were summarized using a descriptive scale of 1 to 5: 1 = plants had no visual Zn deficiency symptoms, 2 = pale green linear stripes began to appear between the midrib and the margin of new leaves, 3 = white linear stripes appeared between the midrib and the margin of all leaves 4 = white bands appeared between the midrib and the margin of old leaves, 5 = all leaves were small showing linear white bands between their midrib and margin.

The stem height and the stem diameter were measured, on 10 randomly chosen plants per each treatment's replication, at 3 different growing stages: 6-7 leaf stage, 9-10 leaf stage and harvest.

The total leaf area per plant at harvest was measured in 3 randomly chosen plants per each treatment's replication using the formula (1) developed by Montgomery (1911) and cited in Mokhtarpour et al. (2010) for corn:

$$\textit{Total leaf area per plant} = \sum_{i=1}^{j=n}(L \times W \times 0.75) \quad (1)$$

Where L, W, and n are leaf length, leaf greatest width, and last leaf of corn, respectively.

The harvest was done manually on November 20, 2012, approximately at shoot moisture content of 67%. 10 randomly chosen plants per each treatment's replication were cut close to the soil surface, and the fresh weights of the harvest were measured. 3 randomly chosen ears per each treatment's replication were taken in order to determine different kernels yield compounds: i) total number of ovules per ear using equation (2) ii) pollination rate using equation (3), and iii) thousand-kernel dry weight.

$$\textit{Total number of ovules per ear} = \textit{Number of kernels} + \textit{Number of sterilized ovules} \quad (2)$$

$$\textit{Pollination rate (\%)} = \textit{Number of kernels per ear/Total number of ovules per ear} \times 100 \quad (3)$$

In order to evaluate macro and micro nutrients shoot contents at harvest, 3 plants from each of the following treatments (Control, 0.03%, 0.10% and 0.18%) were dried at 70 °C in an oven until constant weight, chopped and mixed. Thereafter, 3 subsamples of 0.6 g per these treatments were digested with a mixture of salicylic and sulfuric acids in order to determine plant N, P, K, Ca and Mg contents. Besides, 3 other subsamples of 2 g were digested with a mixture of nitric, perchloric, and sulfuric acids to determine plant Zn, Mn, Fe and Cu contents.

Concerning K, Ca, Mg, Zn, Fe and Mn, they were analyzed using an atomic absorption spectrophotometer (Varian AA 240 Fast Sequential; air acetylene flame). However, N and P were auto-analyzed using a continuous flow analyzer (Skalar San^{++}).

2.5 Statistical Analysis

All data were subjected to analysis of variance at 5% significance level. Taking into account that Zn concentration treatment is a quantitative variable, an appropriate regression analysis, which was selected on the basis of higher r^2, was used to evaluate the response of each measured parameter. The optimum Zn foliar spray for each parameter was determined by equating the first derivative of the regression equation to zero. These statistical analyses were performed using the program SPSS (Version 17.0).

3. Results

3.1 Visual Zn Deficiency Symptoms

Zn deficiency symptoms appeared at an earlier stage (5 leaf stage) as white linear stripes between the midrib and the margin of leaves for all treatments (score 3). Zn foliar applications up to 0.03% induced a clear disappearance of these symptoms (score ≈ 1). But, some green pale lines persisted on plants during all growing season even with high Zn foliar spray concentration (0.14% and 0.18%). Control plants also showed a slight recovery from Zn deficiency at 8-9 leaf stage (score 2). On the other hand, due to high Zn level concentration (up to 0.10%), an obvious leaf injury with brown leaf areas was observed after one day of each application. This damage concerned only the touched leaves, thereafter, new safety leaves appeared.

3.2 Plant Growth

Just after the first Zn foliar application (6-7 leaf stage), a clear increase on stem height was marked. At 9-10 leaf stage, all Zn foliar treatments induced a significant increase of 14.6% compared to control. While at harvest, a clear positive quadratic response was recorded. Zn foliar application at 0.10% induced the most significant stem height increase of 11.8% compared to control. Moreover, the total leaf area per plant at harvest was significantly raised. Compared to control, all Zn foliar treatments induced a significant rise of 23.40%. Concerning stem diameter, it was significantly increased at 9-10 leaf stage. But at harvest, no significant difference between treatments was recorded. Nonetheless, an obvious growth inhibition was marked with high Zn foliar spray concentrations up to 0.10% (Table 3).

Table 3. Effect of Zn foliar spray concentration (%) on stem height, stem diameter and total leaf area of corn silage

Zn foliar spray concentrations	Stem height (cm)			Stem diameter (cm)			Total leaf area at harvest (dm^2 $plant^{-1}$)
	6-7 leaf stage	9-10 leaf stage	Harvest	6-7 leaf stage	9-10 leaf stage	Harvest	
Control	16.2±3.0	39.8± 12.5	199.7±21.0	1.2±0.2	1.4±0.2	1.4±0.2	30.20±8.76
0.03	17.7±3.9	46.5±10.8	215.2±17.8	1.3±0.2	1.5±0.2	1.5±0.1	36.02±4.77
0.07	16.8±2.7	45.9±9.9	214.9±13.2	1.2±0.2	1.5±0.1	1.4±0.2	36.47±6.42
0.10	16.2±2.8	45.5±8.1	223.3±10.6	1.2±0.2	1.5±0.1	1.5±0.2	38.67±4.69
0.14	16.7±2.4	44.4±7.9	214.4±12.9	1.3±0.2	1.5±0.1	1.4±0.2	37.33±7.11
0.18	17.2±2.9	45.7±10.4	210.4±12.2	1.3±0.2	1.5±0.2	1.4±0.2	37.92±7.94
F test	*	*	*	n. s.	*	n. s.	*
Regression							
r^2	0.03	0.50	0.81	0.06	0.65	0.42	0.86
b_0	16.79	41.60	201.76	1.30	1.47	1.44	31.20
b_1	-2.63	82.71	362.42	-0.44	2.03	1.12	115.87
b_2	20.86	-366.00	-1787.11	2.53	-8.86	-7.33	-455.74

Data are the means ± standard deviation.

* Significant at 5% probability level; n.s.-Not significant at 5% probability level.

3.3 Silage Yield and Shoot Dry Matter Partitioning

Silage yield was significantly influenced with Zn foliar spray. It had a positive quadratic response and the highest yield was recorded with Zn spray at 0.09%. It induced an increase of 49.4% compared to control. However, a clear decline of about 26% in optimal yield was noted with Zn spray concentration up to 0.10% (Figure 1).

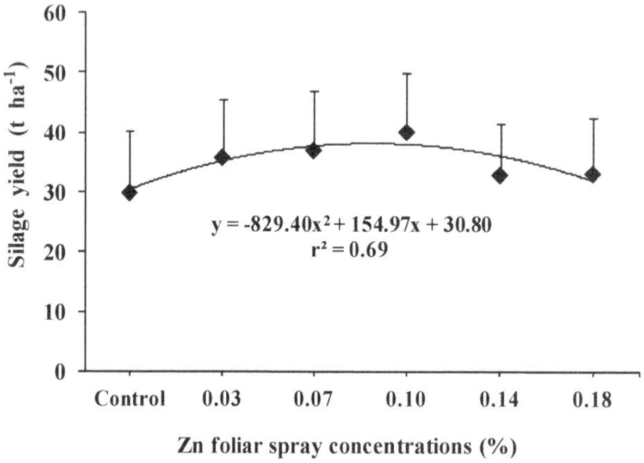

Figure 1. Effect of Zn foliar spray concentration on silage yield. Vertical bars denote standard deviation

The shoot dry matter partitioning analysis showed that stem, leaves and ear dry weights were all significantly enhanced by Zn foliar applications and had the same trend of the regression equation describing silage yield (Figure 2a). Within ear, we found that especially kernels dry weight was significantly influenced (Figure 2b).

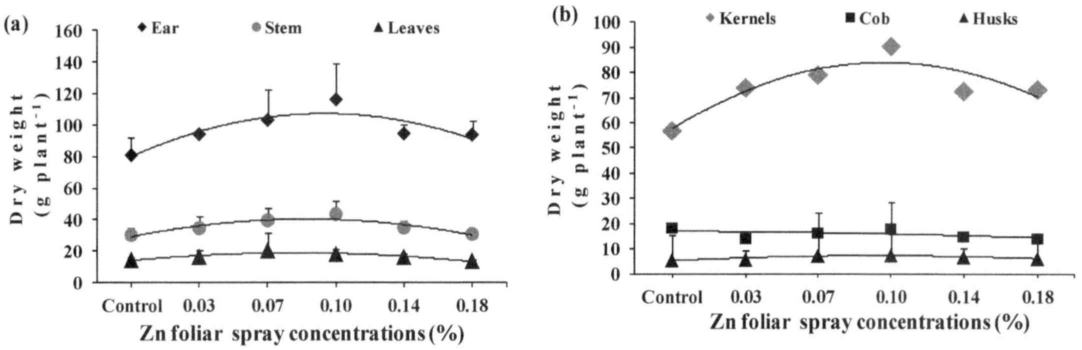

$y=-2648.86x^2+530.36x+80.78$	$r^2=0.75$ for ear	$y=-2374.22x^2+490.52x+58.67$	$r^2=0.77$ for kernels
$y=-1321.71x^2+240.14x+29.46$	$r^2=0.86$ for stem	$y=-196.4x^2+39.69x+5.25$	$r^2=0.83$ for husks
$y=-634x^2+108.2x+14.14$	$r^2=0.85$ for leaves	$y=-78.24x^2+0.14x+16.86$	$r^2=0.26$ for cob

Figure 2. Effect of Zn foliar spray concentration on shoot dry matter partitioning into ear, stem, leaves (a), kernels, cob and husks (b) of corn silage at harvest. Vertical bars denote standard deviation

The analysis of kernels yield components showed no significant effect of Zn foliar spray on the total number of ovules per ear (Figure 3a). But, two main components were significantly influenced: the pollination rate and the thousand-kernel dry weight (Figures 3b and 3c). Both had quadratic responses and the highest response was approximately at 0.10% of Zn. Compared to control, this Zn rate induced significant increases of 15.4% and 19.1% on pollination rate and on thousand-kernel dry weight, respectively.

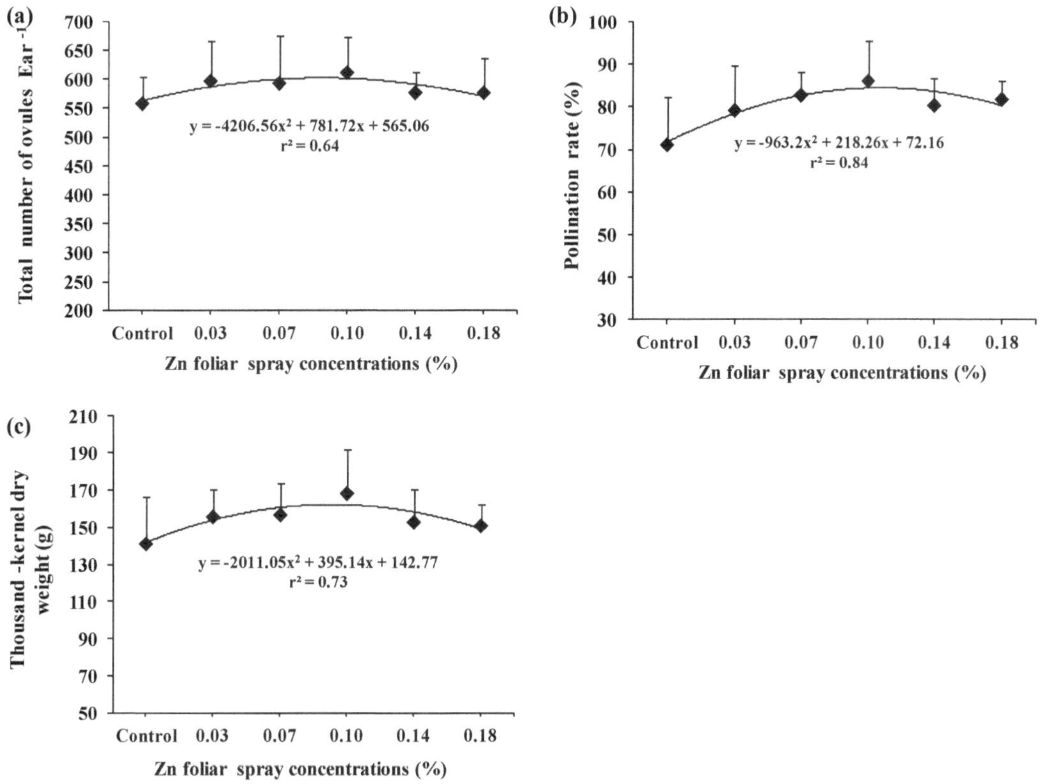

Figure 3. Effect of Zn foliar spray concentration on total number of ovules per ear (a), pollination rate (b) and thousand-kernel dry weight (c) of corn silage at harvest. Vertical bars denote standard deviation

3.4 Mineral Shoot Content

3.4.1 Micro Elements Shoot Content

The chemical analysis of shoot dry matter at harvest showed that Zn foliar spray influenced some micro nutrients plant concentrations (Figures 4a, 4b, 4c and 4d). The increase of Zn foliar spray concentration resulted in significant linear increase in shoot Zn content. Zn spray concentration at 0.09% which induced optimal yielding led to a shoot Zn content of 21.8 mg kg^{-1}. However, plant concentration in Mn, Fe and Cu showed linear significant decrease with increasing Zn foliar spray concentration. Zn foliar applications at 0.09% induced mineral contents of 93, 35.8, 6.7 mg kg^{-1} for Fe, Mn and Cu, respectively.

3.4.2 Macro Elements Shoot Content

Furthermore, shoot concentration in K and Mg had negative linear relationship with Zn concentration spray. At Zn foliar application of 0.09%, we recorded 0.8% as shoot k content and 0.12% as Mg content (Figures 5a and 5b).

In contrast, shoot P content showed a quadratic response towards Zn foliar spray concentration (Figure 5c). Zn applications at 0.09% induced shoot P content of 0.27%. Other macro elements as N and Ca didn't show any significant difference between treatments and were in adequate levels (DATA not shown here).

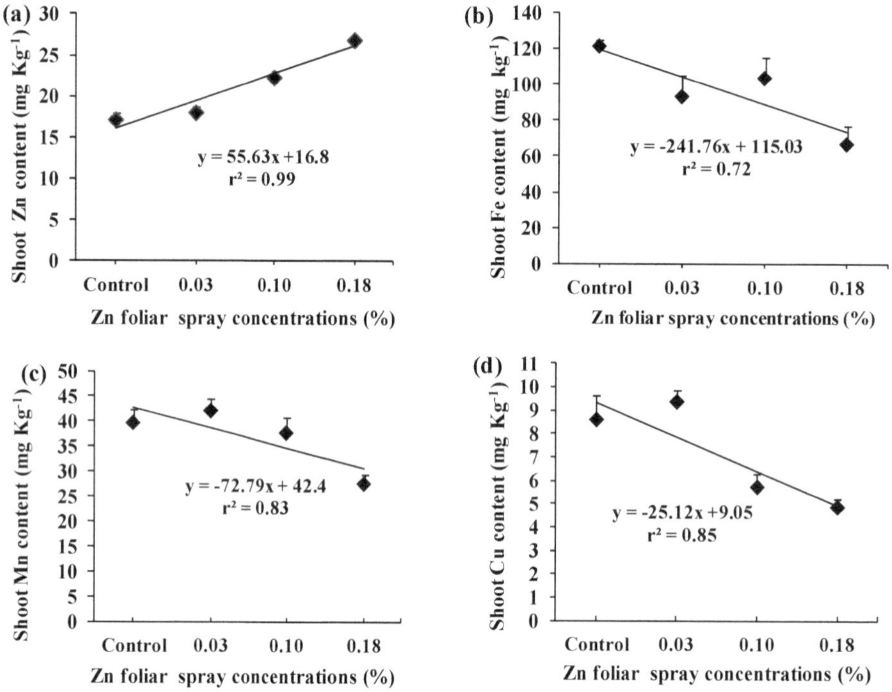

Figure 4. Effect of Zn foliar spray concentration on Zn (a), Fe (b), Mn (c) and Cu (d) contents of the shoot dry matter for corn silage at harvest. Vertical bars denote standard deviation

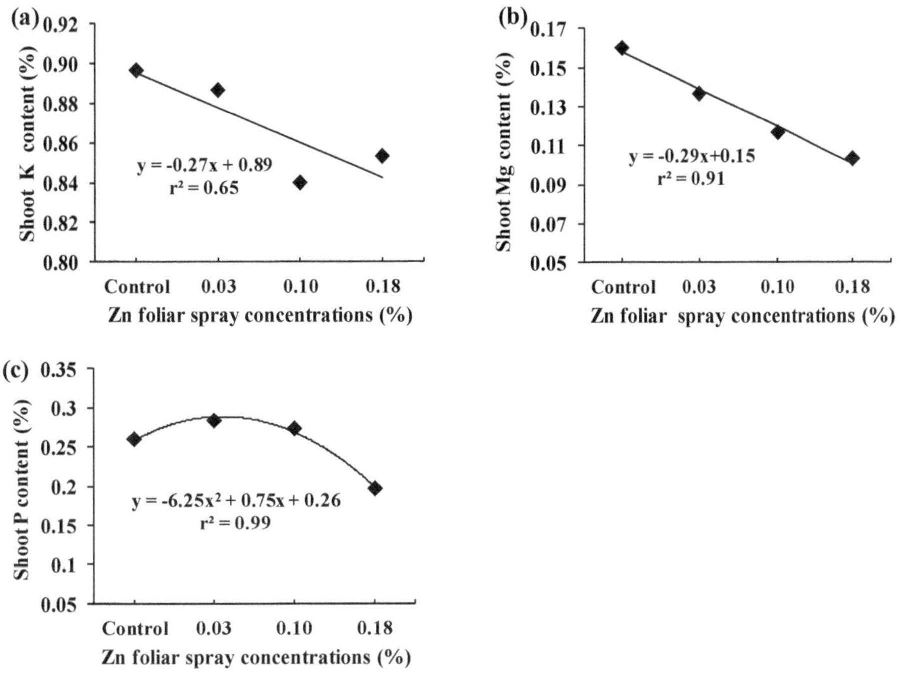

Figure 5. Effect of Zn foliar spray concentration on K (a), Mg (b), and P (c) contents of the shoot dry matter for corn silage at harvest. Vertical bars denote standard deviation

4. Discussion

Results from the present study have shown that Zn foliar spray prevented Zn deficiency stress in corn when it's applied at adequate level. Approximately 0.09% is the optimum Zn foliar spray concentration for corn silage production when grown in Zn deficient sandy soil. Furthermore, adverse effect on plant growth can happen with higher Zn foliar spray concentration.

4.1 Visual Zn Deficiency Symptoms

The appearance of Zn deficiency symptoms on corn was mainly attributable to a low soil born Zn content (0.13 mg kg^{-1}) which was below the critical level of 0.8 mg kg^{-1} required for corn production (Landsay & Norvel, 1977). Their appearance at an earlier stage (5-6 leaf stage) was reported by many authors (Kuldeep & Banerjee, 1986; Sharma, Chatterjee, & Saharma, 1990; Wang & Jin, 2007). As soon as we increased Zn foliar spray concentration, these symptoms became less severe. The same result was reported by Kaya and Higgs (2002) on tomato. Nevertheless, the slight recovery from Zn deficiency noticed in control plants at 8-9 leaf stage agreed with earlier result reported by Rungruang, Doyle, and Raymond (1978). On the other hand, the clear leaf injury noted at high Zn foliar spray concentration (up to 0.10%) was noticed in rice with Zn foliar spray concentration up to 0.03% (Qiao et al., 2014) and in tomato at 0.015% of Zn (Kaya & Higgs, 2002).

4.2 Plant Growth, Silage Yield and Shoot Dry Matter Partitioning

Synchronization of visual Zn deficiency symptoms and stem height shortening in control and low Zn foliar spray concentrations (0.03% and 0.07%) supported the physiological mediating function of Zn in the metabolism of growth regulators as indole 3 acetic acid and gibberellic acid (Cakmak, Marshner, & Bangerth, 1988; Balal et al., 1998). The leaf area was also enhanced by Zn foliar spray; a result which was reported by Chaab, Savaghebi, and Motesharezadeh (2011) on maize due to Zn soil supply. While no increase in stem diameter at harvest was noted, this result agreed with the finding of Rajabi et al. (2013) on safflower. In contrast, high Zn foliar spray concentration (up to 0.09%) induced a clear inhibition of plant growth and yield decline, which agreed with the result of Kaya and Higgs (2002) on tomato. In this regard, a recent study by Qiao et al. (2014) explained that Zn foliar application at high concentration on rice, exceeding 0.03%, induced a significant decrease in phosynthesis rate. They explained this decrease by the inhibition of chlorophyll synthesis and by inactivation of β-carbonic anhydrase enzyme which contains Zn and plays crucial role in photosynthesis.

Many authors reported the importance of Zn foliar applications at adequate levels to enhance yields of many crops such as corn (Potarzycki & Grzebisz, 2009), wheat grain (Haslett, Reid, & Rengel, 2001), tomato fruit (Kaya & Higgs, 2002) and safflower (Rajabi et al., 2013). From this study, the highest silage yield was recorded with Zn foliar applications at 0.09% of Zn, equivalent to 0.40% of $ZnSO_4.7H_2O$, applied at 5-6 and 9-10 leaf stages. Similarly, Lindsay (1972) reported in his review that Zn foliar applications at 0.5% of $ZnSO_4{\cdot}7H_2O$, applied at 3 and at 5 leaf stages corrected Zn deficiency on corn. In contrast, Potarzycki and Grzebisz (2009) reported that optimal maize grain yield was recorded with high Zn foliar spray concentration ranging between 0.25% and 0.37% of Zn. This latter was applied as oxysulfate at 5 leaf stage and was equivalent to 1.11-1.64% of $ZnSO_4.7H_2O$. Dry matter partitioning analysis showed that stem, leaves and ear biomass were all influenced by Zn foliar applications. The marked positive effect of Zn foliar spray on biomass can be explained through the role of Zn in increasing both the carbonic anhydrase enzyme activity (Qiao et al., 2014; Sasaki, Hirose, Watanabe, & Ohsugi, 1998) and chlorophyll synthesis (Qiao et al., 2014; Cakmak & Marshner, 1993), which improved photosynthesis rate (Qiao et al., 2014). As kernels presented about 50% of the total aerial dry matter (Figure 6), the silage yield response was especially linked to kernels weight. This latter was enhanced mainly through two components: thousand-kernel dry weight and pollination rate. The increase of thousand-kernel dry weight was reported by Potarzycki and Grzebisz (2009) on corn, and by Rajabi et al. (2013) on safflowers. It could be attributed mainly to photosynthesis activation mentioned above. Regarding pollination rate, its increase with adequate Zn foliar application can be related to male fertility enhancement found by Sharma et al. (1990).

4.3 Mineral Shoot Content

Control plants showed insufficient shoot Zn content at harvest (15.8 mg Kg^{-1}) which was below the critical level required in corn 22 mg kg^{-1} (Singh et al., 2005). This content was linearly increased with raising Zn foliar spray concentration, the same result was reported by Cakmak et al. (2010) on wheat and by Kaya and Higgs (2002) on tomato. This linear positive relationship illustrates the luxurious consumption character of corn silage towards Zn.

Zn foliar supplementations at 0.09%, which induced the highest silage yield, provided an adequate shoot Zn content (21.8 mg kg^{-1}). This was approximately equal to the critical level cited above (22 mg kg^{-1}). Adequate Zn level in corn can both enhance silage nutritional quality and suppress or minimize Zn feed additives supply used in dairy cattle feeding. Even if we remarked negative relationships between shoot Zn content and Mg ($r^2 = 0.73$), K ($r^2 = 0.59$), Mn ($r^2 = 0.63$), Cu ($r^2 = 0.68$) and Fe ($r^2 = 0.59$) at harvest, Zn foliar spray at 0.09% resulted in adequate plant content of k (0.8%) which was approximately equal to the critical level ranging between 0.7% and 0.9% (Arnon, 1975). Also, it induced adequate levels of Mn, Cu and Fe (Rashid & Ryan, 2008). The plant Mg content (0.12%) was lower than the adequate concentration of 0.20% requested for corn silage by Fox and Piekielek (1984). In the case of this study, no Mg deficiency symptoms were noticed. In this context, Arnon (1975) reported that Mg deficiency symptoms in corn appeared at Mg concentration below 0.06%, and content around 0.12% and 0.24% induced optimal yields.

The negative interactions between Zn and other nutrients were in line with the finding of Kaya and Higgs (2002) who reported that Zn foliar applications on tomato induced antagonism between Zn and K, Mg and Fe. The same relationship was found between Zn and Cu (Kumar et al., 2009) and between Zn and Fe (Ai-Qing et al., 2011) on wheat. This kind of interaction, which is presumably related to an inhibition of elements uptake due to Zn foliar applications, hasn't been well explained in the literature yet, and gives a reason for further investigation.

The shoot P content behaved differently from the other elements; it was enhanced with slight Zn concentration applications (below or equal to 0.06%) and decreased with higher ones (up to 0.06%). At an earlier study, Orabi, Mashadi, Abdllah, and Morsy (1981) found a positive relationship between P and Zn on corn. However, Zhu, Smith, and Smith (2001) reported that high soil P supply induced a low Zn content on wheat. Interaction between P and Zn is widely reported in the literature for different crops, but little is known about the specific mechanism involved either under Zn foliar spray or soil Zn supply. On the other hand, the optimal silage yield recorded at 0.09% had an adequate P content (0.27%) which was around the required level (0.1-0.5%) (Arnon, 1975).

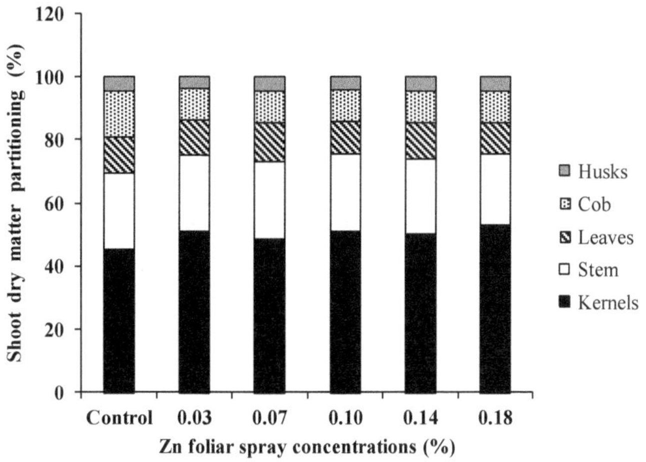

Figure 6. Effect of Zn foliar spray concentration on ratio of kernels, stem, leaves, cob and husks dry weights to total shoot dry weight

5. Conclusion

Results have indicated that 0.09% of Zn, using as source Zn sulfate ($ZnSO_4{\cdot}7H_2O$), is the adequate concentration to spray at 5-6 and 8-9 leaf stages of corn silage in order to cure Zn deficiency. Such level of concentration proved to increase plant Zn content without any disruption in plant mineral composition, and to enhance plant growth, thereby increasing silage yield. However, caution must be taken with Zn foliar spray at concentration up to 0.10% which induced leaf injury, disrupted mineral plant composition and inhibited plant growth.

Acknowledgements

The authors are grateful to Mr. Loultiti M. M. and Mr. Coquant J. M. for their financial support. We wish to thank Mrs. Mouhatdi R., Mrs. Zehri B., Mr. Ziti S., Mr. Soulaimani A., Mr. Touhami D. and Mr. Daoudi M. for their valuable help during laboratory analysis. We also acknowledge Dr. Yacholti and Mr. El Gharradi Y. O. for language correction.

References

Ai-Qing, Z., Qiong-Li, B., Xiao-Hong, T., Xin-Chun, L., & Gale, W. J. (2011). Combined effect of iron and zinc on micronutrient levels in wheat (*Triticum aestivum* L.). *Journal of Environemental Biology, 32*, 235-239. Retrieved from http://www.jeb.co.in/journal_issues/201103_mar11/paper_16.pdf

Alloway, J. B. (2008). *Zinc in soils and crop nutrition* (2nd ed.). International Zinc Association Communications. IZA Publications, Brussel. Retrieved from http://www.zinc.org/general/Zinc_in_Soils_and_Crop_Nutrition_ALLOWAY.pdf

Arnon, I. (1975). *Mineral nutrition of maize.* International Potash Institute, Bern. Retrieved from http://www.ipipotash.org/udocs/mineral_ nutrition_ of_ maize.pdf

Balal, H., Norihiro, H., Yoshitaka, N., Masahito, S., & Hiroshi, T. (1998). Zinc nutrition and levels of endogenous Indole-3-acetic acid in radish shoots. *Journal of Plant Nutrition, 21*, 1113-1128. http://dx.doi.org/ 10.1080/01904169809365470

Cakmak, I., Kalayci, M., Kaya, Y., Torun, A. A., Aydin, N., Wang, Y., ... Horst, W. J. (2010). Biofortification and localization of zinc in wheat grain. *Journal of Agricultural and Food Chemistry, 58*, 9092-9102. http://dx.doi.org/10.1021/jf101197h

Cakmak, I. (2008). Enrichment of cereal grains with zinc: Agronomic or genetic biofortification? *Plant and Soil, 302*, 1-17. http://dx.doi.org/10.1007/s11104-007-9466-3

Cakmak, I., & Marshner, H. (1993). Effect of zinc nutritional status on activities of superoxide radical and hydrogen peroxide scavenging enzymes in bean leaves. *Plant and Soil, 155-156*, 1127-1130. http://dx.doi.org/ 10.1007/BF00025000

Cakmak, I., Marshner, H., & Bangerth, F. (1988). Effect of zinc nutritional status on growth, protein metabolism and levels of indole 3 acetique acid and other phytohormones in bean (*Phaseolus vulgaris*). *Journal of*

Experimental Botany, 40, 405-412. http://dx.doi.org/10.1093/jxb/40.3.405

Chaab, A., Savaghebi, Gh. R., & Motesharezadeh, B. (2011). Differences in the zinc efficiency among and within maize cultivars in a calcareous soil. *Asian Journal of Agricultural Sciences, 3*, 26-31. Retrieved from http://www.maxwellsci.com/print/ajas/v3-26-31.pdf

De Vasconcelos, A. C. F., Clístenes, W. A. N., & Fernando, F. C. F. (2011). Distribution of zinc in maize plants as a function of soil and foliar Zn supply. *International Research Journal of Agricultural Science, 1*, 1-5. Retrieved from http://interesjournal.org/IRJAS/Pdf/2011/March/de%20Vasconcelos%20et%20al.pdf

Fox, R. H., & Piekielek, W. P. (1984). Soil magnesium level, corn (*Zea mays* L.) yield, and magnesium uptake. *Communications in Soil Science and Plant Analysis, 15*, 109-123. http://dx.doi.org/10.1080/00103628409367459

Gupta, U. C., Kening, W. U., & Siyuan, L. (2008). Micronutrients in soils, crops and livestock. *Earth Science Frontiers, 15*, 110-125. http://dx.doi.org/10.1016/S1872-5791(09)60003-8

Hansch, R., & Mendel, R. R. (2009). Physiological functions of mineral micronutrients (Cu, Zn, Mn, Fe, Ni, Mo, B, Cl). *Current Opinion in Plant Biology, 12*, 259-266. http://dx.doi.org/10.1016/j.pbi.2009.05.006

Haslett, B. S., Reid, R. J., & Rengel, Z. (2001). Zinc mobility in wheat: Uptake and distribution of zinc applied to leaves or roots. *Annals of Botany, 87*, 379-386. http://dx.doi.org/10.1006/anbo.2000.1349

Kaya, C., & Higgs, D. (2002). Response of tomao (*Lycopersicon esculentum* L.) cultivars to foliar application of zinc when grown in sand culture at low zinc. *Scientia Horticulturae, 93*, 53-64. http://dx.doi.org/10.1016/S0304-4238(01)00310-7

Knoche, M., Petracek, P. D., Bukovac, M. J., & Shafer, W. E. (1994). Urea penetration of isolated tomato fruit cuticles. *Journal of the American Society for Horticultural Science, 119*, 761-764. Retrieved from http://journal. ashspublications.org/content/119/4/761.full.pdf

Kuldeep, S., & Banerjee, N. K. (1986). Growth and zinc content of maize (*Zea mays* L.) as related to soil-applied zinc. *Field Crops Research, 13*, 55-61. http://dx.doi.org/10.1016/0378-4290(86)90007-9

Kumar, R., Mehrotra, N. K., Nautiyal, B. D., Kumar, P., & Singh, P. K. (2009). Effect of copper on growth, yield and concentration of Fe, Mn, Zn and Cu in wheat plants (*Triticum aestivum* L.). *Journal of Environemental Biology, 30*, 485-488. Retrieved from http://www.jeb.co.in/journal_issues/200907_jul09/paper_03.pdf

Lindsay, W. L., & Norvell, W. A. (1977). Development of a DTPA soil test for zinc, iron, manganese, and copper. *Soil Science Society of America Journal, 42*, 421-428. http://dx.doi.org/%2010.2136/sssaj1978.03615995004200030009x

Lindsay, W. L. (1972). Zinc in soils and plant nutrition. *Advances in Agronomy, 24*, 147-186. http://dx.doi.org/10.1016/S0065-2113(08)60635-5

Mokhtarpour, H., The, C. B. S., Saleh, G., Selamat, A. B., Asadi, M. E., & Kamkar, B. (2010). Non destructive estimation of maize leaf area, fresh weight, and dry weight using leaf length and leaf width. *Communications in Biometry and Crop Science, 5*, 19-26. Retrieved from http://agrobiol.sggw.waw.pl/~cbcs/articles/CBCS_5_1_4.pdf

Mousavi, S. R., Galavi, M., & Rezaei, M. (2013). Zinc (Zn) importance for crop production – A review-. *International Journal of Agronomy and Plant Production, 4*, 64-68. Retrieved from http://ijagcs.com/wp-content/uploads/2013/01/1881-18841.pdf

Orabi, A. A., Mashadi, H., Abdllah, A., & Morsy, M. (1981). Effect of zinc and phosphorus on the grain yield of corn (*Zea mays* L.) grown on a calcareous soil. *Plant and Soil, 63*, 291-294. http://dx.doi.org/10.1007/BF02374607

Potarzycki, J., & Grzebisz, W. (2009). Effect of zinc foliar application on grain yield of maize and its yielding components. *Plant Soil Environment, 55*, 519-527. Retrieved from http://www.agriculturejournals.cz/publicFiles/13721.pdf

Qiao, X., He, Y., Wang, Z., Li, X., Zhang, K., & Zeng, H. (2014). Effect of foliar spray of zinc on chloroplast b-carbonic anhydrase expression and enzyme activity in rice (*Oryza sativa* L.) leaves. *Acta Physiologia Plantarum, 36*, 263-272. http://dx.doi.org/10.1007/s11738-013-1407-6

Rajabi, M., Fetri, M., Eghbal, G. M., Faraji, A. M. H., & Asadian, G. (2013). Foliar application of Zn and Mn fertilizers on yield and yield components of safflower (*Carthamus tinctorius* L.) cultivars. *International*

Journal of Agriculture and Crop Sciences, 5, 718-722. Retrieved from http://ijagcs.com/wp-content/uploads/2013/ 04/718-722.pdf

Rashid, A., & Ryan, J. (2008). Micronutrient constraints to crop production in the Near East. Potential significance and management strategies. In V. B. Alloway (Ed.), *Micronutrient Deficiencies in Global Crop Production* (pp. 149-180). Springer Netherlands. http://dx.doi.org/10.1007/978-1-4020-6860-7_6

Rungruang, I., Doyle, P., & Raymond, L. (1978). Utilization and redistribution of Zn during vegetative growth of corn. *Agronomy Journal, 70*, 243-246. http://dx.doi.org/10.2134/agronj1978.00021962007000020008x

Sasaki, H., Hirose, T., Watanabe, Y., & Ohsugi, R. (1998). Carbonic anhydrase activity and CO_2 transfer resistance in Zn deficient rice leaves. *Plant Physiology, 118*, 929-934. http://dx.doi.org/10.1104/pp.118.3.929

Sharma, P. N., Chatterjee, A. S. C., & Saharma, C. P. (1990). Zinc deficiency and pollen fertility in maize (*Zea mays*). *Plant and Soil, 124*, 221-225. http://dx.doi.org/10.1007/BF00009263

Singh, B., Natesan, S. K. A., Singh, B. K., & Usha, K. (2005). Improving zinc efficiency of cereals under zinc deficiency. *Current Science, 88*, 36-44. Retrieved from http://www.iisc.ernet.in/currsci/jan102005/36.pdf

Takkar, P. N., & Walker, C. D. (1993).The distribution and correction of zinc deficiency. In A. D. Robson (Ed.), *Zinc in soil and plants* (pp. 151-166). Kluwer Academic Publishers, Dordrecht, the Netherlands. http://dx.doi.org /10.1007 /978-94-011-0878-2_11

Van Biljon, J. J., Wright, C. A., Fouche, D. S., & Botha, A. D. P. (2010). A new optimum value for zinc in the main maize producing sandy soils of South Africa. *South African Journal of Plant and Soil, 27*, 252-255. http://dx.doi.org/10.1080/02571862.2010.10639994

Wang, H., & Jin, Ji-Y. (2007). Effects of zinc deficiency and drought on plant growth and metabolism of reactive oxygen species in maize (*Zea mays* L.). *Agricultural Sciences in China, 6*, 988-995. http://dx.doi.org/10.1016/S1671-2927(07)60138-2

Wojcik, P. (2004). Uptake of mineral nutrients from foliar fertilization. *Journal of Fruit and Ornamental Plant Research, 12*, 201-218. Retrieved from http://www.insad.pl/files/journal_pdf/journal_2004spec/full2004-24spec.pdf

Zhu, Y. G., Smith, S. E., & Smith, F. A. (2001). Zinc (Zn)-phosphorus (P) interactions in two cultivars of spring wheat (*Triticum aestivum* L.) differing in P uptake efficiency. *Annals of Botany, 88*, 941-945. http://dx.doi.org/ 10.1006/ anbo.2001.1522

5

Stability Analyses for Seed Yield of Chickpea (*Cicer arietinum* L.) Genotypes in the Western Cold Zone of Iran

Homayoun Kanouni[1], Yadolah Farayedi[2], Ali Saeid[3] & Sayyed Hossein Sabaghpour[4]

[1] Seed and Plant Unit, Agriculture and Natural Resources Research Center of Kurdistan, Sanandaj, Iran

[2] Food Legume Research Department, Dryland Agricultural Research Institute (DARI), Maragheh, Iran

[3] Seed and Plant Unit, Agriculture and Natural Resources Research Center of West-Azerbaijan, Urmia, Iran

[4] Seed and Plant Unit, Agriculture and Natural Resources Research Center of Hamedan, Hamedan, Iran

Correspondence: Homayoun Kanouni, Seed and Plant Unit, Agriculture and Natural Resources Research Center of Kurdistan, P. O. Box 714, Sanandaj, Iran. E-mail: hkanouni@yahoo.com

Abstract

Different methods have been developed to estimate of genotype by environment interaction (GEI) in crop plants. In this study, 14 kabuli type chickpea genotypes were assessed for seed yield in four stations over three successive years (2010-2013) at west highlands of Iran. Randomized complete block design was used in all test environments with four replicates. Combined analysis of variance for seed yield revealed significant differences between genotypes, locations, and interaction between these two sources. The mean seed yield of genotypes averaged over environments showed that V4 and V2 had the highest (1163.58 kg ha^{-1}) and the lowest seed yield (759.07 kg ha^{-1}), respectively. Significant GE interaction implied that chickpea genotypes had various responses to different environments and, the stability analysis could be performed. To investigate GEI and identify the best performing stable genotypes, several stability parameters were employed. According to Wricke's ecovalance, stability variance, Plaisted method, and genotypic stability V5, V8 and V3 were the most stable genotypes. Based on CV, regression coefficient and MS(GE), V1 and V5 found to be stable and adapted to diverse environments, and the other genotypes distributed among stability statistics. Based on the AMMI biplot, 12 test environments divided into two mega environments. These mega environments included very cold districts like Maragheh and similar areas, and relatively softened regions of Kurdistan and similar environments. For these two mega environments, V6 and V4 showed more adaptability, respectively. In conclusion, the two genotypes, V4 (FLIP 00-39C) and V6 (FLIP 99-26C) could be recommended as new cultivars to chickpea farmers for autumn sowing in west areas of Iran.

Keywords: chickpea, G×E interaction, stability parameters, seed yield

1. Introduction

Chickpea is one of the most important components of agricultural systems in Iran (Kanouni et al., 2006). Iran ranks seventh in chickpea growth and production after India, Australia, Turkey, Myanmar, Pakistan and Ethiopia (FAOSATA, 2012). Chickpea (*Cicer arietinum* L.), which is the unique agronomic species in genus *Cicer*, provides a non-expensive and high quality protein rich source for poor societies. Chickpea has two Desi and Kabuli types, which are sowing in different parts of the world, and one has preference over another based on consumer's taste. During 2011-12, production and the area under sowing of chickpea in Iran were 300 thousand tone and 550 thousand ha, respectively (Danyali et al., 2012). Average seed yield of chickpea in Iran is 350 kg ha^{-1}, which is lower than world's average. Whereas, potential yield of 4000 kg ha^{-1} has been recorded in a farm in Syria (Singh & Saxena, 1999). This gap between average and potential yield of chickpea could be due to susceptibility of this crop to the abiotic constraints.

In a field trial, a genotype may be known as superior variety when have high potential to produce seed yield at different environmental conditions, as well as high level of phenotypic stability, simultaneously. When GEI has importance to plant breeders, stability and adaptability parameters of yield will help them to find superior varieties and adaptation of these varieties to a divers set of environmental conditions. However, breeding for yield increment has highly depending on selection of favorable genotypes to the target environment (Bakhsh et al., 2009; Raffi et al., 2004).

From three components of phenotypic variation (Genotype, environment and genotype by environment interaction), so far the highest attention allocated to genotype (Halluer & Miranda, 1981). Between other two components, environment and GEI, during last decades GEI has investigated more (Lin et al., 1986). In this regard, different regression models have been applied widely, and several univariate and multivariate parametric and non-parametric statistics have been suggested to identifying response of genotypes to the environmental variations.

Phenotypic stability has been studied widely by biometricians, who have developed several methods for its analysis (Eberhart & Rusell, 1966; Lin & Binns, 1988). Lin et al. (1986) have classified stability parameters in three groups namely type I, II and III. By type I, a genotype has named stable when its environmental variance is small. Environmental variance S_i^2 (Romer, 1917; cited from Becker & Léon, 1988), regression coefficient (Finley & Wilkinson, 1963), and coefficient of variation (Francies & Kanenberg, 1978) are belonging to this group. In type II, which stability of Shukla (Shukla, 1972), Wricke's ecovalance (Wricke, 1965), variance mean of Plaisted and Peterson (1959), Plaisted (1960) variance, and regression coefficient of Finlay and Wilkinson (1963) classified in this group, a genotype is stable when its response to environment is parallel with average response of all experimental genotypes. In the stability type III, which includes squared deviations from the regression (Eberhart & Russell, 1966) and coefficient of determination (Pinthus, 1976), a genotype considers stable when mean of residual squares is smaller than regression models on environmental index. Later Lin and Binns (1988) proposed a fourth type. In type IV, a variety considers to be stable if the year mean square in locations is small. Wricke (1962) suggested that contribution of a genotype in interaction's sum of square could be used as an estimate of stability (w_i). Shukla (1972), to obtain an unbiased estimate of GEI variance for each genotype, made changes in ecovalance (w_i) and called it stability variance. Francis and Kannenberg (1978) used coefficient of variation of each genotype as an estimate of stability. Lin and Binns have developed a priority index (P_i) or superiority measure of genotypic performance, which defined as the distance mean square between the cultivar's response and the maximum response averaged over all locations. Kang (1988) suggested a rank-sum method that provides selection for yield and the stability variance statistic, simultaneously. Rank-sum obtains via summation rank of mean yield and rank of Shukla's stability variance. However, lower rank-sum indicates greater stability. Based on Tai's stability parameters (Tai, 1971), a perfectly stable cultivar has $\lambda = 1$ and $\alpha = -1$. Average stability can identify by $(\alpha, \lambda) = (1, 0)$. In higher and lower than average, α is smaller and larger than zero, respectively.

Multivariate stability statistics have been proposed to eliminate noise from the data, and to reveal the structure in the data (Akter et al., 2014; Zobel et al., 1982). One of the most applicable multivariate parameters, which have a high scientific validity, is additive main effects and multiplicative interactions (AMMI) method. AMMI model increases the probability of high yielding genotypes selection (Zobel et al., 1982). Furthermore, this model has integrated analysis of variance (ANOVA) and principal component analysis (PCA) which applies to estimate the main effect of genotypes as well as genotypic and environmental components of interaction, respectively. However, the AMMI model does not make provision for a quantitative stability measure, such a measure is provided by AMMI stability values (ASV) which quantify and rank genotypes according to their yield stability (Purchase et al., 2000).

There is little information on the yield stability of chickpea in Iran. Hence, the objectives of this study were: (1) to identify high and stable yielding chickpea genotypes across environmental variation and (2) to study the measure of genotype × environment interaction in chickpea genotypes.

2. Material and Methods

Data analyzed in this study obtained via a set of yield trials conducted during three consecutive years, 2010 to 20 13, at four experimental stations of west of Iran including Kurdistan, Maragheh, Urmia and Hamedan. More details about these locations has presented in Table1. At each environments (year by location integration), 13 elite chickpea genotypes which were developed by International Center for Agricultural Research in the Dry Areas (ICARDA), as well as one local check, Djam (Table 2), studied in randomized complete block design (RCBD) with four replicates. The experiments were conducted on 1st week of October and plant population density was maintained with 35 plants per m^2 in all locations.

Table 1. Code and climatic properties of experimental environments, west of Iran (2010-13)

Environments				Latitude & Longitude	Altitude (m)	Temp. (°C)		Annual Rainfall (mm)
Location	Year	Code	Env. Index			Min.	Max.	
Maragheh	2010-11	M1	727.48	37°24′ N	1720	-15.28	36.25	498.10
	2011-12	M2	-181.82	47°15′E		-18.11	34.52	351.40
	2012-13	M3	14.33			-21.53	35.77	263.20
Kurdistan	2010-11	K1	393.86	35°40′ N	2120	-16.05	36.66	442.30
	2011-12	K2	-102.07	47°07′E		-20.41	30.18	305.11
	2012-13	K3	-615.80			-18.22	32.80	352.80
Urmia	2010-11	U1	925.12	37°32′ N	1880	-17.22	33.23	291.22
	2011-12	U2	-12.47	45°5′E		-17.93	29.61	303.20
	2012-13	U3	-327.84			-19.32	30.44	300.16
Hamedan	2010-11	H1	55.20	37°32′ N	1610	-15.64	32.09	298.93
	2011-12	H2	-627.08	45°5′E		-16.19	31.38	309.37
	2012-13	H3	-248.91			-14.24	31.44	333.05

Table 2. Name, origin, and code of chickpea genotypes used in the experiment

Genotype Name	Origin	Code	Genotype Name	Origin	Code
FLIP 01-40C	ICARDA	V1	FLIP 02-512C	ICARDA	V8
SEL 99 TH150454	ICARDA	V2	FLIP 00-84C	ICARDA	V9
FLIP 97-85C	ICARDA	V3	FLIP 01-9C	ICARDA	V10
FLIP 00-39C	ICARDA	V4	FLIP 01-18C	ICARDA	V11
FLIP 97-230C	ICARDA	V5	FLIP 98-15C	ICARDA	V12
FLIP 99-26C	ICARDA	V6	FLIP 99-45C	ICARDA	V13
FLIP 02-84	ICARDA	V7	DJAM (Check)	Iran	V14

For each location every year, analysis of variance was performed. At the next step, Bartlett's homogeneity test of variances was accomplished. Subsequently, referring to result of Bartlett's test, combined analysis of variance was done (Crossa, 1990). Stability statistics including coefficient of variation (Franscis & Kannenberg, 1978), Wricke's ecovalance (Wricke, 1962), Shukla stability variance (Shukla, 1972), regression line slop (Finlay & Wilkinson, 1963), genotypic stability (Hanson, 1970), Plaisted stability parameter (1960), mean square of GEI (Lin & Binns, 1988), priority index (Lin & Binns, 1988), and Tai's alpha and lambda (1971) were performed for each genotype. Rank-sum method (Kang, 1988), as one non-parametric statistic was measured for experimental genotypes. Among multivariate stability parameters, AMMI model was used to investigate of GE interaction and could be indicated as below (Thillainathan & Fernandez, 2001):

$$Y_{ge} = \mu + \alpha_g + \beta_e + \sum_{n=1}^{N} \lambda_n \gamma_{gn}^2 \delta_{en} + \theta_{ge} \quad (1)$$

Where,

Y_{ge}isyield of genotype by environment, μ is grand mean, α_g is genotype mean deviations, $\beta_e$is environmental mean deviations, λ_n is eigen value of PCA axis n, γ_{gn}^2 and δ_{en} are genotype and environment PCA scores for PCA axis n, N is number of PCA axes retained in the model and θ_{ge} is residuals.

To employ of AMMI stability value (ASV), the following equation was proposed by Purchase et al. (2000):

$$\text{AMMI Stability Value (ASV)} = \sqrt{\frac{IPC1\ sum\ of\ squares}{IPC2\ sum\ of\ squares} IPC1^2 + IPC2^2} \quad (2)$$

The ASV is the distance from zero in a two dimensional plot of IPCA1 (Interaction Principal Component Analysis axis 1) scores against IPCA2 scores. Since the IPCA1 score contributes more to G × E sum of squares, it has to be weighted by the proportional difference between IPCA1 and IPCA2 scores to compensate for the relative contribution of IPCA1 and IPCA2 total G × E sum of squares.

A detailed explanation on mathematical basis of AMMI analysis and its interpretations has been presented by Hussein et al. (2000). All analyzes performed using SAS GLM procedures (SAS Institute, 1999).

3. Results and Discussion

3.1 Analysis of Variance

Prior to combined analysis of variance, homogeneity of error variances of twelve experiments was determined by Bartlett's test. Result of this test revealed that environments were not heterogeneous. However, combined analysis of variance of environment, genotype, and GE interaction were significant for seed yield (Table 3). Significant GE interaction implied that chickpea genotypes had various responses to different environments and, the stability analysis could be performed to estimate the adaptation of the genotypes. More than 70% of the total sum of square explained by environment, whereas, only 3% of variation attributed to genotypes and 11% of that variation explained via GE interaction. Gauch and Zobel (1988) believe that, low variance of genotypic effect may be due to proximity of genetic potential of the test genotypes.

Table 3. Combined ANOVA of seed yield for 14 chickpea genotypes at 12 environments of Iran (2010-13)

S.O.V.	d.f.	SS	MS	F	Explained%
Environment (E)	11	141509118	12864465	102.21**	70.61
Replication/E	36	4530979	125861	2.31**	2.26
Genotype(G)	13	6423185	494091	3.14**	3.20
G × E	143	22473269	157156	2.89**	11.21
Error	468	25477431	54439		12.71
Total	671	200413982			
CV%		23.16			

Note. S.O.V. = source of variation; d.f. = degree of freedom; SS = sum of square; MS = mean square; ** = Significant at 1% probability level.

Mean seed yield of chickpea genotypes varied between 1163.5 kg ha^{-1} for V4 and 756.02 kg ha^{-1} for V2 (Table 5). Grand mean of seed yield over 12 locations and mean yield of check variety (Djam) were 1007.26 and 968.91 kg ha^{-1}, respectively. The least significant difference (LSD) value for means comparison was189.87 kg ha^{-1}. The LSD value shows that there were significant differences among genotypes ($P \leq 0.05$).

3.2 Stability Analysis

Regression of location on seed yield for 14 chickpea genotypes across different location and year combinations analyzed and results have shown in Table 4. These results imply that, genotype by environment interaction (linear) is significant ($P < 0.05$). This significant effect identified linear relationship between yield of genotypes and environmental index. In other words, the slope of the regression line for test genotypes is not same for all.

Table 4. Stability analysis for seed yield of chickpea genotypes with regression of location on yield in Iran (2010-13)

S.O.V.	d.f.	SS	MS	F	P value
Genotype (G)	13	0.1605E+07	123523.1		
Environment (E)	11	0.353773E+07	0.321612E+07		
Regression	1	0.141509E+09	0.141509E+09		
Deviation	10	-0.106132E+09	-0.106132E+08		
G × E	143	0.561832E+07	39288.9		
G × E (Linear)	13	0.1032E+07	79454.3	2.253	0.011
Deviation	130	0.4585E+07	35272.4		
Total	167	1.07611+07			

The genotypes with significantly higher ecovalance (W_i) ($P < 0.05$) are presented in Table 5. Corresponding to Wrick's ecovalance (W_i), genotypes of V5, V8, and V3 had less ecovalance than others, respectively and considered more stable genotypes across environments. Based on Plaisted method ($\theta_{(i)}$), genotypes V5, V8, and V3 estimated higher value of $\theta_{(i)}$ and were considered more stable than others. Regression coefficient (b_i) of varieties on the environmental index, ranged from 0.597 to 1.333. According to this parameter (Fransis & Kanenberg, 1978), genotypes with line slop close to one have general adaptability. Regards to this parameter, all genotypes in this study, unless V2, V6 and V11 were not significant. In fact, these genotypes have potential to increase their seed yield with improvement of environmental conditions. According to the of coefficient of variation proposed by Fransis and Kanenberg (CV), genotypes V9 and V12 with less CV and high seed yield seems to be more stable. In line with Lin and Binns statistics (P_i) and MS(GE), it could be seen that, V4, V6, and V9 have less P_i values and are considered relatively stable . Though, based on MS(GE), genotypes V1, V5, and V7 were stable. Lin & Binns (1988) tested both Pi and MS(GE) against the MS of deviation from regression. Based on stability variance (σ_i^2) of Shukla (1972), residual variance for each genotype was calculated and explored that, genotypes of V5, V8, and V3 had less variance. Becker and Léon (1988) suggested that ecovalance (Wricke, 1962) and stability variance (Shukla, 1972) give same results for ranking of genotypes. Hanson (1970) has proposed a genotypic stability (D_i^2), which based on this statistic genotypes V3, V5, and V1 were most stable. Considering regression procedure of Tai (1971), the highest stability was obtained by genotypes V6, V9 and V12 (Figure 1). As shown in Figure 1, a three-dimension plot presents the genotypes that have alpha and lambda amounts closer to -1 and 1, respectively. Based on this procedure, V3 and V5 could be selected as stable genotypes. Non-parametric statistic of rank-sum (Kang, 1988) showed that V1, V5, and V8 had desirable stability. All these genotypes had lower performance than grand mean. Using this parameter, V2 with lowest seed yield identified as the most non-stable genotype.

It is common that, it will be difficult to find a unique strategy for evaluating the stability of all varieties across all environments. Some of scientists believe that, should use all stability analysis methods including parametric and non-parametric, univariate and multivariate methods and then take a general conclusion (Kang, 1988; Yan & Kang, 2003).

Table 5. Genotypic mean (kg ha^{-1}) and stability parameters for 14 kabuli type chickpea genotypes across different environments in Iran (2010-13).

Code	Mean (kg ha^{-1})	Parameters										
		W_i	$\theta_{(i)}$	b_i	CV	P_i	MS(GE)	σ_i^2	D_i^2	α_i	λ_i	RS
V_1	982.33	284709.31	40163.10	0.951	48.95	101460.55*	20961.10	26928.30	589338.37	-0.0540	1.996	14
V_2	756.02	1128033.57*	33282.84	0.597*	50.61	291019.47*	94103.30*	116371.78*	713999.62	-0.4089*	5.104*	28
V_3	945.08	177507.76	41037.71	0.859	44.96	136411.93*	40273.82	15558.43	299529.88	-0.1442*	0.903	16
V_4	1163.5	625514.91*	37382.64	1.159	51.34	47658.47	23448.75	63074.34*	1351139.10	0.1539	4.073*	14
V_5	983.07	87201.27	41774.47	1.022	50.52	98735.12*	18534.17	5980.47	540664.00	0.0195	0.620	9
V_6	1141.2	522411.01*	38223.81	1.333*	57.52	52500.67	23128.21	52139.08*	1619695.82	0.3374*	1.708	13
V_7	1015.7	274090.68	40249.74	0.958	47.66	90677.39*	23001.65	25802.08	597114.59	-0.0449	1.935*	10
V_8	974.21	175058.83	41057.69	1.038	52.89	109313.69*	25524.78	15298.70	676486.33	0.0432	1.225	12
V_9	1111.3	405369.20*	39178.70	0.956	44.62	72583.56	35512.70	39725.56*	726464.29	-0.0459	2.878*	13
V_{10}	994.64	333955.32*	39761.33	0.987	50.57	110661.15*	35023.99	32151.36*	715657.05	-0.0159	2.397*	14
V_{11}	1069.5	523688.95*	38213.39	1.255*	59.39	80867.64	31533.59	52274.62*	1498139.25	0.2768	2.388*	16
V_{12}	1051.0	370263.02*	39465.11	0.937	45.96	80549.32	25254.70	36002.17*	649905.87	-0.0664	2.584*	13
V_{13}	945.14	300070.18	40037.78	0.963	51.77	132808.17*	36693.26	28557.48	634974.11	-0.0391	2.131*	18
V_{14}	968.91	400265.65*	39220.34	0.985	52.57	156555.22*	48621.98	39184.27*	792029.92	-0.0083	2.878*	20

Note. * Significant at 5% probability level; LSD 5% = 189.87 kg ha^{-1}; Mean = Mean yield; Wi = Wricke's ecovalance, $\theta_{(i)}$ = Plaisted's stability parameter; bi = coefficient of regression; CV = coefficient of variation; Pi = priority index, MS(GE) = mean squares of genotype by environment interactions; σ_i^2 = stability variance; D_i^2 = genotypic stability, α_i & λ_i = Tai's alpha and Lambda, RS = rank-sum.

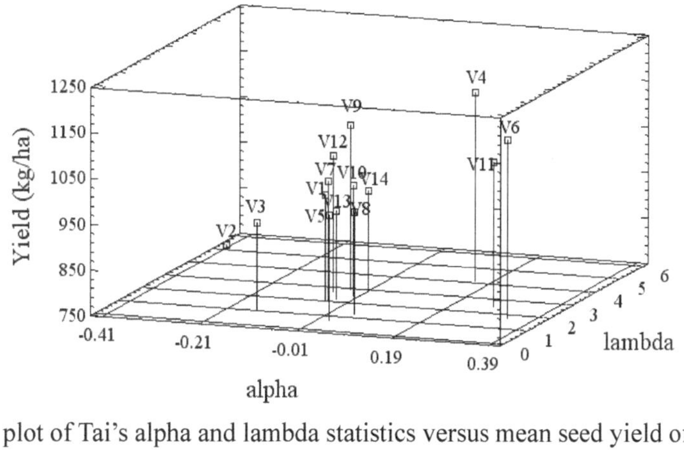

Figure 1. Scatter (3-D) plot of Tai's alpha and lambda statistics versus mean seed yield of 14 chickpea genotypes

Analysis of additive main effects and multiplicative interactions (AMMI) for seed yield of 14 elite chickpea genotypes at 12 environments are presented in Table 6. As can be seen, 82.2% of the total sum of squares is attributed to the environmental effects, and only 4.6% has allocated to genotype and 13.2% to genotype by environment interaction. Large amount of environment SS imply that experimental environments are varied, and great differences between mean of environments has created main portion of variations in seed yield of experimental genotypes.

Table 6. AMMI analysis for seed yield of 14 chickpea genotypes at 12 environments of Iran

S.O.V.	d.f.	SS	MS	F	Prob. level	Explained%
Genotype (G)	13	0.160 E+07	123523.1			4.6
Environment (E)	11	0.353E+07	0.321E+07			82.2
G×E	143	0.561E+07	39288.9			13.1
AMMI1	23	0.225E+07	97949.9	3.493	0.000	43.2
AMMI2	21	0.125E+07	59775.3	2.804	0.000	23.5
AMMI3	19	744988.0	39209.9	2.298	0.005	11.3
AMMI4	17	501151.1	29479.4	2.149	0.015	8.8
Residual G×E	63	864050.0	13715.1			13.2
Total	167					

Results of this analysis also showed that the first principal component of interaction (first component of AMMI) accounted for approximately 43% and second principal component accounted for about 24% of the GEI sum of squares, and cumulatively both AMMI1 and 2 explained approximately 67% of the total interaction of GE. Primary evaluations showed that, all estimated principal components are significant at 1% level. Meanwhile, residual (noise) included about 13% of GE sum of square. In this study, the best estimated model has obtained via two IPC1 and IPC2 components. Although some authors suggest use of four main components as necessity to estimate the AMMI model (Thillainathan & Fernandez, 2001); however, review of the literatures suggests that factors such as the type of crop plant, germplasm diversity, and range of environmental conditions are effective on the degree of complexity of the estimated model (Malhotra & Singh, 1991; Atta & Shah, 2009; Acikgoz et al., 2009).

In order to observe the pattern of interaction between genotype and environment and interpret the results, biplot polygons were used (Yan & Kang, 2003). Corresponding to GGE biplot method and through symmetric scoring, biplot of IPC1 against IPC2 was drawn and given in Figure 2. In this biplot, 65 percent of variation explained belongs to GE interaction and genotypes with highest distance to the origin of coordinate have attached together with dash line and made biplot polygon. Among these genotypes, entries such as V6, V11, and V4, which are vertices of the polygon, produced the highest seed yield in this study. Based on the biplot, five rays divided the biplot to the five sections and environments fell in three of them and vertex genotype for this section is V6, indicating high yielding genotype for theses environments is V6. Similarly, environments U2 and H1 fell into section 2 and there is no vertex genotype. Another section including K1, K2, K3, H3, and U3 and vertex genotype for this section is V4.

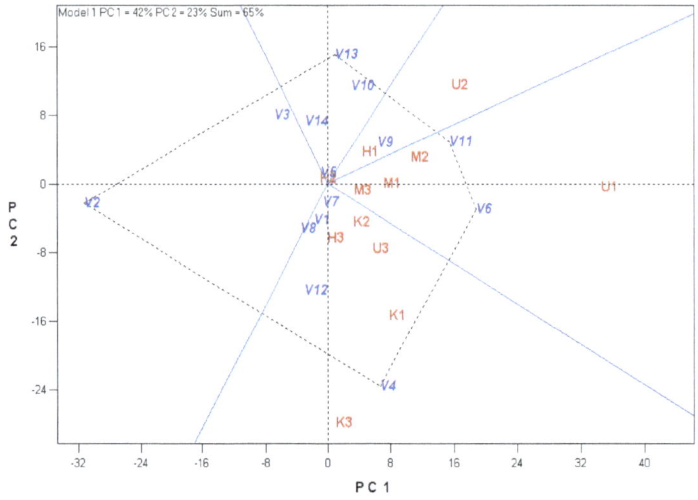

Figure 2. Biplot of first interaction principal components (IPC1) and second interaction principal components (IPC2) scores of 14 chickpea genotypes in 12 environments

Genotypes, such as V1, V7, V8, and V5 which were close to the origin of coordinate produce average yield at all of experimental environments. Figure 2 implies that there are two main environments for cold highlands of West Iran. First one corresponding with very cold regions, Maragheh and similar environments, where identified with genotype V6, and second one is corresponding with Kurdistan and same environments, where has relatively milder temperature and determined with V4. On the other hand, both AMMI1 and 3 explain more than 53% of variation, which cannot be ignored. Here, we presented the second biplot, which confirms somewhat the former biplot and show the consistent situation of V6 and V2 (Figure 3). This biplot suggests that genotype V6 with high IPC1and low IPC3 can be considered as more stable genotypes for most of environments.

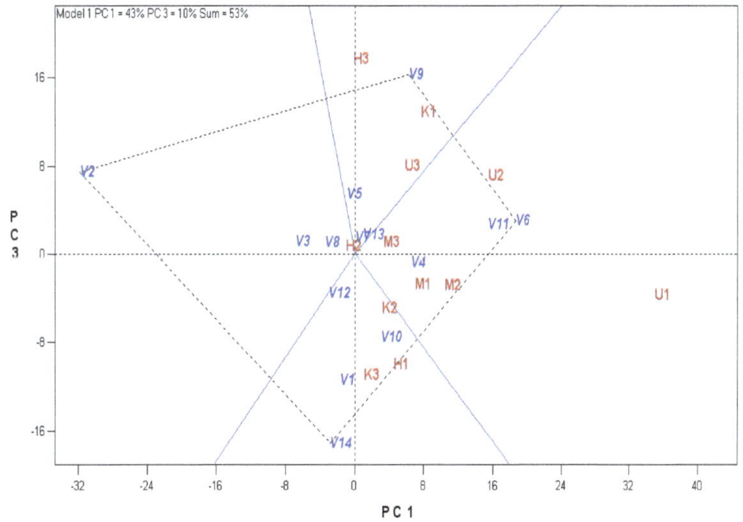

Figure 3. Biplot of first interaction principal components (IPC1) and third interaction principal components (IPC3) scores of 14 chickpea genotypes in 12 environments

The AMMI stability values (ASV) were calculated according to Purchase et al. (2000). The results showed that the first two PCs explained 66.7% of variance of data set (Table 7). According to the ASV, V5, V7, and V1 are most stable genotypes, respectively. This parameter is comparable with other stability statistics (Purchase et al., 2000).

Table 7. First two principal components and AMMI stability values for 14 chickpea genotypes

Gen	IPC1	IPC2	ASV
V1	2.47	5.75	6.63
V2	25.19	-8.17	34.72
V3	1.814	-11.03	11.29
V4	9.48	17.44	21.56
V5	-0.31	-0.82	0.92
V6	-13.66	11.79	21.75
V7	1.674	3.88	4.48
V8	4.288	6.06	8.35
V9	-3.687	-11.43	12.45
V10	-8.37	-8.47	14.05
V11	-16.37	6.39	22.84
V12	10.44	5.78	15.14
V13	-9.74	-9.39	16.08
V14	-2.28	-6.30	7.01
% of variance	43.2	23.5	-

3.3 Comparison of Stability Parameters

Table 8 indicates the ranking orders for 14 chickpea genotypes, according to the different stability parameters. Based on Shukla's stability variance (σ^2_i), Wricke's ecovalence (W_i), Mean square of GEI (MS(GE)), Kang's Rank sum method (RS), and AMMI stability value (ASV) the most stable genotype was V5. According to mean seed yield of genotypes and Lin and Binns priority index (P_i), the superior genotype was V4.

Spearman's coefficient of rank correlation was employed for each of pair wise comparisons of the ranks of stability parameters (Table 9). The mean of seed yield was correlated to the stability parameters Pi ($r = 0.96$**), D^2_i ($r = -0.57$*), and Tai's α_i ($r = -0.55$*). Ecovalence (W^2_i), stability variance (σi^2), D^2_i, λi and AMMI stability value (ASV) were highly correlated ($P \leq 0.01$). These results showed that one of these parameters could be used as a substitute for the others in GE interaction study of chickpea. Plaisted stability parameter ($\theta_{(i)}$) had high negative correlation with stability variance (σ^2_i), D^2_i, λ_i and Purchase's AMMI stability value (ASV). Regression coefficient (b_i) had positive correlation with ASV at 5% probability level. Correlation between coefficient of variation (CV), D^2_i and α_i was significant at 5 and 1 percent probability levels, respectively. P_i had positive and significant correlation with the procedures of MS(GE) and RS. Desirability index (D^2i) correlation with Tai's stability parameter (λ_i) and ASV was positive and significant ($P \leq 0.01$). Meanwhile, rank sum (RS) was correlated with ASV at 5% probability level. Purchase et al. (2000) found same findings in a wheat study.

Table 8. Ranking orders of mean yield and various stability parameters of 14 chickpea genotypes evaluated across 12 environments in the chickpea growing areas west of Iran

Gen. code	Mean	W_i	$\theta_{(i)}$	b_i	CV	P_i	MS(GE)	σ_i^2	D_i^2	αi	λi	RS	ASV
V1	9	5	10	6	5	8	2	5	3	4	7	8	3
V2	14	14	1	14	8	14	14	14	8	1	14	14	14
V3	13	3	12	9	2	12	12	3	1	2	1	10.5	6
V4	1	13	2	11	9	1	5	13	12	12	13	8	11
V5	8	1	14	8	6	7	1	1	2	10	3	1	1
V6	2	11	4	13	13	2	4	11	14	14	5	5	12
V7	6	4	11	4	4	6	3	4	4	6	6	2	2
V8	10	2	13	10	12	9	7	2	7	11	4	3	5
V9	3	10	5	5	1	3	10	10	10	5	11.5	5	7
V10	7	7	8	1	7	10	9	7	9	8	10	8	8
V11	4	12	3	12	14	5	8	12	13	13	9	10.5	13
V12	5	8	7	7	3	4	6	8	6	3	11	5	9
V13	12	6	9	3	10	11	11	6	5	7	8	12	10
V14	11	9	6	2	11	13	13	9	11	9	11.5	13	4

Table 9. Rank correlation coefficients among mean yield and stability parameters of 14 chickpea genotypes in 12 environments

Parameter	Mean	W_i	$\theta_{(i)}$	b_i	CV	P_i	MS(GE)	σ_i^2	D_i^2	αi	λi	RS
W_i	-0.36											
$\theta_{(i)}$	0.36	-1.00										
b_i	-0.14	0.39	-0.39									
CV	-0.06	0.31	-0.31	0.34								
P_i	0.96**	-0.25	0.25	-0.25	0.02							
MS(GE)	0.53*	0.35	-0.35	-0.08	0.06	0.63*						
σ_i^2	-0.36	1.00**	-1.00**	0.39	0.31	-0.25	0.35					
D_i^2	-0.57*	0.79**	-0.79**	0.27	0.61*	-0.40	0.18	0.79**				
αi	-0.55*	0.12	-0.12	0.20	0.75**	-0.47	-0.36	0.12	0.61			
λi	-0.20	0.79**	-0.79**	-0.06	0.06	-0.07	0.38	0.79**	0.55*	-0.13		
RS	0.49	0.51	-0.51	0.03	0.26	0.58*	0.78**	0.51	0.21	-0.26	0.45	
ASV	-0.19	0.82**	-0.82**	0.54*	0.42	-0.14	0.41	0.82**	0.63*	0.13	0.50	0.52*

Note. * and ** = Significant at 5 and 1 percent probability levels, respectively.

4. Conclusions

An ideal variety should be adaptable to a wide range of growing conditions, with above average yield and below average variance across different environments. In the present study, AMMI stability value (ASV) had correlations with a number of other procedures, which imply that AMMI analysis was more efficient than univariate stability parameters and ASV procedure can be useful to rank genotypes. AMMI model, according to two first principal components, has graphical ability to determine chickpea genotypes with high seed yield, stable performance, and broad adaptation. The genotype V5 (FLIP 97-230C) presents high seed yield stability; however, its mean performance was not desirable. Overall, genotypes V4 (FLIP 00-39C) and V6 (FLIP 99-26C) can be recommended to western cold zone farmers of Iran to autumn planting of chickpea. Further studies are warranted to understand why in same environment a genotype is more stable than the others which could be helpful for breeding high yielding stable chickpea cultivars in the region.

References

Acikgoz, E., Ustun, A., Gul, I., Anlarsal, E., Tekeli, A. S., Nizam, I., & Yavuz, M. (2009). Genotype × environment interaction and stability analysis for dry matter and seed yield in field pea (*Pisum sativum* L.). *Spanish Journal of Agricultural Research, 7*(1), 96-106.

Akter, A., Jamil, H., M., Umma, K. M., Islam, M. R., Hossain, K., & Mamunur, R. M. (2014). AMMI Biplot Analysis for Stability of Grain Yield in Hybrid Rice (*Oryza sativa* L.). *Journal of Rice Research, 2*, 126. http://dx.doi.org/10.4172/jrr.1000126

Atta, B. M., & Shah, T. M. (2009). Stability analysis of elite chickpea genotypes tested under diverse environments. *Australian Journal of Crop Science, 3*(5), 249-256.

Becker, H. C., & Léon, J. (1988). Stability Analysis in Plant Breeding. *Plant Breeding, 101*, 1-23. http://dx.doi.org/10.1111/j.1439-0523.1988.tb00261.x

Bakhsh, A., Akhtar, L. H., Malik, S. R., Masood, A., Iqbal, Sh. M., & Qurashi, R. (2011). Grain yield stability in chickpea (*Cicer arietinum* L.) across environments. *Pakistan Journal of Botany, 43*(5), 2947-2951.

Crossa, J. (1990). Statistical analyses of multi-location trials. *Advances in Agronomy, 44*, 55-85. http://dx.doi.org/10.1016/S0065-2113(08)60818-4

Danyali, S. F., Razavi, F., Ebadi-Segherloo, A., Dehghani, H., & Sabaghpour, S. H. (2012). Yield Stability in Chickpea (*Cicer arietinum* L.) and Study Relationship among the univariate and multivariate stability Parameters. *Research in Plant Biology, 2*(3), 46-61.

Eberhart, S. A., & Russel, W. A. (1966). Stability Parameters for Comparing Varieties. *Crop Science, 6*, 36-40.

http://dx.doi.org/10.2135/cropsci1966.0011183X000600010011x

FAOSTAT. (2012). *FAO Statistical Yearbook*. Food and Agricultural Organization of the United Nations, Rome, Italy. Retrieved from http://faostat.fao.org/site/291/default.aspx

Finlay, K. W., & Wilkinson, G. M. (1963). The analysis of adaptation in plant breeding programs. *Australian Journal of Agricultural Research, 14*, 742-754.

Francis, T. R., & Kannenberg, L. W. (1978). Yield stability studies in short-season maize I .A descriptive method for grouping genotypes. *Canadian Journal of Plant Science, 58*, 1029-1034.

Gauch, H. G., & Zobel, R. W. (1988). Predictive and postdictive success of statistical analyses of yield trials. *Theoretical and Applied Genetics, 76*, 1-10. http://dx.doi.org/10.1007/ BF00288824

Halluer, R., & Miranda, J. B. (1981). *Quantitative genetics in Maize breeding*. Iowa State University Press, Ames.

Hanson, W. D. (1970). Genotypic stability. *Theoretical and Applied Genetics, 40*, 226-231. http://dx.doi.org/10.1007/BF00285245

Hussein, M. A., Bjornstad, Å., Aastveit, A. H., & Berg, T. (2000). SASG × ESTAB - A SAS program for computing genotype × environment stability statistics. *Agronomy Journal, 92*, 454-459. http://dx.doi.org/10.2134/agronj2000.923454x

Kang, M. S. (1988). A rank-sum method for selecting high-yielding, stable corn genotypes. *Cereal Research Communication, 16*, 113-115.

Kanouni, H., Taleei, A. R., & Khalily, M. (2006). Stability analysis of seed yield and one hundred seeds weight in desi type chickpea genotypes under dryland conditions. *Seed and Plant, 23*, 297-310.

Lin, C. S., & Binns, M. R. (1988). A superiority measure of cultivar performance for cultivar × location data. *Canadian Journal of Plant Science, 68*, 193-198. http://dx.doi.org/10.4141/cjps88 -018

Lin, C. S., Binns, M. R., & Lefkovitch, P. (1986). Stability analysis: Where do we stand? *Crop Science, 26*, 894-899. http://dx.doi.org/10.2135/cropsci1986.0011183X002600050012x

Malhotra, R. S., & Singh, K. B. (1991). Classification of chickpea growing environments to control genotype by environment interaction. *Euphytica, 58*, 5-12. http://dx.doi.org/10.1007/BF00035334

Pinthus, J. M. (1973). Estimate of genotype value: a proposed method. *Euphytica, 22*, 121-123. http://dx.doi.org/10.1007/BF00021563

Plaisted, R. L. (1960). A shorter method of evaluating the ability of selection to yield consistently over seasons. *American Potato Journal, 37*, 166-172.

Plaisted, R. L., & Peterson, L. C. A. (1959). Technique for evaluating the ability of selections and yield consistency in different locations or seasons. *American Potato Journal, 36*, 381-385.

Purchase, J. L., Hattingh, H., & Van Deventer, C. S. (2000). Genotype × environment interaction of winter wheat (*Triticum aestivum* L.) in South Africa: II. Stability analysis of yield performance. *South African Journal of Plant and Soil, 17*, 101-107. http://dx.doi.org/10.1080/02571862.2000.10634878

Raffi, S. A., Newaz, M. A., & Khan, N. (2004). Stability analysis for pod and seed production in dry bean (*Phaseolus vulgaris* L.). *Asian Journal of Plant Science, 3*, 239-242.

SAS institute. (1996). *SAS/STAT User's Guide* (2nd ed.). SAS institute Inc., Cary, NC.

Shukla, G. K. (1972). Some aspects of partitioning genotype × environmental components of variability. *Heredity, 28*, 237-245. http://dx.doi.org/10.1038/hdy.1972.87

Singh, K. B., & Saxena, M. C. (1999). Chickpeas. In R. Coste (Ed.), *The Tropical Agriculturist*. Macmillan, London.

Tai, G. C. C. (1971). Genotypic stability analysis and application to potato regional trials. *Crop Science, 11*, 184-190. http://dx.doi.org/10.2135/cropsci1971. 0011183X001100020006x

Thillainathan, M., & Fernandez, G. C. J. (2001). SAS applications for Tai's stability analysis and AMMI model in genotype × environmental interaction (GEI) effects. *Journal of Heredity, 92*, 367-371. http://dx.doi.org/10.1093/jhered/92.4.367

Wricke, G. (1962). Uber eine Methode zurr Efassung der okologischen Streubretite in Feldversuchen. *Z Pflazenzuecht, 47*, 92-96.

Yadav, S. S., Redden, R. J., Chen, W., & Sharma, B. (2007). *Chickpea Breeding and Management* (p. 638). Wallingford, Oxon, UK: CAB International.

Yan, W., & Kang, M. S. (2003). *GGE biplot analysis: A graphical tool for breeders, geneticists and agronomist* (p. 271). Boca Raton, FL: CRC Press.

Zobel, R. W., Wright, M. J., & Gauch, H. G. (1988). Statistical analysis of a yield trial. *Agronomy Journal, 80*, 388-393. http://dx.doi.org/10.2134/agronj2006.0282

6

Effect of Organic Manure in the Induction of Resistance of Citrus to *Phytophthora parasitica*

Luciana Faldoni[1,2], Mariângela Cristofani-Yaly[1], Leonardo Pires Boava[1], Evandro Henrique Schinor[1] & Kátia Cristina Kupper[1,2]

[1] Centro APTA Citros Sylvio Moreira, Cordeirópolis-SP., Brazil

[2] Universidade Federal de São Carlos, Araras-SP, Brazil

Correspondence: Kátia Cristina Kupper, Centro APTA Citros Sylvio Moreira, CEP 13490-970, Cordeirópolis-SP., Brazil. E-mail: katia@centrodecitricultura.br

Abstract

Gummosis and root rot caused by *Phytophthora* are among the most economically important diseases in *Citrus*. Chemical control has been used; however, it is essential to search for new alternatives for its control. This study aimed to: (i) produce an organic compound from the anaerobic fermentation of cattle manure, (ii) determine the microbial diversity present in the compound, (iii) determine the effect of compound on mycelial growth of *Phytophthora parasitica* (*P. nicotianae*), (iv) evaluate its effect in inducing resistance to the pathogen in five *Citrus* rootstocks and, (v) evaluate the expression profile of some defense-related genes in resistant plants of *Poncirus trifoliata* and susceptible plants of Caipira sweet orange. The largest microbial population occurred between 25 and 35 days after production of organic manure with predominance of *Bacillus* spp. The organic manure at concentration of 20% inhibited about 80% of the mycelial growth of *P. parasitica*. Plants of Caipira sweet orange (susceptible to *P. parasitica*) when treated with organic manure at all concentrations (10, 20 and 50%), had lower disease severity when compared with the respective control. The analysis of gene expression suggested that the combination of treatment followed by inoculation with the pathogen in plants of Caipira sweet orange, the susceptible genotype, was able to increase the expression level of the genes encoding chitinase (*CHI*), lipoxygenase (*LOX*), peroxidase (*POX*), β-1,3-glucanase and chalcone synthase (*CHS*). It is believed that they should be involved with the resistance of *Citrus* plants to *P. parasitica*.

Keywords: gummosis, root rot, gene expression, rootstock

1. Introduction

Among the problems faced by the Brazilian citriculture, gummosis and root rot caused mainly by *Phytophthora parasitica* Dastur (= *P. nicotianae* Breda de Haan var. *parasitica* (Dastur) Waterhouse) are among the most economically important diseases in *Citrus*, occurring in nearly all producing regions (Leoni & Ghini, 2006).

The oomycete infects the main scaffold branches, leading to canker development and gum exudation, lesion spreads upward, affecting secondary branches, and downward, affecting the trunk and roots (Alvarez et al., 2009). Chemicals, such as metalaxyl (a systemic benzenoid fungicide) and fosetyl Al (aluminium ethylhydrogenphosphonate) have been successfully used to control both gummosis and root rot. However, they are not always desirable due to the high costs of application, potential hazards to the environment, and the development of fungicide-resistant strains. Selection and breeding for resistance to *Phytophthora* in *Citrus* species is considered the most efficient approach to control the disease, since there are varying degrees of resistance within the genus *Citrus* and its relatives (Boava et al., 2011).

An alternative approach to control plant diseases is the application of aqueous extracts of organic matter and biofertilizers. The advantages of using organic matter and biofertilizers are the low cost and easy availability of the product, since the farmer can produce it using the available organic material. It can provide solutions to both objectives of safe waste management and sustainable agriculture by converting all the societal waste organics into nutritive organic manure which can be used by farmers to grow more food for society (Sinha et al., 2010).

Induced resistance has been demonstrated in many plant species, and numerous studies have evaluated the expression of defence responses against several important causal agents of plant diseases (Jiang et al., 2008).

Induced resistance is triggered by the activation of genes that encode a series of pathogenesis-related (PR) proteins which include protein families of varied characteristics, such as chitinases (*CHI*), β-1,3-glucanases (*GLU*), lysozymes, peroxidases (*POX*), besides enzymes involved in lignin and phytoalexin synthesis. A key gene in the route of the biosynthesis of, phytoalexins is Chalcone synthase (*CHS*). All the above mentioned genes are related to defense processes during pathogenesis (Van Loon et al., 2006).

Given this scenario, this study aimed to: (i) produce an organic compound from the anaerobic fermentation of cattle manure, (ii) determine the microbial diversity present in the organic manure, (iii) determine the effect of the prepared manure on mycelial growth of *P. parasitica*, (iv) evaluate its effect in inducing resistance to *P. parasitica* in five *Citrus* rootstocks and (v) evaluate the expression profile of some defense-related genes in resistant plants of *Poncirus trifoliata* and susceptible plants of Caipira sweet orange.

2. Materials and Methods

2.1 Preparation of Organic Manure

The organic manure was produced according to Kupper et al. (2006), using cattle manure supplemented with macro and micronutrients to stimulate the fermentation (6.5 g/L of N; 0.68 g/L of P_2O_5; 2.5 g/L of K_2O; 0.975 g/L of Ca; 0.66 g/L of Mg; 4.75 g/L of S; 1.8 g/L of Fe; 0.49 g/L of Mn; 0.29 g/L of Cu; 2.9 g/L of Zn; 0.35 g/L of B; 1.20 g/L of Na), 6 L of cattle manure, 2 L of sugarcane bagasse and water in the ratio of 50% (v/v) in a closed plastic recipient completed to a 20 L volume with water, under anaerobic conditions for a period determined by microbiological analysis.

2.2 Determination of Culturable Microbial Community in the Organic Manure

The microbial community composition was determined in filtered samples of the organic manure, diluted up to 10^{-4} taken every five days for 60 days. After homogenizing, 100 μL of each dilution were spread over the surface of an agar medium in Petri dishes, with a Drigalski loop and the incubation of cultures took place on B.O.D incubator at temperature of 25 ± 2 °C in the dark. It was used different culture media to verify the microbial community: a) Martin´s medium, to count the total population of fungal colonies; b) Nutrient Agar (NA) for determining the number of bacterial colony-forming units and; c) Potato Dextrose Agar (PDA) for the determination of the colonies of *Bacillus* spp., an aliquot of the organic manure sample was heat treated at 80 °C/10min and the plated on the medium (PDA) (Kupper et al., 2006). Three plates were used for each dilution. The evaluation was performed daily and we proceeded counting and isolation of morphologically distinct microorganisms in each plate. The values were converted into colony-forming units per milliliter (c.f.u./mL). To compare the means it was used the Tukey test at 5% probability.

2.3 Effect of Organic Manure on the Mycelial Growth of P. parasitica

To determine the effect of organic manure on growth of *P. parasitica*, the product obtained was filtered using gauze and added to the culture medium carrot agar (CA), at the following concentrations: 0, 1.0, 2.5, 5.0, 10.0 and 20.0%. After the preparation of culture medium, 16 g/L of agar were added to the medium and then autoclaved for 20 minutes at 100 kPa. Subsequently, the medium were poured into Petri dishes. Discs of 5 mm diameter removed from a 7 days-old culture of *P. parasitica* (isolate IAC/95) were transferred to the center of the plates. Incubation of the cultures was carried out in a B.O.D incubator at 25 °C ± 2 °C in the dark. Determination of growth was realized when the colony of the pathogen in the control treatment reached the edge of the plate. The experimental design was a completely randomized set-up, with three replicates. The means were compared by Tukey's test ($P \leq 0.05$).

2.4 Effect of Organic Manure in the Control of P. parasitica in Different Citrus Rootstocks

To evaluate the effect of organic manure in controlling the disease, *Poncirus trifoliata* cv Rubidoux (highly tolerant or resistant), Sunki mandarin (*Citrus sunki* Hort. Ex Tan) (susceptible), Rangpur lime (*C. limonia* Osbeck) (moderately resistant), Sour orange (*C. aurantium*) (highly tolerant or resistant) and Caipira sweet orange [*C. sinensis* (L.) Osbeck] (susceptible) plants were sprayed with the aqueous compound at concentrations equivalent to the treatments applied to the soil, for instance plants that have the organic manure incorporated into the soil at a concentration of 10% have leaves sprayed with organic manure in 10% of concentration, and so on. Inoculation with *P. parasitica* was carried out seven days after spraying (DAS) with the organic manure. This period was based on studies with eucalyptus (*Eucalyptus* spp.) in the control of rust (*Puccinia psidii*), according to Teixeira et al. (2005). Stems of *Citrus* plants were inoculated with a mycelial disk (3 mm) taken from a 7 days-old culture of *P. parasitica* (isolate IAC/95). Mycelial blocks were placed onto the center of a small cut made in plant stem. The inoculation site was protected with moistened cotton and parafilm. The plants were placed in an environment with a photoperiod of 16 hours at 25 °C and relative humidity (RH) of 85%. This experiment consisted of four treatments: 1) plants that were only treated with organic manure (OM), 2) plants inoculated with *P. parasitica* (IP), 3) plants treated with organic manure and inoculated with *P. parasitica* (OM + IP) and 4) control plants, untreated with organic manure and non-inoculated. The evaluation was made by measuring the length of lesions in the inoculated stems, 60 days after inoculation of the pathogen. The experimental design was randomized blocks with four replications, with one plant per plot. The means were compared by Tukey's test ($P \leq 0.05$).

2.5 Analysis of Gene Expression

In an attempt to elucidate the potential of the organic manure in the induction of resistance, we performed a study of gene expression using a reverse transcription quantitative real-time PCR (qPCR), in which levels of expression of genes encoding peroxidase (*POX*), lipoxygenase (*LOX*), β-1,3-glucanase, chitinase (*CHI*) and chalcone synthase (*CHS*), were evaluated in *Poncirus trifoliata* and Caipira sweet orange genotypes. As the plants were small, samples of leaves were collected from all parts of the plants of the treatment with 0 and 20% of organic manure concentration of the experiment carried out to evaluate the effect of biofertilizer in the control of *P. parasitica* in different *Citrus* rootstocks. The leaves were harvested 48 hours after inoculation with *P. parasitica,* flash-frozen in liquid nitrogen, and then stored at -80 °C prior to RNA isolation. The choice of the genes and of the time to collect the leaves were based on previous studies focusing on the expression of local and systemic defense responses activated during the *Citrus-Phytophthora* interaction, detailed by Boava et al. (2011). The RNA extraction was performed using Trizol (Invitrogen Life Technologies), following the manufacturer's recommendations. After extraction, we proceeded to quantification in NanoDrop spectrophotometer ND - 1000 (Thermo Fischer Scientific, Massachusetts, USA). Then the treatment with RNase-free DNase I (Qiagen, Maryland, USA) was carried out at a concentration of 1 U per 1,000 ng of RNA, to eliminate any contamination by DNA. The reaction was incubated for 30 min at 37 °C and stopped by heat shock at 65 °C for 5 min. The reverse transcription reactions (synthesis of cDNA first strand) were performed according to manufacturer's instructions of SuperScript First Strand Synthesis System (Invitrogen) with oligo (dT) primer (DT12-18, Invitrogen). Specific primers were designed based on their predicted function retrieved from *Citrus* Expressed Sequence Tags - CitEST database (http://biotecnologia.centrodecitricultura.br) (Table 1) using the Primer Express 2.0 software (Applied Biosystems) and synthesized by Integrated DNA Technologies (Prodimol Biotech A / S). The tests were conducted in the ABI Prism 7000 Sequence Detection System and 7500 Fast Real Time PCR System (Applied Biosystems, USA) by the system SYBR-green of detection (Wittwer, 1997).

Table 1. Oligonucleotide primers used for qPCR analysis

	Description	CitEST	Forward/Reverse	Amp
PR2	β-1,3-glucanases	CA-4000443	GCCGGCCTTGGAAACC/ GCCGGCCTTGGAAACC	97
PR3	Chitinase class I	CAS-CR-204551	ACAGAATTGTGGAAGCGG/ AGCAAGTCTTCAAACATCTCC	107
CHS	Chalcone synthase	CAS-CS-207154	AGTCTGAGGAAGCGAAA/ CAAGCTTTGATGGGGACACT	102
LOX	Lipoxygenase	CAS-CS-111108	AATTCGACTTGGGTCTGTTC CCGTAAGCGGATTGATAGG	94
POX	Peroxidase	CAS-CR-201851	GATCTTCGTGCTCGTGTTCA TGCGAATGTTTTGCTGTCTC	96
UBQ	Ubiquitin	CAS-PT-300961	TTCGTCAGTTGACTAATCCT/ GTTGCTGTGTTGACTGTG	95
TUB	Tubulin	CAS-LT-602290	TTTGTAAGATCCCTCCGA/ TCACCCTCCTGAACATTT	87

The expression levels were evaluated based on the number of PCR cycles required to reach a fixed threshold (threshold cycle - Ct) in the exponential phase of PCR. Ct values were analyzed using version 4.3.6 of the software Genex (http://www.multid.se/). For relative quantification, the method $2^{-\Delta\Delta CT}$ for the conditions for qPCR was applied (Livak & Schmittgen, 2001). For plants treated with organic manure, plants inoculated with *P.*

parasitica and plants treated with organic manure and inoculated, the levels of transcription of the genes were compared with the corresponding control (untreated plants with organic manure and non-inoculated), including Ubiquitin and β-tubulin gene as the endogenous control for reference. The standard deviations were used to show significant differences ($P < 0.05$).

3. Results and Discussion

3.1 Determination of Microbial Community in the Organic Manure

The results showed that the highest microbial populations was present in organic manure between 25 and 35 days after production, in which it was observed predominance of bacteria and *Bacillus* and a low total amount of fungi (Figure 1). These data are consistent with Kupper et al. (2006) which found that during fermentation, the fungus population is lower, whereas the total number of bacteria and *Bacillus* was significantly higher. However, it is believed that the total population of microorganisms in organic manure is higher than that indicated from plated samples.

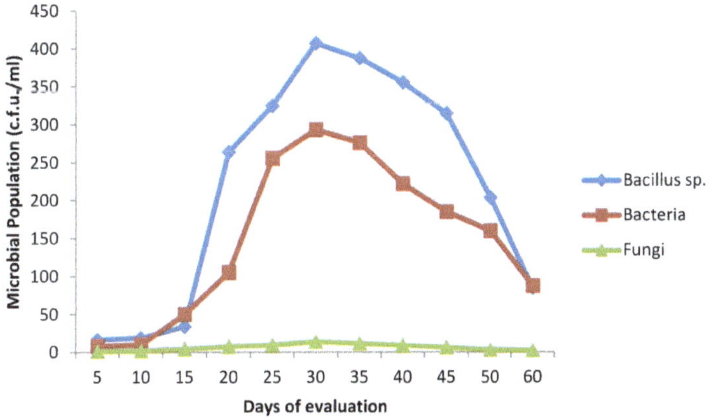

Figure 1. Microbial community present in the organic manure determined every five days for 60 days. Different culture media were used to count the total population of yeast, the number of bacterial colonies formed and the number of colonies of *Bacillus* spp. The values were converted to colony forming units per milliliter (c.f.u./mL)

3.2 Effect of Organic Manure on the Mycelial Growth of P. parasitica

The evaluations of the effect of organic manure showed that all concentrations affected significantly the mycelial growth of *P. parasitica*. But at concentrations of 20%, the organic manure inhibited about 80% of the mycelial growth of the pathogen (Figure 2). It was observed that the colony diameter was inversely proportional to the organic manure concentration. For Kupper et al. (2006), concentrations higher than 10% of biofertilizer, produced from anaerobic fermentation of cattle manure, caused a reduction in colony size of *Phyllosticta citricarpa* (teleomorph: *Guignardia citricarpa*) and the inhibition of the mycelial growth of pathogens was due the metabolites of the biofertilizer. Thus, it is believed that inhibition of colony growth of *P. parasitica* by organic manure was due to toxic metabolites (thermally stable) produced by microorganisms.

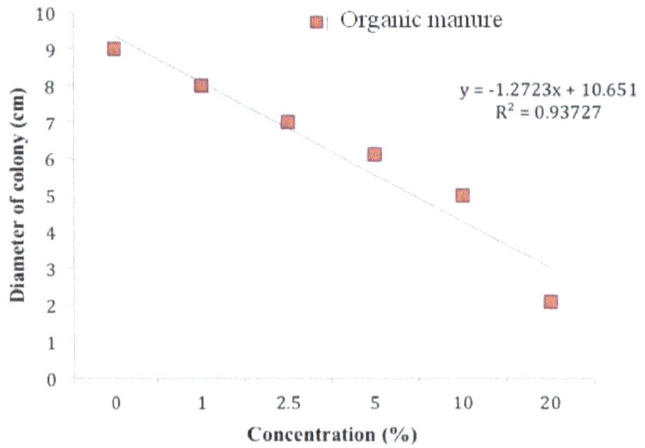

Figure 2. Effect of organic manure on *Phytophthora parasitica* colony diameter (cm), the product obtained was filtered using gauze and added to the culture medium carrot agar (CA), at the following concentrations: 0, 1.0, 2.5, 5.0, 10.0 and 20.0%

3.3 Effect of Organic Manure in the Control of P. parasitica in Different Citrus Rootstocks

When the plants were inoculated with *P. parasitica* but not treated with organic manure the results showed that the susceptible genotypes Caipira sweet orange, Sunki mandarin and Rangpur lime had the highest average length lesions 2.43 cm, 2.25 cm and 1.96 cm respectively (Table 2). These results confirm that *Poncirus trifoliata* is highly resistant, the Sour orange as resistant, the Rangpur lime as moderately resistant and Caipira sweet orange and Sunki mandarin as susceptible (Alvarez et al., 2009). According Siviero et al. (2006), the lesions length is more suitable to characterize the rootstocks and new genotypes resistance to *Phytophthora* spp. when compared with measurements of the width of injury.

Table 2. Mean longitudinal length of the lesion (cm) caused by *Phytophthora parasitica*, in five different genotypes of citrus rootstocks, *Poncirus trifoliata* cv Rubidoux Sunki mandarin (*Citrus sunki* Hort. Ex Tan) Rangpur lime (*Citrus limonia* Osbeck) Sour orange (*Citrus aurantium*) and Caipira sweet orange [*C. sinensis* (L.) Osbeck]

Varieties	Concentrations of organic manure (%)			
	0	10	20	50
Trifoliata	0.91* a A	0,83 a A	1.03 a A	1.03 a A
Sour orange	1.36 a A	1.87 b A	1.42 a A	1.67 b A
Rangpur lime	1.96 b A	1.56 b A	1.67 a A	1.95 b A
Caipira sweet orange	2,43 b B	1.70 b A	1.56 a A	1.65 b A
Sunki mandarin	2.25 b B	1.90 b A	1.98 a A	1.81 b A

Note. * Means followed by the same letter, lowercase in columns and capital in rows, are not significantly different to according to the Tukey's test ($P \leq 0.05$).

When the plants were treated with organic manure and inoculated with *P. parasitica* (Table 2), the susceptible genotypes Caipira sweet orange and Sunki mandarin expressed a decrease in lesion size at all concentrations of the organic manure tested, when compared to inoculated plants that were not treated with organic manure. This suggests that organic manure increases the resistance of susceptible rootstocks. However, at the concentration of 20%, the lesion size among the genotypes did not differ significantly. McDonald et al. (2001) have also demonstrated the effect of biofertilizer in the increase of resistance of plants to pathogens.

3.4 Analysis of Expression Gene

Based on results of the effect of organic manure in the control of *P. parasitica* in different *Citrus* rootstocks we selected the genotypes *Poncirus trifoliata* (resistant) and Caipira sweet orange (susceptible) to verify if the resistance was associated with the induction of some selected genes. According to Hammerschmidt and Dann

(1997), induced resistance can be activated as plant natural defences by certain types of infection or in response to treatment with biotic or abiotic agents.

The gene expression for pathogenesis-related genes, such as PR1 (function unknown), PR2 (b-1,3-endoglucanase), PR3 (chitinase class I), PR5 (Thaumatin-like protein), chalcone synthase (*CHS*), PAL, lipoxygenases (*LOX*) and peroxidases (*POX*), was significantly higher in the resistant as compared to the susceptible interaction, as it was demonstrated by Boava et al. (2011) when they compared the expression of defense-related genes in *P. trifoliata* and *C. sunki*, resistant and susceptible, respectively, infected with *P. parasitica*. It is noteworthy, in this work that, all genes expression evaluated were significantly upregulated ($P \leq 0.05$) in relation to the respective controls (non-inoculated and not treated with organic manure) in the susceptible genotype, Caipira sweet orange, after organic manure treatment followed by inoculation with *P. parasitica* suggesting that this combination was more effective to increase the resistance of Caipira sweet orange (Figure 3). According to Boava et al. (2009), metabolism alterations are more intense in an elicitor-induced plant challenged with a pathogen than in an only challenged or only induced plant suggesting that the former is more capable of responding to the presence of the pathogen.

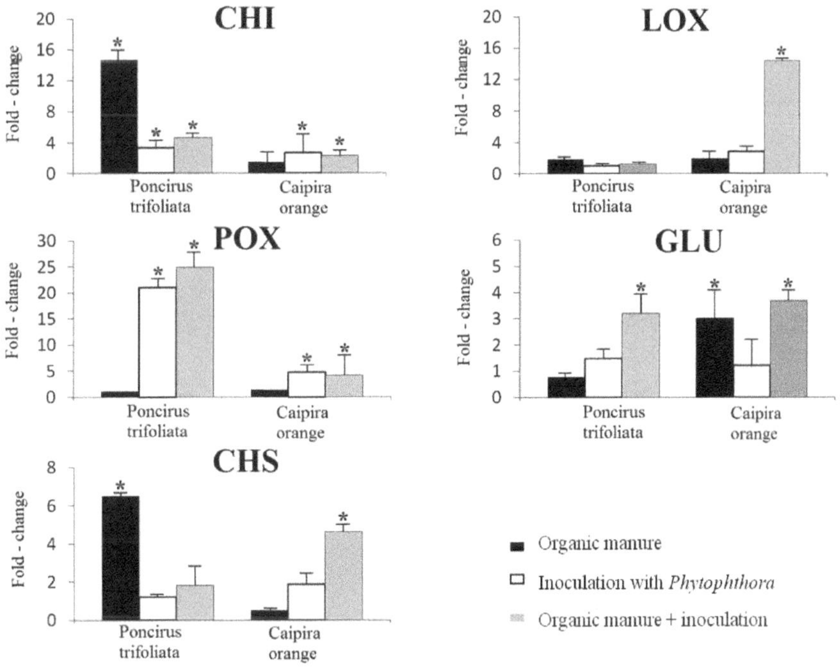

Figure 3. Expression pattern of defense and stress related genes in resistant *P. trifoliata* and susceptible Caipira sweet orange plants in response to effect of organic manure (at 20%) and *Phytophthora parasitica* infection. Transcripts levels were represented as fold change in relation to non-inoculated control, including *β-tubulin* and *Ubiquitin* as reference genes. Vertical bars represent standard deviations of the means of three replicates, *indicates that transcripts ratio was significantly (p-value ≤ 0.05) greater than 2.0-fold up or less than 0.5-fold down in relation to non-inoculated controls. *CHI* = chitinase; *LOX* = lipoxygenase; *POX* = Peroxidase, *GLU* = β-1,3-glucanase and *CHS* = chalcone synthase

When the plants were treated only with the organic manure with no inoculation, *CHI* and *CHS* genes were highly expressed in *P. trifoliata* in relation to respective control, and β-1,3-glucanase gene was highly expressed in Caipira sweet orange in relation to respective control. Thus, we may suggest that it is possible that the expression of these genes can be triggered mainly in response to the organic manure, especially if one considers the presence of the microbial community in the compound. Nandakumar et al. (2001) induced resistance in rice by *Pseudomonas fluorescens*, a growth promoting rhizobacteria, and found a reduction in the severity of sheath rot caused by *Rhizoctonia solani*, correlated with an increase in chitinase (*CHI*), which has antibacterial activity in reason of its lytic action on the cell wall and increased chalcone synthase (*CHS*), which is key in the route of the biosynthesis of phenylpropanoids, such as phytoalexins and β-1,3-glucanase (*GLU*), widely distributed in higher plants and induced during the hypersensitive response of plants to pathogens or some elicitors (Bucher et al.,

2001).

The results revealed that the gene *POX* was significantly upregulated ($P \leq 0.05$) in *Poncirus trifoliata* and Caipira sweet orange in the treatments IP and OM + IP in relation to the respective controls (non-inoculated and not treated with organic manure) in both genotypes (Figure 3). Thus, we suggest that it is possible that expression of this gene can be triggered mainly in response to infection, and can be associated with enhance the resistance of the resistant genotype to infection by *P. parasitica*, since it is well documented that the susceptibility or resistance is not only determined by the presence or absence of resistance genes, but by the level of gene expression. Thus, it is believed that the smallest expression of *POX* gene in Caipira orange may not have been sufficient to contribute to the inhibition of colonization by *P. parasitica*, as noted by differences in lesion size between the genotypes *P. trifoliata* and Caipira sweet orange. This is in agreement with what was observed by Boava et al. (2011), where the expression of *POX* gene was significantly upregulated in resistant and susceptible plants at 48 h after *P. parasitica* inoculation. But in resistant plants the increase in gene expression was higher than in susceptible plants. According to Kawaoka et al. (2003) the peroxidase in plants includes several isoenzymes and isoforms. The catalytic reactions of peroxidase are diverse, and it plays an important role in plants. Aboshosha et al. (2008) found consistent results on the association of peroxidase with the resistance of plants. According to the authors, the peroxidase enzyme is one of the first to respond with increased activity as a specific reaction of resistance against plant pathogens. Previous studies suggest that peroxidase activity can lead to oxidation of different phenolic acids to antifungal compounds or direct inhibition of fungi. Thus, during the interaction between genotype and pathogen, peroxidase activity may not be produced at adequate quantities or on the correct time and site, which could justify the major severity of the disease affecting Caipira sweet orange, relative to *P. trifoliata*. This result also suggests that the induction of POX expression was most likely due to the effect of the inoculation of *P. parasitica* than the induction by the organic manure treatment.

When the plants were treated with the organic manure and followed by inoculation with *P. parasitica* (PI), in *P. trifoliata CHI*, *POX* and β-1,3-glucanase genes, were more expressed in relation to control plants and in Caipira sweet orange *CHI*, *LOX*, *POX*, *CHS* and β-1,3-glucanase genes were more expressed in relation to its control. Conrath (2009) related that the induced plants are sensitized and respond faster than the control. Aziz et al. (2003), observed an increase of chitinase and β-1,3-glucanase after the treatment of vine plants with a natural compound, thereby reducing the infection of *Botryts cinerea* and *Plasmopara viticola* in 55 and 75% respectively. It is noteworthy in this work that, the major changes in metabolism for *LOX* in the susceptible genotype, Caipira sweet orange, occurred after organic manure treatment followed by inoculation with *P. parasitica* suggesting that this combination was more effective to increase the resistance of Caipira sweet orange. Boava et al. (2011) observed that the expression of *LOX* in the resistant genotype occurred transiently in the early stages of pathogen infection, showing significant upregulation ($P \leq 0.05$) at 24 h but not after 48 h post pathogen inoculation. However, the expression of *LOX* in susceptible plants was significantly upregulated ($P \leq 0.05$) at all stages for inoculated and wound plants treatments.

4. Conclusions

In the period between 25 and 35 days of production the organic manure had the highest microbial population and the analysis of gene expression suggested that the combination of organic manure treatment followed by inoculation with *P. parasitica* activated genes encoding chitinase, lipoxygenase, peroxidase, β-1,3-glucanase and chalcone synthase and it is believed that they can be involved with the resistance of *Citrus* plants to *P. parasitica*.

References

Aboshosha, S. S., Atta Alla, S. I., El-Korany, A. E., & El-Argawy, E. (2008). Protein analysis and peroxidase isozymes as molecular markers for resistance and susceptibility of sunflower to *Macrophomina phaseolina*. *International Journal of Agriculture and Biology, 10*, 28-34. http://dx.doi.org/114/AKP/2008/10-1-28-34

Alvarez, L. A., Gramaje, D., Abad-Campos, P., & García, J. J. (2009). Seasonal susceptibility of citrus scions to *Phytophthora citrophthora* and *P. nicotianae* and the influence of environmental and host-linked factors on infection development. *European Journal of Plant Pathology, 124*, 621-635. http://dx.doi.org/10.1007/s10658-009-9449-8

Aziz, A., Poinssot, B., Daire, X., Adrian, M., Lambert, B., Joubert, J. M., & Pugin, A. (2003). Laminarin elicits defense responses in grapevine and induces protection against *Botrytis cinerea* and *Plasmopora viticola*. *Molecular Plant-Microbe Interactions, 16*, 1118-1128. http://dx.doi.org/10.1094/MPMI.2003.16.12.1118

Boava, L. P., Cristofani-Yaly, M., Stuart, R. M., & Machado, M. A. (2011). Expression of defense-related genes in response to mechanical wounding and *Phytophthora parasitica* infection in *Poncirus trifoliata* and *Citrus sunki*. *Physiological and Molecular Plant Pathology, 77*, 1-7. http://dx.doi.org/10.1016/j.pmpp.2011.07.004

Boava, L. P., Kuhn, K. J., Pascholati, S. F., Di Piero, R., & Furtado, E. L. (2009). Effect of acibenzolar-S-methyl and *Saccharomyces cerevisiae* on the activation of Eucalyptus defences against rust. *Australasian Plant Pathology, 38*, 594-602. http://dx.doi.org/10.1071/AP09045

Bucher, M., Rausch, C., & Daram, P. (2001). Molecular and biochemical mechanisms of phosphorus uptake into plants. *Journal of Plant Nutrition and Soil Science, 164*, 209-217. http://dx.doi.org/10.1002/1522-2624(200104)

Conrath, U. (2009). *Priming of Induced Plant Defense Responses* (pp. 362-395). In Advances in Botanical Research (Ed.). Elsevier Ltda.

Hammerschmidt, H., & Dann, E. K. (1992). Induced resistance to disease. In N. A. Rechcigl & J. E. Rechcigl (Eds.), *Environmentally Safe Approaches to Crop Disease Control* (Vol. 8, pp. 177-199). Boca Raton: CRC - Lewis Publishers.

Jiang, S., Park, P., & Ishii, H. (2008). Ultrastructural study on acibenzolar-S-methyl induced scab resistance epidermal pectin layers of Japanese pear leaves. *Phytopathology, 98*, 585-591. http://dx.doi.org/10.1094/PHYTO-98-5-0585

Kawaoka, A., Matsunaga, E., Endo, S., Kondo, S., Yoshida, K., Shinmyo, A., & Ebinuma, H. (2003). Ectopic Expression of a Horseradish Peroxidase Enhances Growth Rate and Increases Oxidative Stress Resistance in Hybrid Aspen. *Plant Physiology, 132*, 1177-1185. http://dx.doi.org/10.1104/pp.102.019794

Kupper, K. C., Bettiol, W., De Goes, A., Souza, P. S., & Bellotte, J. A. M. (2006). Biofertilizer for control of *Guignardia citricarpa*, the causal agent of *Citrus* black spot. *Crop Protection, 25*, 569-573. http://dx.doi.org/10.1016/j.cropro.2005.09.002

Leoni, C., & Ghini, R. (2006). Sewage sludge effect on management of *Phytophthora nicotianae* in *Citrus*. *Crop Protection, 25*, 10-22. http://dx.doi.org/10.1016/j.cropro.2005.03.004

Livak, K. J., & Schmittgen, T. D. (2001). Analysis of relative gene expression data using real-time quantitative PCR and the 2-ΔΔCT method. *Methods, 25*, 402-408. http://dx.doi.org/10.1006/meth.2001.1262

McDonald, A. E., Grant, B. R., & Plaxton, W. C. (2001). Phosphite (phosphorous acid): its relevance in the environment and agriculture and influence on plant phosphate starvation response. *Journal Plant Nutrition, 24*, 1505-1519. http://dx.doi.org/10.1081/PLN-100106017

Nandakumar, R., Babu, S., Viswanathan, R., Raguchander, T., & Samiyappan, R. (2001). Induction of systemic resistance in rice against sheath blight disease by *Pseudomonas fluorescens*. *Soil Biology & Biochemistry, 33*, 603-612. http://dx.doi.org/10.1016/S0038-0717(00)00202-9

Sinha, R. K., Valani, D., Chauhan, K., & Agarwal, S. (2010). Embarking on a second green revolution for sustainable agriculture by vermiclture biotechnology using earthworms: Reviving the dreams of Sir Charles Darwin. *Journal of Agricultural Biotechnology and Sustainable Development, 2*(7), 113-128. http://dx.doi.org/10.5897/JABSD

Siviero, A., Cristofani, M., Furtado, E. L., Garcia, A. A. F., Coelho, A. S. G., & Machado, M. A. (2006). Identification of QTLs associated with citrus resistance to *Phytophthora* gummosis. *Journal of Applied Genetics, 47*, 23-28. http://dx.doi.org/10.1007/BF03194595

Teixeira, D. A., Alfenas, A. C., Mafia, R. G., Maffia, L. A., & Ferreira, E. M. (2005). Evidence of induction of systemic resistance to eucalyptus rust by plant growth promoting rhizobacteria. *Fitopatologia Brasileira, 30*, 350-356. http://dx.doi.org/10.1590/S0100-41582005000400003

Van Loon, L. C., Rep, M., & Pieterse, C. M. J. (2006). Significance of inducible defense-related proteins in infected plants. *Annual Review of Phytopathology, 44*, 135-162. http://dx.doi.org/10.1146/annurev.phyto.44.070505.143425

Wittwer, C. T. (1997). The Light Cycler: A microvolume multisample fluorimeter with rapid temperature control. *BioTechniques Natick, 22*, 176-181.

Matching Nutrient Requirements to Hay Produced in Tennessee: A Modeling Study

Brian T. Campbell[1], John C. Waller[2], William M. Backus[3], Arnold M. Saxton[2] & Deborah K. Joines[4]

[1] Virginia Tech Southern Piedmont AREC, Blackstone, VA, USA

[2] Department of Animal Science, The University of Tennessee, Knoxville, TN, USA

[3] Research and Extension Center at Ames Plantation, The University of Tennessee, Grand Junction, TN, USA

[4] The Tennessee Soil, Plan, and Pest Center, Nashville, TN, USA

Correspondence: Brian T. Campbell, Virginia Tech Southern Piedmont AREC, Blackstone, VA, USA. E-mail: brian3@vt.edu

Abstract

Winter hay feeding is generally the most expensive cost of maintaining a cow for the year. Because of this it is beneficial for a producer to maximize the use of their hay supply. A modeling study was completed that compared the nutrient requirements of three fall calving herds and three spring calving herds to determine differences in nutrient requirements between the herds. Hay feeding should be tailored to the operation and the calving season for the most efficient use of hay.

Keywords: winter feed, hay, cattle, calving season

1. Introduction

Tennessee's beef cow herd of approximately 1,000,000 cows grazing approximately 3 million hectares of Tall Fescue (*Lolium Arundinacea*) throughout the year. When the forage does not meet the requirements of these cows due to dormancy or low quality, hay or supplemental nutrients must be supplied. Hay is the most prevalent and least expensive way to supplement cattle during the winter months. Winter hay feeding generally lasts from mid-November to mid to late-march.

Beef cattle production in Tennessee is primarily cow/calf operations (Waller et al., 2013). The traditional calving season is a spring calving season beginning in mid-January and ending in mid-March. Some producers maintain a fall calving herd, and the fall calving cows will begin calving in mid-September and calve through mid-November.

Feed is the most expensive component of raising any animal and for beef cattle winter is generally when the majority of supplemental feed is provided to the animals (Waller et al., 2013). With this in mind most farmers are interested in lowering the cost of winter feeding (Volesky et al., 2002). The hay feeding period generally runs from when the supply of stockpiled tall fescue has been depleted until April when the spring growth begins. Nutrient requirements of cows change depending on the size of the cow, the milk production of the cow and the stage of production. A computer model was developed to estimate the requirements of cows in 6 herds (3 fall calving and 3 spring calving) to determine the percentages of Tennessee hay that met the requirements for the herds.Because of the time of year when hay is produced and the weather conditions that exist in the south east at that time making high quality hay is often a challenge due to weather and the time needed for hay to properly dry in the field.. The objective of this study was to compare the nutrient composition of hays produced in Tennessee and their ability to meet the nutrient requirements of beef cattle during the winter.

2. Materials and Methods

Forage samples were obtained from the Tennessee forage testing laboratory in Nashville, TN. Samples used were [bermudagrass (*Cynodon dactylon*) – 392; tall fescue (*Lolium arundinacea*) – 607; grass/clover – 200; mixed grass – 793; and orchardgrass (*Dactylis glomerata*) – 84] for the analysis (Table 1).

Table 1. Type of hay, number of samples and the average nutrient content of the samples

Hay Type	Number of Samples	% TDN[1]	% CP[2]
Bermudagrass	392	62.8	12.4
Fescue	607	62.8	12.0
Grass/Clover	200	63.7	14.4
Mixed Grass	793	59.36	13.2
Orchardgrass	84	59	12.2
Total	2076		

Note. [1]TDN = Total digestible nutrients; [2]CP = Crude protein.

These forage samples were submitted from Jan. of 2007 – Dec. 2009. They were analyzed for CP and TDN using NIR spectroscopy. The cow herds were modeled based on herd records from the Ames Plantation herds. The calving distribution was determined from eight years of data and was modeled on a 60-d calving season. The spring calving season began Feb. 1stand ended April 1st and the fall calving season began Oct. 1st and ended Dec. 1st. The herds were modeled as three different size herds based on average mature weight a small, medium, and large herd being 454 ± 45 kg, 544 ± 45 kg, and 635 ± 45 kg respectively. Milking ability was modeled for each herd with an average of 9 kg ± 4.5 kg and ranged from 4.5 to 14.5 kg.

The distribution of calving dates was obtained from herd records maintained for the research and education center at Ames Plantation. Six cow herds were modeled in this study. The herds were three spring and three fall calving beef herds. The herds were based on mature weight of the animal and had an average of 453.6, 544.3, and 635.0 kg ± 45.4 kg respectively and an average milk production of 9.1 kg ± 4.5 kg. A 60 d calving season was modeled, with fall calving starting on October first and a spring calving season starting February first. Data were analyzed using SAS 9.2 in a system that checked each hay sample as to when it would meet the cow's requirements. For a herd averaging 544.3 kg tall fescue hay will meet the requirements of 80% of the cows between 56 and 85% of the time depending on their calving date. The requirements of fall calving cows will decrease during the winter feeding period while the requirements of the spring calving cows will increase.

Hay feeding was modeled to begin on Nov. 15 and ended Mar. 15. All hays were compared with individual cow requirements in each 10d period. Tall fescue comparisons are presented because this is the predominant hay fed in Tennessee and the mid-south United States. The amount of hay that needed to be fed was based on published dry matter intake estimates.

Data were analyzed using SAS 9.2 (Cary, NC). The data were analyzed as a completely random design using the Mixed Model analysis of variance. This was used to make the comparison of each hay sample with each cow for all periods. Each hay sample was compared to each cow for CP and TDN resulting in over 12 million comparisons.

3. Results and Discussion

The forage samples are expected to be the higher quality hay samples that are produced in the state, due to better producers being the ones that will submit hay samples for testing. There was more variation with in forage species than there was between species suggesting that it does not matter what forage type is used as long as it is produced well.

The spring calving herd requirements for quality increased during the period of winter feeding with a dramatic increase at calving (Figure 1).

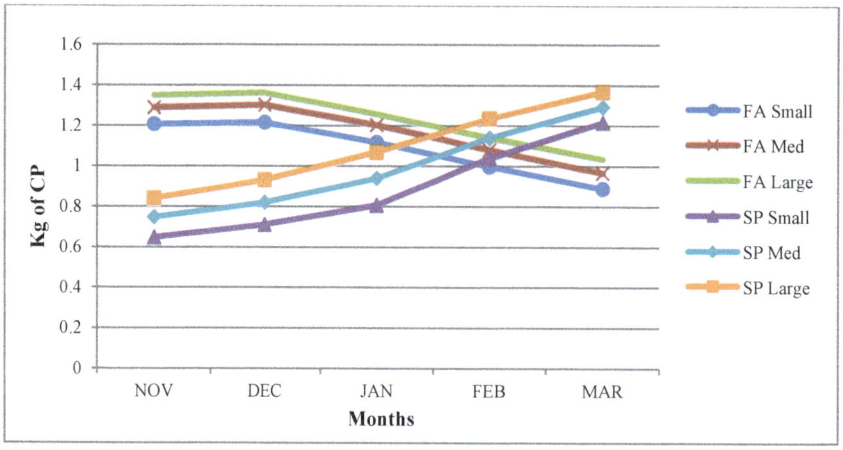

Figure 1. Kilograms of protein required per cow per day during winter feeding

In the month of November for the herd averaging 543 kg, 99% of the tall fescue samples met the requirements for CP and 93% met the requirements for TDN for 80% of the herd (Figure 2).

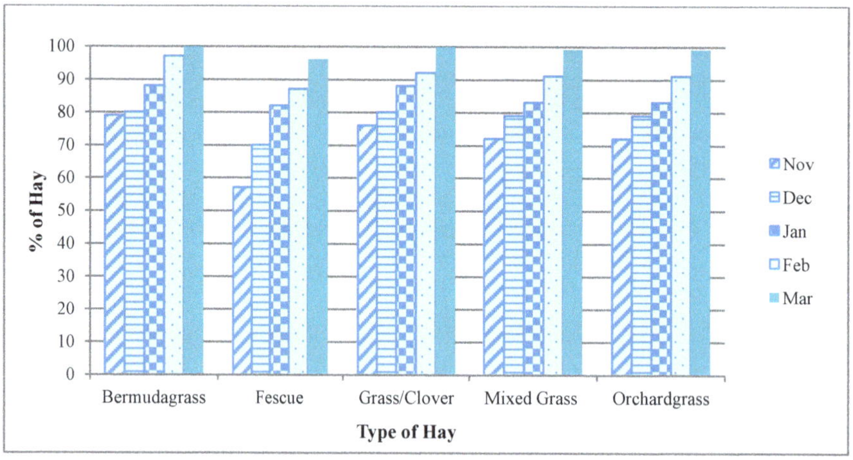

Figure 2. Percentage of hay that meets the TDN requirements of a 636 kg cow milking 9.1 kg/day

This is due to the relatively low metabolic needs of a dry gestating mature cow. In March only 57% and 28% of the samples met the requirements for CP and TDN of 80% of the herd respectively. There is a dramatic increase in the nutrient requirements of the cows at calving. As milk production peaks the nutrients required are greatest and the fewest hays will meet the requirements of the cow. For producers with spring calving cows this would mean they should feed their poorest quality hay first and feed higher quality hay when the demand for nutrients is greater.

The fall calving herd, which is the less prevalent management plan in Tennessee, requires higher nutrient quality hay at the beginning of the winter period. In November only 65% of hay samples met CP requirements and 27% of the hay met the TDN requirements for 80% of the herd. These cows have recently calved and are milking heavily. In March 89% of the hay met the requirement for CP and 74% met the requirement for TDN for 80% of the herd. Milk production in a fall calving herd is decreasing throughout the winter feeding period because of this producers with a fall calving herd should feed their highest quality hay first.

When nutrients are compared on a per head basis there were several facts that became apparent. Cow size has an impact on the amount of feed that is required, but also on the quality of feed that must be provided. A larger cow can have its nutrient requirements met by poorer quality hay due to a higher level of consumption. Smaller cows while they do eat less require a higher quality feed to meet their nutrient requirements.

When cows were compared across sizes and levels of milk productions it became apparent that levels of milk production had a larger impact on nutrients required than did the size of the cow. So if nutrients and feed resources are a limiting factor for a beef operation it is better to select for cows with a lower milk production level rather than for smaller framed cows.

4. Conclusions

It is believed that these samples represent the highest quality hay being produced in Tennessee as the lower quality hay is not tested. This means that forages are cut, dried, baled, and called hay. When nutrient analysis is done on these hays they are more closely resemble straw than hay. Much of this is due to weather and the ability to cut and bale hay at the correct stage of maturity without rain fall. Most of the herds in Tennessee would benefit from a supplementation during their peak lactation, and because Tall Fescue is the primary forage an energy supplement rather than a protein supplement would be the best option for supplementation, and it is important to remember that milk production, not size has the greatest impact on the nutrient requirements of cattle.

References

NRC. (1996). *Nutrient Requirement of Beef Cattle* (7th ed.). National Research Council. Nat. Acad. Sci., Washington, DC.

Garrand, R. M., & Glaze, J. B. Jr. (2008). Supplement Allows Ranchers to Eliminate Hay Feeding and Reduce Winter Feed Costs. *Journal of Extension, 46*(3), Article 3IAW6. Retrieved from http://www.joe.org/joe/2008june/iw6.php

Volesky, J. D., Adams, D. C., & Clark, R. T. (2002). Windrow grazing and baled-hay feeding strategies for wintering calves. *J. Range Management, 55*(1). http://dx.doi.org/10.2307/4003259

Waller, J. C., Fisher, A. E., Bates, G. E., Schrick, F. N., & Campbell, B. T. (2013). Tall Fescue Based Forage Systems Supplemented with Winter Annuals for Stocker Cattle. *Journal of Agriculture Science, 1*, 61-66.

8

Effects of Land-Use Intensification on Distribution and Diversity of *Fusarium* Species in Machakos County, Kenya

P. K. Maina[1], P. M. Wachira[1], S. A. Okoth[1], J. W. Kimenju[2], M. Otipa[3] & J. W. Kiarie[1]

[1] School of Biological Sciences, University of Nairobi, Nairobi, Kenya

[2] Department of Plant Science and Crop Protection, University of Nairobi, Nairobi, Kenya

[3] Kenya Agricultural and Livestock Research Organization, Nairobi, Kenya

Correspondence: P. K. Maina, School of Biological Sciences, University of Nairobi, Kenya. E-mail: wmaina89@yahoo.com

Abstract

Land-use intensification has a significant influence on occurrence of soil microorganisms. The effect of this phenomenon on *Fusarium* species is poorly characterized. One hundred soil samples were obtained from 3 replicated land- use types (LUT) in Mwala and Kauti irrigation regions in Machakos County. These included two intensive land-uses under irrigation and rain-fed agriculture and undisturbed lands. Mwala irrigated lands were divided into four blocks based on history of cultivation. Using soil dilution plate technique, 1,546 isolates of *Fusarium* were recovered and identified into twelve species namely; *F. oxysporum*, *F. solani*, *F. nygamai*, *F. equiseti*, *F. chlamydosporum*, *F. beomiforme*, *F. verticillioides*, *F. proliferatum*, *F. acuminatum*, *F. compactum*, *F. semitectum*, and *F. merismoides*. *Fusarium oxysporum* was the most abundant and diverse *Fusarium* species. *Fusarium semitectum*, *F. compactum* and *F .merismoides* had the least distribution being isolated from only one LUT. *Fusarium beomiforme* and *F. acuminatum* were recovered from irrigated farmlands only while *F. verticillioides*, *F. proliferatum* and *F. acuminatum* were restricted to disturbed lands only. The difference in abundance of *Fusarium* between the three LUTs was significant ($P = 0.047$) with irrigated lands having the highest abundance. Mwala block A had the highest abundance, richness and diversity of *Fusarium*. Lands with a higher intensity of disturbance had a higher abundance and richness of *Fusarium* than the less undisturbed lands. This may have severe implication on crop production as most species of *Fusarium* isolated are pathogenic. Sustainable ways of controlling these potential crop pathogens should be sought.

Keywords: biodiversity, distribution *Fusarium*, intensification, land-use

1. Introduction

Fusarium is a large and diverse genus of filamentous fungi classified in the Order Hypocreales of the Phylum Ascomycetes. This form-genus was first described by Link in1809. Later, the genus *Fusarium* was defined by Wollenweber and Reinking (1935) based on morphological and cultural characteristics. However, *Fusarium* is a polyphyletic genus which includes over 50 species and has cosmopolitan distribution in the air, soil and in association with many plants (Alexopoulos et al., 1996; Leslie et al., 2006). Many *Fusarium* species cause diseases of economic importance in crops. Besides causing air-borne respiratory infections in human, some species of *Fusarium* may also cause a range of opportunistic infections such as keratomycosis (Mselle, 1999), onychomycosis (Godoy et al., 2004), pulmonary infections (Gorman et al., 2006) and endophthalmitis (Goldblum et al., 2000) in immune-compromised humans. *Fusarium* species also produce a wide array of mycotoxins. These are fumonisins, beauvericin, moniliformin, zearalenone, nivalenol, deoxynivalenol, fusaproliferin, and trichothecene in cereals (Blaney & Dodman, 2002). These mycotoxins are responsible for allergies, growth defects and cancer in humans and domestic animals (Broomhead et al., 2002). It is known that *Fusarium* is a cosmopolitan genus and a natural soil fungus (Nurhazrati et al., 2012). In agricultural soils, *Fusarium* is a typical genus. Previous studies in Kenya on *Fusarium* prevalence in soils indicate that it is a highly distributed and diverse genus (Kedera et al., 1999; Maina et al., 2009). This fungus was found to be prevalent in agricultural soils of Malaysia, with *Fusarium oxysporum* being the most abundant species (Siti et al., 2012).

The predominant interest in the genus *Fusarium* is its role as plant pathogens (Booth, 1971). *Fusarium* species cause a variety of serious plants diseases. These include vascular wilts, cankers, rots of seed, fruit, root and stem, and blights in a wide range of economically important crops. This is due to the fact that *Fusarium* chlamydospores, conidia and hyphae are distributed widely in cultivated soil and soil debris (Bolkan et al., 1979). These propagules gain entry into the plant through cut surfaces of seeds, damaged roots and stem tissues of young and stressed plants. Infection through wounds caused by insects can also act as point of entry, therefore, causing diseases to susceptible plants (Leslie et al., 2006).

Fusarium species are able to survive in the soil for long periods of time as chlamydospores (Vakalounakis & Chalkias, 2004). These are resting spores of *Fusarium* species produced in the soil during periods of unfavorable conditions (Leslie et al., 2006). Many *Fusarium* species also exist as harmless saprobes in the soil while others establish long-term associations with crop plants as endophytes (Bacon & Hinton, 1996). A high diversity index of soil *Fusarium* species is important for their involvement in soil structure formation; decomposition of organic matter; toxin removal; and the cycling of carbon, nitrogen, phosphorus, and sulphur (Garbeva et al., 2004). However, the presence of diverse species of *Fusarium* in agricultural soil may lead to disease infection of susceptible plants (Silvestro et al., 2013). Apparently the cultivated soil is the most important source of *Fusarium* spores. Moreover, *Fusarium* inoculum is spread by air, infected debris and soils (Summerell et al., 2010).

The rapid increase in human population in Kenya has resulted in increased demand for food leading to rapid conversion of natural vegetation lands to agricultural lands. Some expansions have involved clearing natural forests and grasslands. Being largely semi-arid, Kenya experiences shortage of arable land leading to increased fragmentation and intensification of the available arable land. These land-use practices have major effects on soil biodiversity including increase in soil plant pathogenic species especially of *Fusarium* (Foley, 2005; Luque et al., 2005; Siti et al., 2012; Okoth et al., 2013). Silvestro et al. (2013) reported that the prevalence of *Fusarium* in agricultural soils is largely determined by plant species and soil depth. Prevalence of *Fusarium* in agricultural soil is a threat to the productivity of agricultural crops as all crops have a certain *Fusarium* species that infect them. Therefore, this study sought to determine the effect of land-use intensification on the occurrence, distribution and diversity of *Fusarium* species in Kenyan soils. The inventory obtained will form basis for comparison studies and possible policy formulations in an endeavor to produce food sustainably.

2. Materials and Methods

2.1 Study Area

The study was carried out in Kabaa irrigation scheme, Mbiuni location, Mwala Sub-county of Machakos County that experiences arid and semi-arid type of climate and also the temperature varies from 6 °C to 29 °C. Two regions were selected: Mwala and Kauti irrigation schemes. These areas were selected due to variations in land-use which includes intensive irrigated and rain-fed agriculture as well as undisturbed lands in the vicinity for comparison. In each region, soils were sampled from the 3 replicated LUTs. Mwala is located in lower midland zone 4, with cambisol soils. The main crops cultivated are French bean, tomato, onion, banana, kale, cabbage and passion fruit. The scheme is divided into four blocks (A to D) with A being the oldest block while D is the most recent. Kauti irrigation scheme lies in the upper midland zone 4 with alfisols and acrisols soils. The main crops grown here are French bean, kale, tomato, maize and coffee.

2.2 Soil Sampling

Soil samples were collected from 3 major LUTs; from irrigated agricultural and rain-fed agricultural farmlands and from undisturbed lands. This was replicated in the two regions of Mwala and Kauti irrigation schemes. The main crops grown under irrigation include French bean, maize, tomatoes, common bean and kales. Under rain-fed agriculture, diverse crops are grown including maize, common beans, sorghum, cassava, kales, millet among other crops. The undisturbed lands are composed of native grasses and deciduous thorn bushes made of *Commiphora/Sanseveria* association. At each sampling point, 12 subsamples were collected from the top 20 cm of soil (Figure 1).

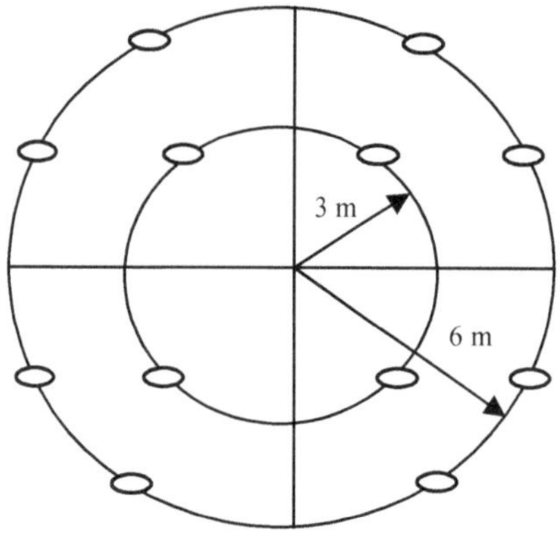

Figure 1. Schematic representation of the twelve soil sampling points which comprised one main sampling point

Soil samples were collected using a sterilised soil auger and mixed thoroughly before obtaining 100 g from this sample. To avoid cross-contamination between sampling points, the auger was sterilized between sampling points by using cotton wool soaked in 70% ethanol. The samples were put in paper bags and transported in the laboratory where they were stored at 5 °C awaiting further analysis. The soil samples were air-dried at room temperature (27±1 °C) for 5 days and ground and then sieved with 0.5 mm sieve to separate larger particles. These soil samples were used for isolation and quantification of *Fusarium*.

2.3 Isolation of Fusarium from the Soil

Soils were assayed using serial dilution plating method as described by Burgess et al. (1988) with 0.1% Tap Water Agar (Brayford, 1993). *Fusarium*-selective PCNB-Peptone Agar (PPA) media were used to recover *Fusarium* isolates from the soil. From each of the of soil sample, 10 g of air-dried soil were removed and added to 90 ml of sterile 0.1% Tap Water Agar. The mixture was vigorously agitated (200 rpm for 60 s on a Lab-Line Orbital shaker, Melrose Park, IL) and 10 ml of the resulting suspension were pipetted into a flask containing 90 ml of sterile distilled water. This procedure was repeated up to the third ten-fold dilution. Then, 1-ml aliquots from second and third dilutions, in three replicates, were aseptically pipetted on to the Petri dishes containing *Fusarium*-selective PPA media (Burgess et al., 1988). The Petri dishes were then incubated in an alternating temperature regime, 25 °C day/20 °C night, at about 65% relative humidity, under cool white fluorescent lights (Philips TL 40W/80 RS F40BLB) with a 12 h photoperiod for 7 to 10 days. Observations were made from the third day onwards for developing colonies.

2.4 Sub Culturing, Purification and Identification of Fusarium Species

Colonies from PPA media were transferred to Spezieller Nahrstoffarmer Agar (SNA) media to form distinctive colonies and sporulation and incubated at 25 °C for 5 days (Leslie & Summerell, 2006). Subsequently, in order to obtain monosporic cultures of each colony formed on SNA, from which identification was based, very dilute inocula, of 5 to 10 spores per drop of suspension (when viewed at low power magnification), were prepared and spread on 2% plates. Germlings on 2% Tap Water Agar (TWA) were then sub cultured on SNA, Carnation-Leaf-Agar (CLA) and Potato-Dextrose-Agar (PDA) media, for growth and identification (Leslie & Summerell, 2006). PDA cultures were used to assess pigmentation and gross colony morphology. Cultures grown on SNA were evaluated for microconidia and for chlamydospores.

All the pure isolates sub cultured on PDA, CLA and SNA were incubated for ten to twenty days at 25 °C under fluorescent lamps (Sylvania cool white tubes) with a 12 h photoperiod. *Fusarium* species were identified by morphological characteristics (Brayford, 1993; Burgess et al., 1988; Leslie & Summerell, 2006; Nelson et al., 1983). After identification, the single spore cultures were preserved in agar slants of SNA in screw cap bottles at 4 °C and also in sterilized soil in screw cap bottles.

2.5 Statistical Analyses

The data obtained showing occurrence, richness and diversity of *Fusarium* species in the soil in relation to LUTs were analyzed using GenStat computer package, discovery edition. Data were first standardized before analyses to take into account the difference in sample numbers between the different LUTs. Data obtained on abundance were transformed using logarithm for normality. Analysis of variance was used to determine the effects of different land uses on occurrence, richness and diversity of *Fusarium* species. Diversity was calculated using

Shannon`s diversity index (H'). Significance was evaluated at $P < 0.05$ for all analyses. Means found to be significantly different were separated using Tukey test at $P < 0.05$.

3. Results

A total of 1,546 *Fusarium* isolates were recovered during the study. The identification of these isolates resulted into 12 *Fusarium* species (Table 1). *Fusarium oxysporum* was the most abundant *Fusarium* species, with a frequency of isolation of 60.9%, followed by *F. solani* at 16.3%. The two accounted for 77.2% of all *Fusarium* isolates recovered. *Fusarium merismoides* had the least isolation frequency at 0.1% and was only isolated from Mwala rainfed farms (Table 1). Results further revealed that *F. oxysporum* and *F. solani* accounted for 97.4% of all fusaria isolated from the Mwala irrigation blocks.

Table 1. Frequency of isolation of *Fusarium* species from different Land Use Types

Fusarium species	Proportion (%) per land use type						Overall proportion (%)
	Mwala irrigated	Mwala rainfed	Mwala undisturbed	Kauti irrigated	Kauti rainfed	Kauti undisturbed	
F. oxysporum	75.9	38.9	40	22.2	27.3	18.2	**60.9**
F. solani	21.5	7.7	0	0	18.2	18.2	**16.3**
F. nygamai	0	24.0	13.3	2.2	9.1	9.1	**6.8**
F. equiseti	0	16.8	0	2.2	4.5	36.3	**5.0**
F. chlamydosporum	0	9.7	6.7	2.2	22.7	18.2	**3.5**
F. beomiforme	0.4	0	0	51.1	0	0	**3.0**
F. verticillioides	0.8	2.4	0	6.8	0	0	**1.5**
F. proliferatum	1.0	0	0	2.2	18.2	0	**1.3**
F. acuminatum	0.4	0	0	11.1	0	0	**0.9**
F. compactum	0	0	26.7	0	0	0	**0.5**
F. semitectum	0	0	13.3	0	0	0	**0.2**
F. merismoides	0	0.5	0	0	0	0	**0.1**
Frequency of isolation (%)	**63.4**	**25.3**	**1.8**	**5.5**	**2.7**	**1.3**	**100%**

The difference in abundance of *Fusarium* between the LUTs was significant ($P = 0.047$). The *Fusarium* Log (Abundance +1) for the irrigated, rainfed and undisturbed were, 3.49, 2.8 and 1.96, respectively (Figure 2). Lands with a higher intensity of disturbance had a higher abundance of soil *Fusarium* than the undisturbed farmlands as this was recorded in both Mwala and Kauti regions (Figures 3 and 4).

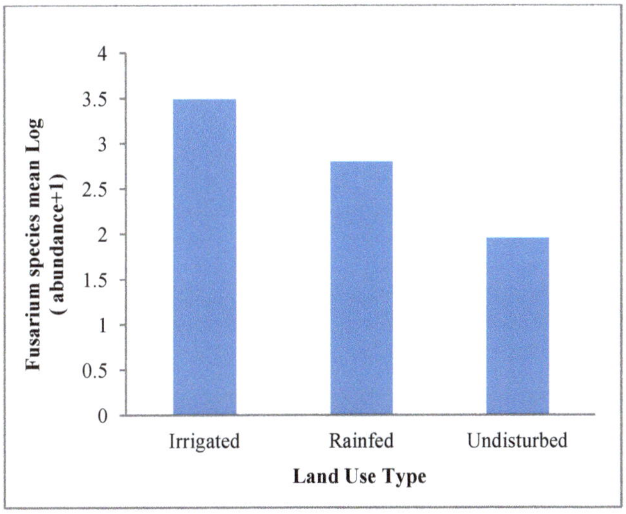

Figure 2. *Fusarium* mean Log (Abundance+1) for different LUTs

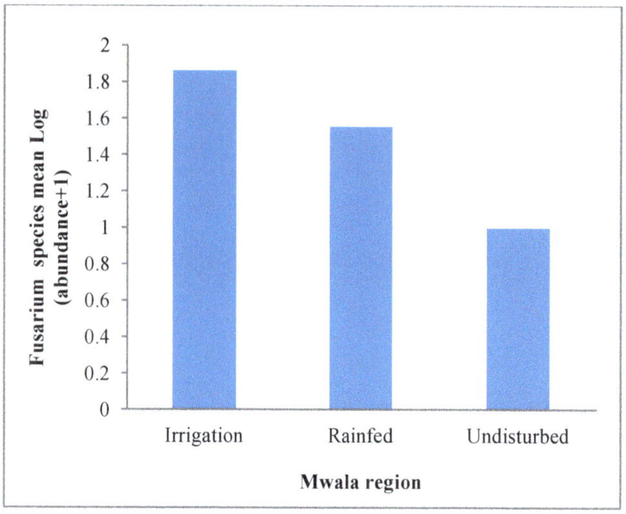

Figure 3. *Fusarium* mean Log (Abundance +1) within Mwala region

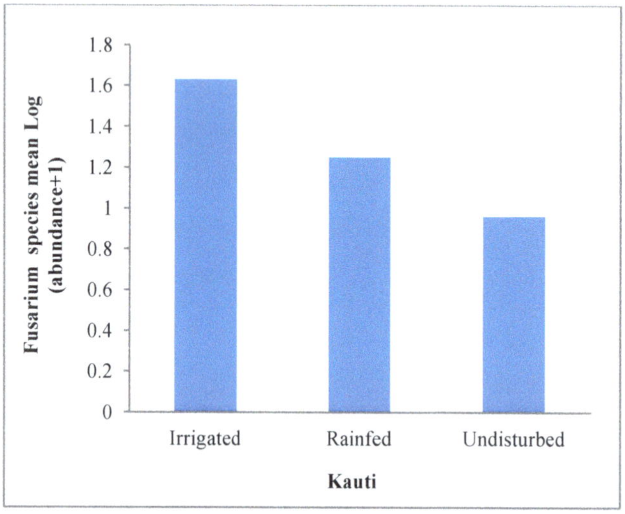

Figure 4. *Fusarium* mean Log (Abundance +1) within Kauti region

The relationship between LUT and abundance of soil *Fusarium* was clear. This was evidenced by the greater numbers of *Fusarium* recovered from irrigated LUT, followed by rainfed LUT, with undisturbed having least soil *Fusarium* in both schemes. Specifically, the difference in abundance of *Fusarium* species within Mwala irrigation blocks were not significant ($P = 0.356$). Block A had a higher abundance of *Fusarium* compared to the rest (Figure 5). Block A has a longer history of disturbance than the other blocks and it accounted for 63.6% of *Fusarium* isolates from Mwala irrigated lands (Table 2). Block D which is a relatively recent irrigation block had the least isolation frequency of 2.1%.

Table 2. Frequency of isolation of *Fusarium* species in Mwala irrigation blocks

***Fusarium* spp.**	**% of isolation in the irrigation blocks**				
	A	**B**	**C**	**D**	**%**
F. oxysporum	72.3	82.9	82.1	72.7	**75.9**
F. solani	25	13.8	16.1	3.3	**21.4**
F. nygamai	0	0	0	0	**0**
F .equiseti	0	0	0	0	**0**
F. chlamydosporum	0	0	0	0	**0**
F. beomiforme	0.3	0.8	0	0	**0.4**
F. verticillioides	0.9	0.8	0	0	**0.8**
F. proliferatum	0.9	1.6	1.8	0	**1.1**
F. acuminatum	0.6	0	0	0	**0.4**
F. compactum	0	0	0	0	**0**
F. semitectum	0	0	0	0	**0**
F. merismoides	0	0	0	0	**0**
%	**63.6**	**23.6**	**10.7**	**2.1**	**100**

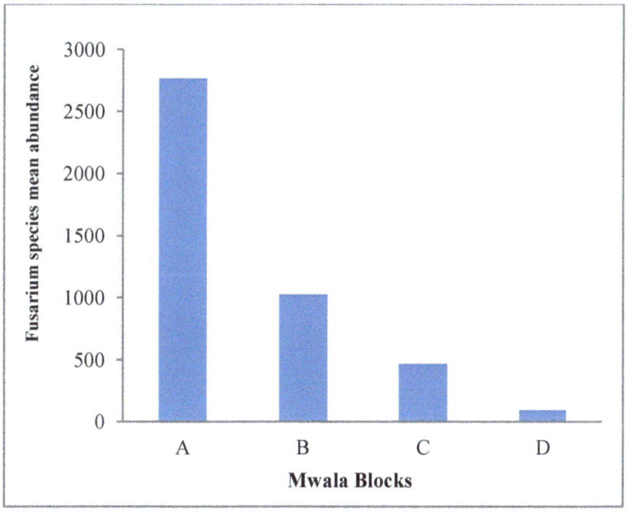

Figure 5. *Fusarium* mean Log (Abundance +1) within Mwala irrigation blocks

Among the *Fusarium* species, *F. oxysporum* had the widest distribution occurring in all the LUTs under study. *Fusarium nygamai*, *F. equiseti* and *F. chlamydosporum* also had a wide distribution. *Fusarium beomiforme* and *F. acuminatum* were only recovered from irrigated farmlands while *F. semitectum* and *F. compactum* were only isolated from undisturbed soils. Isolates of *F. beomiforme*, *F. verticillioides*, *F. proliferatum* and *F. acuminatum* were only isolated from disturbed land use during this study while *Fusarium merismoides* was only isolated from land under rain-fed land use.

Results indicated that irrigated farmlands had the highest *Fusarium* species mean richness while undisturbed lands had the least richness. This indicated that intensity in land use had an influence on *Fusarium* species richness. Although there is no significant difference in *Fusarium* species richness across the LUTs ($P = 0.825$), the intensively cultivated lands had higher species richness than the undisturbed farmlands (Figure 6).

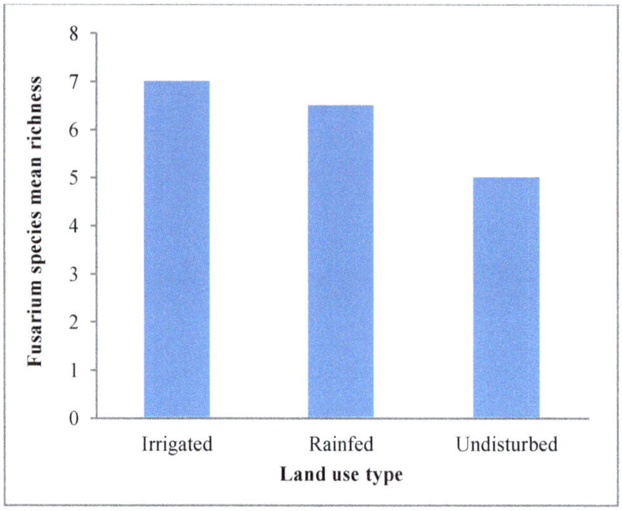

Figure 6. *Fusarium* species mean richness for different LUTs

The undisturbed lands had the lowest *Fusarium* species mean richness in both Mwala and Kauti regions (Table 3). The irrigated land use in Kauti had the highest mean richness followed by Mwala rain-fed land use (Table 3). Mwala region had a lower *Fusarium* mean richness than Kauti. The relationship on history of cultivation and richness was clear among Mwala irrigation blocks, with block A having the highest species richness, followed by block B. Block D, being the most recent irrigation block had the lowest richness (Figure 7). However, the difference in species richness between these irrigation blocks of Mwala was not significant ($P = 0.353$).

Table 3. *Fusarium* species mean richness for different land use types.

Land use type	*Fusarium* species mean richness	
	Mwala	**Kauti**
Irrigated	6	8
Rainfed	7	6
Undisturbed	5	5

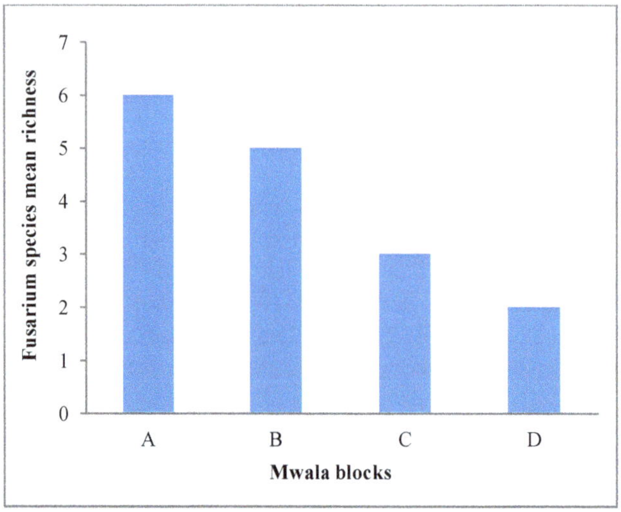

Figure 7. *Fusarium* species mean richness within Mwala irrigation blocks

During this study, the diversity of *Fusarium* species in different LUTs did not show any significant difference ($P = 0.063$), neither was there a relationship observed between the intensity of land-use and *Fusarium* species diversity. Overall, rainfed land-use had the highest diversity followed by undisturbed lands (Figure 8). This trend was evident in both Mwala and Kauti regions. In comparison, Kauti rain-fed land-use had the greatest *Fusarium* species mean index (H' = 1.669), followed by Mwala rain-fed land-use (H' = 1.548) while Mwala irrigated land-use had the least mean Shannon index (H' = 0.671).

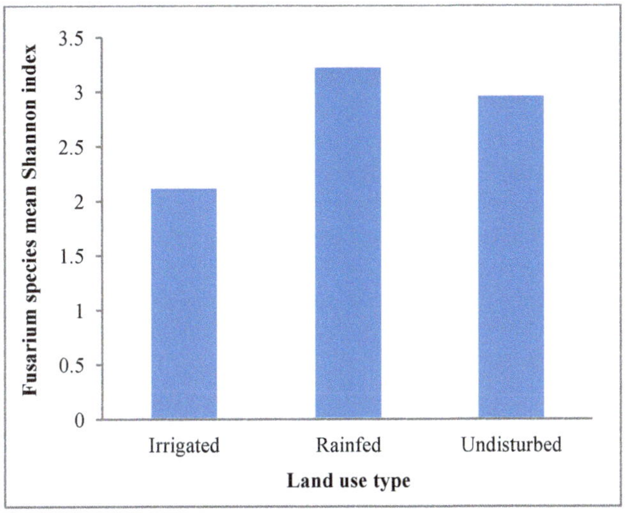

Figure 8. *Fusarium* species mean Shannon indices for different LUTs

Results further indicated that the history of cultivation had no influence on diversity of *Fusarium* species. Within Mwala irrigation blocks, block A had the highest diversity, followed by D while C was least diverse (Figure 9).

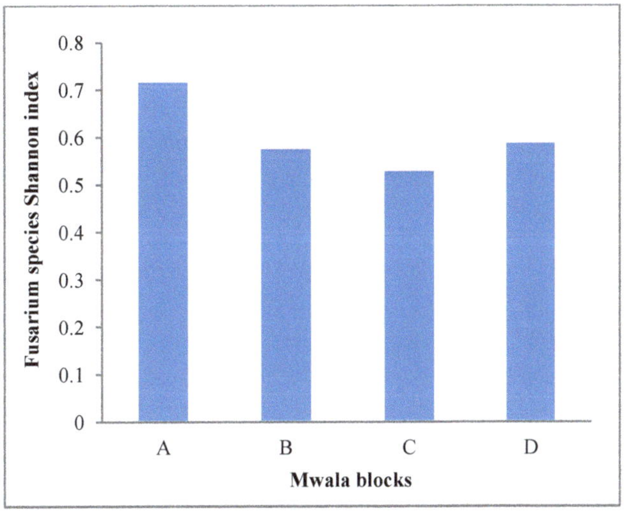

Figure 9. *Fusarium* species mean Shannon indices within Mwala irrigation blocks

4. Discussion

This study has revealed prevalence of *Fusarium* species in the soils of Kabaa irrigation scheme where a total of 1,546 isolates of *Fusarium* species were obtained and characterized into 12 *Fusarium* species. These species are widely distributed in the irrigated, rainfed and disturbed soils in the study area. The results are in iIn agreement with this findings Silvestro et al. (2013) noted that *Fusarium* spp. is a cosmopolitan genus and a natural soil fungus. Members of this genus are ubiquitous in soil and have been isolated from various soil types in tropical and temperate regions, from desert soils to artic and alpine soils (Latiffah et al., 2007; Summerell et al., 2010). However, the majority of *Fusarium* species are recovered in cultivated soils especially near the soil surface. Wakelin et al. (2008) reported that *Fusarium* species occur widely in cultivated soils and is often associated with plant roots either as parasites or saprophytes. The abundance of these species generally resembles that reported in South Africa (Rheeder & Marasas, 1998).

Earlier studies of *Fusarium* in Africa and Asia showed variable numbers and species composition. In their study Latiffah et al. (2007) identified four *Fusarium* species, namely, *F. solani*, *F. equiseti*, *F. semitectum* and *F. oxysporum*. In other studies in Lesotho, Nigeria and Zimbabwe have shown diverse species occurrence: *F. oxysporum*, *F. equiseti*, *F. solani*, *F. moniliforme*, *F. compactum*, *F. nygamai*, and *F. chlamydosporum* (Fawole & Olowonihi, 2005; Nwanwa & Nelson, 1993). A similar study of soils of South Africa showed the presence of nineteen species. These were *F. chlamydosporum*, *F. merismoides*, *F. lateritium*, *F. culmorum*, *F. compactum*, *F. dlamini*, *F. poae*, *F. proliferatum*, *F. verticillioides*, *F. scirpi*, *F. polyphialidicum*, *F. graminearum*, *F. sambucinum*, *F. napiforme*, *F. oxysporum*, *F. equiseti*, *F. semitectum*, *F. nygamai* and *F. solani*. Of these *F. oxysporum*, *F. equiseti*, *F. semitectum*, *F. nygamai* and *F. solani* were the most frequently isolated (Jeschke et al., 1990). *Fusarium semitectum* and *F. equiseti* were commonly isolated from soil as reported by Burgess et al. (1988) and Leslie et al. (1990, 2006), and most probably exist as soil inhabitants. Both species are not regarded as important plant pathogens although the species have been reported to cause diseases on several crops (Elmer, 1996). A previous study in soils of Kenya showed the occurrence of 26 species of *Fusarium* under different land use systems (Maina et al., 2009). The study revealed that *Fusarium oxysporum* was the most abundant and distributed species. These results are similar to those of the current study. This species could probably be highly adapted for competition as well as having a wide host range (Edel et al., 2001; Nelson et al., 1981). The difference in composition and abundance of soil *Fusarium* could be attributed to differences in type of plants grown, soil characteristics, as well as land management practices (Chehri, 2011).

This study indicated a high prevalence of *Fusarium oxysporum* being isolated from all LUTs. In their study, Maina et al. (2009) reported similar results, where *F. oxysporum* was the most prevalent species. Siti et al. (2012) also reported high prevalence of *Fusarium* species in Malaysia. In their study on diversity of *Fusarium* in agricultural lands, it was revealed that *F. oxysporum* was the most prevalent species at 26.2% followed by *F. semitectum* and *F. solani* at 24 and 21.3%, respectively. The prevalence of *F. oxysporum* fungus especially in cultivated lands in this study is an indicative of status of soil health. Members of this fungus have been reported as a serious pathogen on various plants and the prevalence of this fungus in the study area is a cause for concern. The plant pathogenic representatives of *Fusarium oxysporum* are involved in a variety of plant diseases such as vascular wilts, damping-off, crown rots and root rots (Jarvis & Shoemaker, 1978; Summerell & Rugg, 1992). In addition, *F. solani* was the second most prevalent species isolated in this study and was recovered in all the LUTs. It is a common soil-borne fungi and a pathogen to many agriculture crops such as citrus, pepper, and beans, (Fletcher & Karen, 2007; Nemec, 1987). Several studies have revealed that this species is widely distributed in numerous native soils such as in sub tropical, semi-arid and grassland soils (Burgess & Summerell, 1992) and desert soil (El Gindy & Saad, 1990). Therefore, in this study, it was not surprising that *F. solani* was the second most frequent species isolated from the study area. However, in a study covering Malaysian highlands, *F. solani* was the most abundant species at 66.1% of all isolates followed by *F. graminearum* and *F. oxysporum* at 8.5 and 7.8%, respectively (Nurhazrati et al., 2012). This can be explained by the low temperatures of the highlands as well as the type of plants growing in these areas.

In this study, *F. merismoides* had the lowest abundance and the most restricted distribution. These results are in agreement with those of Silvestro et al. (2013) who reported that this species had a restricted distribution being isolated only from 5-10 cm soil layer from the rotation of winter agriculture. Jeschke et al. (1990) also observed low frequency of *F. merismoides* in soil samples collected from different altitudes in Southern Africa. The reason for low abundance and distribution of this species could probably be due to unfavovarable soil conditions. Leslie and Summerell (2006) suggested that *F. merismoides* is a fungal soil saprophyte, but has the potential to cause infections in plants if the environmental conditions are appropriate. Bateman and Murray (2001) observed that the frequency of this species in the soil increased under dry conditions, suggesting that the presence of a mucilaginous matrix produced by this fungal species may favor its survival and ability to compete in dry conditions (Louis & Cooke, 1983).

Results of this study further revealed that the soil abundance of *Fusarium* was related to the LUT. The abundance of soil *Fusarium* was positively influenced by intensity of disturbance. Farmlands with a higher intensity of disturbance had higher abundances of *Fusarium* while the undisturbed farmlands had the least abundance. These observations were replicated for both regions under study. These results are similar to those reported by Maina et al. (2009) who reported a higher abundance of *Fusarium* in disturbed lands. These results also correlate well with the survey carried out by Marasas et al. (1988) which revealed very low populations of *Fusarium* species in undisturbed indigenous sites. Bushula (2008) isolated only 122 *Fusarium* isolates from native soils of Western Cape that were grouped into only 3 species. The results of the current study agree with previous reports from other authors, who observed that *Fusarium* genus is one of the most abundant fungi in agricultural soils and less abundant in undisturbed lands (Marasas et al., 1988; Thorn, 1997).

Disturbance of the soil through tillage leads to fragmentation of fungal hyphae which acts as propagules. Tillage may also help in dispersal of *Fusarium* propagules (Steinkellner & Langer, 2005). It also opens up the soil to aeration and hence supports physiological activities of *Fusarium* in the soil. It creates spaces through which water may percolate, hence aiding in dispersal of propagules. This can explain the high number of fusaria in cultivated under irrigation. According to Salas and Stack (1991), the method of tillage may not only affect the composition of soil *Fusarium* but also the abundance of species. That soil characteristics affect the abundance of soil fusaria has been confirmed (Maina et al., 2009; Wei et al., 2012). Cultivation of crops in the study area may involve use of organic and inorganic fertilizers, which may have an impact on abundance of soil fusaria (Maina et al., 2009). Long-term application of these elements in the soil will therefore have a bearing on abundance of soil fusaria. This may also account for higher abundance of this fungal group in agricultural land than the undisturbed soils. This study also indicated that there is low occurrence of *Fusarium* species in undisturbed soils. This could be as a result of suppression by other predominant saprophytic fungi present or that the fungus is selected against through lack of suitable host in these ecosystems.

The abundance of soil *Fusarium* was also positively influenced by the length in time of disturbance. This trend was clearly shown in Mwala irrigated blocks (Figure 5). Longer history of cultivation may favor proliferation of soil *Fusarium* due to prevalence of suitable host and conducive environmental conditions. Continuous cultivation ensures suitable substrates in the soil for the fungus. Wakelin et al. (2008) suggested that repeated incorporation of stubble into the soil after crop harvest could account for sustained higher populations of *Fusarium* species in the soil. *Fusarium* is a cellulolytic fungus and the stubble incorporation could increase fungal development and favor the breakdown of carbon and nitrogen source incorporated into the soil (Collins et al., 1990). Similar results were observed by Marasas et al. (1988) and Fernández et al. (2008) who showed that the presence of *F. acuminatum*, *F. equiseti* and *F. poae* could be related with the repeated incorporation of oilseed crops in the field sequence. They concluded that change in the structure of *Fusarium* community could be produced by changing type of crop stubble incorporated in the soil.

The results further revealed that *Fusarium* richness was positively related to disturbance. The most disturbed lands had the highest *Fusarium* richness. This was clearly demonstrated in both Mwala and Kauti regions. The results are in agreement with those of Nesci et al. (2006). Disturbance increases and sustains a higher richness of fungal species due to improved sources of nutrients and suitable growth environments. Richness also correlated positively with history of disturbance as reflected in Mwala irrigation blocks. The oldest block had the highest *Fusarium* richness. The results of this study further indicated that diversity between studied farms did not differ significantly. However, rain-fed land-use type had the highest diversity. This could be explained by the diversity of plants grown, soil characteristics and the effect of disturbance. Difference in soil properties and the crop types grown have been shown to affect the occurrence and recovery of *Fusarium* species (Burgess et al., 1988; Inam-Ul-Haq et al., 2009; Larkin & Fravel, 2002; Senthilkumar et al., 2011).

5. Conclusion

The study provides the baseline information on the occurrence, diversity and distribution of *Fusarium* in various land-use systems. The influence of land-use on the occurrence of this fungus is evident. The more disturbed farmlands had the highest abundance, richness and diversity of *Fusarium* species. Similarly, the lands with a longer history of cultivation had the higher abundance, richness and diversity of *Fusarium* species. Differences in abundance and distribution among *Fusarium* species were observed with *F. oxysporum* being the most predominant and with a cosmopolitan distribution. Enhanced population and diversity of *Fusarium* species in agricultural soils could impact negatively on crop health and consequently on production of food for humans and animals. Sustainable ways of managing this highly crop pathogenic fungus should therefore be explored.

Acknowledgements

The authors would like to acknowledge the following institutions for facilitating this study, The Ministry of Agriculture, The ADB Bank, Kenya Agricultural and Livestock Research Organization and the University of Nairobi.

References

Alabouvette, C. (1999). *Fusarium* wilt suppressive soils: An example of disease suppressive soils. *Australasian Plant Pathology, 28*, 57-64. http://dx.doi.org/10.1071/AP99008

Alexopoulos, C. J., Mims, C. W., & Blackwell, M. (1996). *Introductory Mycology* (4th ed., p. 868). New York: John Wiley & Sons, Inc. Retrieved from http://trove.nla.gov.au/version/46511427

Amir, H., & Alabouvette, C. (1993). Involvement of soil abiotic factors in the mechanisms of soil suppressiveness of Fusarium wilts. *Soil Biology and Biochemistry, 25*, 157-164. http://dx.doi.org/10.1016/0038-0717(93)90022-4

Bacon, C. W., & Hinton, D. M. (1996). Symptomless endophytic colonisation of maize by *Fusarium moniliforme*. *Canadian Journal of Botany, 74*, 1195-1202. http://dx.doi.org/10.1139/b96-144

Bateman, G. L., & Murray, G. (2001). Seasonal variation in populations of *Fusarium* species in wheat-field soil. *Applied Soil Ecology, 18*, 117-128. http://dx.doi.org/10.1016/S0929-1393(01)00158-5

Blaney, B. J., & Dodman, R. L. (2002). Production of zearalenone, deoxynivalenol, nivalenol, and acetylated derivatives by Australian isolates of *Fusarium graminearum* and *F. pseudograminearum*in relation to source and culturing conditions. *Australian Journal of Agricultural Research, 53*, 1317-1326. http://dx.doi.org/10.1071/AR02041

Booth, C. (1971). *The genus Fusarium* (p. 237). Commonwealth Mycological Institute, Kew, Surrey, England. Retrieved from http://trove.nla.gov.au/version/27039116

Bolkan, H. A., Dianese, J. C., & Cupertino, F. P. (1979). Survival and colonization potential of *Fusarium moniliforme* var. *subglutinans* in soil. *Phytopathology, 69*, 1298-1301. http://dx.doi.org/10.1094/Phyto-69-1298

Brayford, D. (1993). *The identification of Fusarium species* (p. 119). CAB International, UK.

Broomhead, J. N., Ledoux, D. R., Bermudez, A. J., & Rottinghaus, G. E. (2002). Chronic effects of moniliformin in broilers and turkeys fed dietary treatments to marketage. *Avian Diseases, 46*, 901-908. http://dx.doi.org/10.1637/0005-2086(2002)046%5B0901:CEOMIB%5D2.0.CO;2

Burgess, L. W., Liddell, C. M., & Summerell, B. A. (1988). Laboratory manual for *Fusarium* research (2nd ed., p. 156). University of Sydney, Sydney.

Burgess, L. W., & Summerell, B. A. (1992). Mycogeography of Fusarium: Survey of Fusarium species from subtropical and semiarid grassland soils from Queensland Australia. *Mycological Research, 96*(9), 780-784. http://dx.doi.org/10.1016/S0953-7562(09)80448-6

Bushula, V. S. (2008). *Native Fusarium species from indigenous Fynbos soils of the Western Cape* (MSc. Thesis, p. 197). Stellenbosch University. Retrieved from http://hdl.handle.net/10019.1/2437

Chehri, K., Salleh, B., Yli-Mattila, T., Reddy, N., & Abbasi, S. (2011). Molecular characterisation of pathogenic *Fusarium* species in cucurbit plants from Kermanshah province, Iran. *Saudi Journal of Biological Sciences, 18*(4), 341-35. http://dx.doi.org/10.1016/j.sjbs.2011.01.007

Edel, V., Steinberg, C., Gautheron, N., Recorbet, G., & Alabouvette, C. (2001). Genetic diversity of *Fusarium oxysporum* populations isolated from different soils on France. *FEMS Microbiology Ecology, 36*, 61-71. http://dx.doi.org/10.1111/j.1574-6941.2001.tb00826.x

El Gindy, A. A., & Saad, R. R. (1990). Fungi of virgin and cultivated soil of Salhiah Desert, Egypt. *Zentralbl Mikrobiol,* 145(7): 547-551.

Elmer, W. H. (1996). *Fusarium* fruit rot of pumpkin in Connecticut. *Plant Disease, 80*, 131-135. http://dx.doi.org/10.1094/PD-80-0131

Fawole, O. B., & Olowonihi, E. T. (2005). Distribution of fungi and bacteria in the soils of the University of Ilorin Teaching and Research Farm. *Journal of Agricultural Research, 4*, 218-227.

Fernández, M. R., Huber, D., Basnyat, P., & Zentner, R. P. (2008). Impact of agronomic practices on populations of *Fusarium* and other fungi in cereal and non cereal crop residues on the Canadian Prairies. *Soil Till. Res., 100*, 60-71. http://dx.doi.org/10.1016/j.still.2008.04.008

Fletcher, J. T., & Karen, L. L. (2007). A stem rot of tomato caused by *Fusarium merismoides*. *Plant Pathology, 34*(3), 443-445. http://dx.doi.org/10.1111/j.1365-3059.1985.tb01387.x

Foley, J. A. (2005). Global consequences of land use. *Science, 309*, 570-574. http://dx.doi.org/10.1126/science.1111772

Garbeva, P, van Veen, J. A., & van Elsas, J. D. (2004). Microbial diversity in soil: selection of microbial populations by plant and soil type and implications for disease suppressiveness. *Annual Review of*

Phytopathology, 42, 243-270. http://dx.doi.org/10.1146/annurev.phyto.42.012604.135455

Godoy, P., Nunes, F., Silva, V., Tomimori-Yamashita, J., Zaror, L., & Fischman, O. (2004). Onychomycosis caused by *Fusarium solani* and *Fusarium oxysporum* in Sao Paulo, Brazil. *Mycopathologia, 157*, 287-290. http://dx.doi.org/10.1023/B:MYCO.0000024186.32367.d4

Gorman, S. R., Magiorakos, A., Zimmerman, S. K., & Craven, D. E. (2006). *Fusarium oxysporum* pneumonia in an immune competent host. *Southern Medical Journal, 99*, 613-616. http://dx.doi.org/10.1097/01.smj.0000217160.63313.63

Goldblum, D., Frueh, B. E., Zimmerli, S., & Bohnke, M. (2000). Treatment of post Keratitis *Fusarium* endophthalmitis with amphotericin B lipid complex. *Cornea, 19*, 853-856. http://dx.doi.org/10.1097/00003226-200011000-00019

Inam-Ul-Haq, M., Javed, N., Ahsan-khan, M., Jaskani, M. J., Khan, M. M., Khan, H. U., ... Gowen, S. R. (2009). Role of temperature, moisture and *Trichoderma* species on the survival of *Fusarium oxysporum ciceri* in the rainfed areas of Pakistan. *Pakistani Journal of Botany, 41*, 1965-1974.

Jarvis, W. R., & Shoemaker, R. A. (1978). Taxonomic status of *Fusarium oxysporum* causing foot and root rot of tomato. *Phytopathology, 68*, 1679-1680. http://dx.doi.org/10.1094/Phyto-68-1679

Jeschke, N., Nelson, P. E., & Marasas, W. F. O. (1990). *Fusarium* spp. isolated from soil samples collected at different altitudes in the Transkei, southern Africa. *Mycologia, 82*, 727-733. http://dx.doi.org/10.2307/3760159

Kedera, C. J., Plattner, R. D., & Desjardins, A. E. (1999). Incidence of *Fusarium* spp. and levels of Fumonisin B in Maize in Western Kenya. *Applied and Environmental Microbiology, 65*, 41-44.

Larkin, R. P., & Fravel, D. R. (2002). Effects of varying environmental conditions on biological control of *Fusarium* wilt of tomato by nonpathogenic *Fusarium* spp. *Phytopathology, 92*, 1160-1166. http://dx.doi.org/10.1094/PHYTO.2002.92.11.1160

Latiffah, Z., Mohd, Z. M., & Baharuddin, S. (2007). Diversity of *Fusarium* species in cultivated soils in Penang. *Malaysian Journal of Microbiology, 3*, 27-30.

Leslie, J. F., Pearson, C. A. S., Nelson, P. E., & Toussoun, T. A. (1990). Fusarium species from corn, sorghum, and soybean fields in the central and eastern United States. *Phytopathology, 80*(4), 343-350. http://dx.doi.org/10.1094/Phyto-80-343

Leslie, J. F., Summerell, B. A., & Bullock, S. (2006). *The Fusarium laboratory manual* (p. 388). New York: Wiley-Blackwell. http://dx.doi.org/10.1002/9780470278376

Louis, I., & Cooke, R. C. (1983). Influence of the conidial matrix of *Sphaerellopsis filum* (Darluca filum) on spore germination. *Trans BR Mycol Soc, 81*, 667-670. http://dx.doi.org/10.1016/S0007-1536(83)80151-X

Luque, A. G., Pioli, R., Bonel, B., & Alvarez, D. P. (2005). Cellulolytic fungi populations in stubble and soil as affected by agricultural management practices. *Bulletin of Agriculture Horticulture, 23*, 121-142. http://dx.doi.org/10.1080/01448765.2005.9755316

Maina, P. K., Okoth, S., & Monda, E. (2009). Impact of land use on distribution an diversity of *Fusarium species* in TaitaTaveta, Kenya. *Tropical and Subtropical Agroecosystem, 11*, 323-335.

Marasas, W. F. O., Burgess, L. W., Anelich, R. Y., Lamprecht, S. C., & Van Schalkwyk, D. J. (1988). Survey of *Fusarium* species associated with plant debris in South African soils. *S Afr J Bot, 54*, 63-71.

Mauseth, J. (2008). *Botany: An Introduction to Plant Biology* (p. 672). New York: Jones and Bartlett Publishers.

McGuire, K. L., Fierer, N., Bateman, C., Treseder, K. K., & Turner, B. L. (2012). Fungal community composition in Neotropical rain forests; influence of tree diversity and precipitation. *Microbial Ecology, 63*, 804-12. http://dx.doi.org/10.1007/s00248-011-9973-x

Mselle, J. (1999). Fungal keratitis as an indicator of HIV infection in Africa. *Tropical Doctor, 29*, 133-135.

Nelson, P. E., Toussoun, T. A., & Marassas, W. F. O. (1983). *Fusarium spp: An illustrated Guide for Identification* (p. 193). The Pennsylvania State University Press, University Park, PA.

Nemec, S. (1987). *Fusarium solani* association with branch and trunk cankers on citrus weakened by cold weather in Florida, USA. *Mycopathologia, 97*, 1443-150. http://dx.doi.org/10.1007/BF00437237

Nesci, A., Barros, G., Castillo, C., & Etcheverry, M. (2006). Soil fungal population in pre-harvest maize ecosystem in different tillage practices in Argentina. *Soil Till Res, 91*, 143-149. http://dx.doi.org/10.1016/j.still.2005.11.014

Nurhazrati, M., Hafizi, R., Nor, A. I., Baharuddin, S., & Latiffah, Z. (2012). Diversity of *Fusarium* Species from Highland Areas in Malaysia. *Tropical Life Sciences Research, 23*(2), 1-15.

Nwanma, B. N. O., & Nelson, P. E. (1993). The distribution of *Fusarium* species in soils planted to millet and sorghum in Lesotho, Nigeria and Zimbabwe. *Mycopathologia, 121*, 105-114. http://dx.doi.org/10.1007/BF01103578

Okoth, P., Okoth, S., & Jefwa, J. M. (2013). The conservation and use of micro-organisms and invertebrates in root crop-based systems: State of knowledge, trends and future prospects. *Background study paper* (No. 63, p. 64). Commission on Genetic Resources for Food and Agriculture.

Rheeder, J. P., & Marasas, W. F. O. (1998). *Fusarium* species from plant debris associate with soils from maize producing areas in the Transkei region of South Africa. *Mycopathologia, 143*(2), 113-119. http://dx.doi.org/10.1023/A:1006975825586

Rousk, J., Brookes, P. C., & Baath, E. (2010). The microbial PLFA composition as affected by pH in an arable soil. *Soil Biology and Biochemistry, 42*, 516-520. http://dx.doi.org/10.1016/j.soilbio.2009.11.026

Salas, B., & Stack, R. W. (1991). Influence of tillage on soil populations of *Fusarium* species in a spring wheat cropping system. *Phytopathology, 81*(10), 1215-1216.

Senthilkumar, G., Madhanraj, P., & Panneerselvam, A. (2011). Studies on saprophytic survival of *Fusarium oxysporum* using precolonized paddy straw bits. *Journal of National Product and Plant Resources, 3*, 15-19.

Silvestro, L. B., Stenglein, S. A., Forjan, H., Dinolfo, M. I., Arambarri, A. M., Manso, L., & Moreno, M. V. (2013). Occurrence and distribution of soil *Fusarium* species under wheat crop in zero tillage. *Spanish Journal of Agricultural Research, 11*(1), 72-79. http://dx.doi.org/10.5424/2013111-3081

Siti, N. M. S., Nur, A. I. M. Z., Nur, A. A., & Salleh, B. (2012). Diversity of *Fusarium* species Isolated from Soil Cultivated with Cucurbits within East Coast, Peninsular Malaysia. *Pertanika J. Trop. Agric. Sci., 35*(2), 381-386.

Steinkellner, S., & Langer, I. (2005). Impact of tillage on the incidence of *Fusarium* spp. in soil. *Plant and Soil, 267*(1-2), 13-22. http://dx.doi.org/10.1007/s11104-005-2574-z

Summerell, B. A., Laurence, M. H., Liew, E. C. Y., & Leslie, J. F. (2010). Biogeography and phylogeography of *Fusarium*: A review. *Fungal Diversity, 44*(1), 3-13. http://dx.doi.org/10.1007/s13225-010-0060-2

Summerell, B. A., & Rugg, C. A. (1992). Vascular wilt of *Helichrysum* species caused by *Fusarium oxysporum. Australasian Plant Pathology, 21*, 18-19. http://dx.doi.org/10.1071/APP9920018

Thorn, R. G. (1997): The fungi in soil. In J. D. Van Elsas, J. T. Trevors & E. M. H. Wellington (Eds.), *Modern soil microbiology* (pp. 63-108). Marcel Dekker, NY.

Vakalounakis, D. J., & Chalkias, J. (2004). Survival of *Fusarium oxysporum* f. sp. *radicis-cucumerinum* in soil. *Crop protection, 23*, 871-873. http://dx.doi.org/10.1016/j.cropro.2004.01.011

Wakelin, S. A., Warren, R. A., Kongand, L., & Harvey, P. R. (2008). Management factors affecting size and structure of soil *Fusarium* communities under irrigated maize in Australia. *Applied Soil Ecology, 39*, 201-209. http://dx.doi.org/10.1016/j.apsoil.2007.12.009

Wei, W., Xu, Y. L., Li, S., Liu, J. B., Han, X. Z., Li, W. B., & Ji, P. (2012). Analysis of *Fusarium* populations in a soybean field under different fertilization management by real-time quantitative PCR and Denaturing Gradient Gel Electrophoresis. *Journal of Plant Pathology, 94*(1), 119-126.

9

The Adoption of Temperate Selected Sesame Accessions in the Tropics: Selected for Japan and Grown in Ghana

Kwame D. Ansah[1], Henry O. Sintim[1], Samuel Awuah[1], Joseph E. Ali[1] & George Oteng[1]

[1] Department of General Agriculture, School of Applied Science & Technology, Sunyani Polytechnic, Sunyani, Ghana

Correspondence: Henry O. Sintim, Department of General Agriculture, School of Applied Science & Technology, Sunyani Polytechnic, P.O. Box 206, Sunyani, Ghana. E-mail: hosintim@gmail.com

Abstract

Sesame is an oilseed crop which can be grown on marginal lands. Selection of stable sesame cultivars that can adapt to local environmental conditions can be a very important food security strategy. A set of 21 high yielding sesame accessions that have been selected for a temperate region were grown in the transitional zone of Ghana during the major season of 2014. The seeds were sown after the first rain in the year in a completely randomized design, with three replications. The objective was to evaluate the effect of the contrasting environment on sesame phenology and to select cultivars with yield potential that can be accepted into local farming systems in the new environment. Morphological, physiological and agronomic traits, leading to yield were recorded in this study. Number of capsules per plant had the strongest association (72%) with seed yield. Five accessions showed a combination of early maturity < 12 weeks with high overall mean seed yield (> 20 g per plant) and good harvest index (0.29). Based on their mean performance these cultivars have been selected as promising exotic cultivars for the new locality.

Keywords: sesame, climate adaptability, cultivar selection

1. Introduction

Sesame (*Sesamum indicum* L.) is one of the world's oldest spice and oilseed crop grown mainly for its seeds (Morris, 2002; Dorothea, 2003; Francis, 2013). The leading world producers are India, China, Mexico, Myanmar and Sudan in Africa. Sesame seeds are unusually high in oil, around 50% of the seed weight and 25% protein, compared to 20% seed oil for example in soybeans.

Sesame are those tiny seeds which is most familiar as a topping for hamburger bread buns. It is an annual self-pollinating plant with an erect pubescent and sometimes branching stem. It is between 0.6 to 1.3 m tall. One flower is produced at each leaf axil and the lower flowers usually bloom 2 to 3 months after planting and continue to bloom until the terminal flowers are opened. The world production is estimated at 3.66 million tons with Asia and Africa producing 2.55 and 0.95 million tons respectively. The crop requires only 500-650 mm of rainfall per annum but it could be sensitive to drought during its vegetative stage (Boureima et al., 2011).

As an arable crop, sesame can be grown without irrigation or amended soil nutrition. It is considered drought tolerant, but needs good soil moisture to get established (Myers, 2002). It has a natural pest suppression mechanism (Sintim et al., 2014), can thrive on marginal soils and can be planted three times a year on the same land (Sintim: personal communication). This early or late planting window compares to no other oilseed crop. These attributes make sesame an excellent candidate for inclusion into low-input sustainable food systems.

The total annual consumption is about 65% for oil extraction and 35% for food (RMRDC Survey, 2004). It is used in baking, and has edible oil extracted from the seed (Obiajunwa et al., 2005). The oil carries a premium relative to other cooking oils and is considered more stable due to the presence of some antioxidants (Bedigian & Harlan, 1986; Yoshida & Kajimoto, 1994; Moazzami et al., 2007; Adebowale et al., 2011). After the oil is extracted from the seed, the remnant cake is high in protein and suitable for feedstock (Jefferson Agricultural Institute, 2003). Although for now sesame oil is used almost exclusively for human food consumption, it has potential for a variety of industrial uses. Sesame plants have also been used to successfully remove organochlorine pesticides from soils (Abhilash & Singh, 2010).

Despite the improved potential for increasing the production and productivity of sesame, there are a number of challenges. Over seven million hectares were grown worldwide in 2010, which produced about four million tons of seeds (FAO, 2012). The average world yield is still low at 0.52 ton ha^{-1} (FAO, 2012) which is from low-yielding dehiscent varieties with poor harvest index values, and have significant yield loss during threshing (Ashri, 1998; Uzun & Cagirgam, 2006). Among these production constraints, the most crucial include the lack of improved cultivars which are high yielding, non-dehiscent and also reliable seed supply (Pham et al., 2010; Tiwari et al., 2011). The average grain yield per hectare has been reported to be relatively low in most producing countries; between 300-350 kg/ha (Phillip, 1977; Saha & Bhargava, 1980; Schilling & Cattan, 1991). The fruit capsules dehisce along the suture lines and release the seeds to the ground. These wild characters prevent mechanized harvesting and the expansion of sesame cultivation (Uzun et al., 2003). However, non- dehiscent sesame varieties with yield potential of over 1 ton ha^{-1} and suitable for mechanical harvest have been developed in the last decade (Langham et al., 2008) confirming the status of sesame as a viable cash crop with tremendous export potential (Oplinger et al., 1990).

Sesame is however catching attention because breeding programs for about 80 years has seen a breakthrough (Anwar et al., 2012) where the crop's major drawback, brittle seed capsules has been solved. The main reason for this delayed success in breeding programs was that it is mainly produced in developing countries and usually by smallholders (Ashri, 2007). For now collections of improved sesame species and cultivars are available in the USA, India, Russia, China, Kenya, South Korea and to a lesser extent Japan, and these provide a valuable gene pool. Ashri (2007) has indicated that, introduced sesame varieties can adapt to new environments within 2-3 seasons of replant. Cultivars to be adopted should be selected based on the local climate, resistance to local pests, and intended use (Sheahan, 2014). Although there is ample evidence indicating that sesame yield depends on the interaction of different climatic parameters such as solar radiation, temperature and humidity (Beech, 1985; Nath et al., 2001) we still had the chance to grow temperate selected sesame cultivars in a tropical area to evaluate its adaptability and yield potential.

2. Materials and Methods

2.1 Location and Plant Materials

Twenty-one sesame accessions (Table 1) were collected from the sesame bank of University of Toyama, Japan and the pericarp size was classified as: medium, big or large with the large being the size found on bread buns were grown under field conditions after the first rain in March 2014 at Sunyani. Sunyani lies 7.3333°N, 2.3333°W in a tropical forest-savanna transitional zone with a bi-modal annual rainfall. The field experiment was carried out at the research fields of Sunyani Polytechnic, where the annual temperature fluctuates between 25 °C and 35 °C.

2.2 Experimental Procedures

The cultivars were sown in 6 m long cultivar-alternating plots which were 100 cm apart with an interplant distance of 40 cm. This was replicated in each of the three blocks. After 2 weeks of growth, plants were thinned to 2 plants per stand and data collection was initiated. Ten mid-row plants of each cultivar per block were randomly tagged as target data subjects.

2.3 Data Collection

Plant height from the base to the terminal leaves; stem height from the base to first capsule; fruiting zone length which was the height between the first basal and terminal capsules were each measured using a graduated wooden measure. The number of branches of five randomly selected tagged plants was counted. The number of capsules per plant from each of the five randomly tagged plants was taken. Plants were each cut at the base and the wet biomass yield/plant was determined (leaves, stem, capsules). Fresh capsule weight from ten plants, seed yield per plant, seeds/capsule, and 1000 seed weight were determined using an electronic balance. Harvest Index was taken as the ratio of seed yield to biomass yield per plant.

2.4 Data Analysis

Analysis of variance was carried out on above mentioned traits using 9th edition of Genstat® (Release 9.2, 2007). Means of treatments were separated using Least Significant Difference (LSD) after Analysis of Variance (ANOVA) in common probability levels.

Table 1. Description of 21 sesame accessions used in the study

No.	Accession	Pericarp colour	Pericarp size
1	965	White	big
2	565	White	medium
3	506	Tan	large
4	5951	White	large
5	999004	White	medium
6	987	White	medium
7	5041	Black	large
8	958	Tan	large
9	6197	White	large
10	5183	Black	large
11	Sudan white	White	large
12	Sudan white (sesamin rich)	White	big
13	5261	Black	medium
14	6295	Tan	big
15	9891	White	big
16	605	White	big
17	52621	Beige	medium
18	603	Tan	big
19	5391	White	medium
20	6071	White	medium
21	5762	khaki	large

Note. These accessions were collected from the sesame bank of University of Toyama, Japan.

*Pericarp size was classified as: medium big or large. Large being the normal size found on commercial buns.

3. Results and Discussion

All the tested sesame varieties had seed colours and a ≈ 12 week maturity period which were true to type as described by the seed breeder or collector. The pericarp sizes were classified to be medium, big or large. The large seed (Table 1) is in reference to commercial seeds used in the bakery industry.

Results obtained from the 21 varieties indicate that there exist wide variations among the varieties. The distance from the first capsule down to the base of the plant was between 22-80 cm (Table 2) and was significant ($p < 0.001$). There was a positive correlation (65%) between the first capsule height from the base and plant height. The tallest variety (5762) measuring 149 cm and the shortest variety (603) measuring 112 cm tall had the first capsule being 63 cm and 28 cm above ground respectively. There was a significant difference ($p < 0.02$) among the plant heights of the varieties. Plant height had a 52% correlation with yield.

The highly branched varieties, with greater than 15 branches per plant (Table 2); as compared to the no branching types could attain a higher fruiting zone length if the branches were considered. Fruiting zone in this experiment was restricted to the main stem leading to plant height. Branching is very variable in sesame and for these varieties was significant ($p < 0.001$, Table 2).

Fruiting zone which for sesame is the distance between the first basal capsule and the terminal capsule was significant among the varieties ($p < 0.001$). The mean fruiting zone length was 87 cm. Results from these varieties indicate that fruiting zone length had a negative correlation 36% with yield. This was especially manifested in the variety (987) with a high above average fruiting zone length of 105 cm producing a seed yield of 13 g/plant. This was in contrast with some varieties e.g. (965) with seed yield 29 g/plant from a fruiting zone of 85 cm. Yield per plant as a consequent was highly significantly different ($p < 0.001$, Table 4).

Table 2. Agronomic parameters of sesame cultivars

No.	Accession	Stem height from base to the first capsule [cm]	Plant height [cm]	Fruiting zone length [cm]*	Number of branches
1	965	44.0	129.0	85.0	1.0
2	565	35.1	142.3	107.2	1.7
3	506	36.4	114.4	78.0	1.3
4	5951	32.4	138.7	106.3	1.0
5	999004	79.9	141.5	61.6	17.7
6	987	29.5	133.9	104.4	0.3
7	5041	28.7	133.7	105.0	0.7
8	958	28.9	117.1	88.2	0.3
9	6197	22.0	112.4	90.4	0.6
10	5183	23.9	117.4	93.5	2.3
11	Sudan white	58.9	141.2	82.3	8.3
12	Sudan white (sesamin rich)	54.7	125.1	70.4	9.3
13	5261	72.4	147.0	74.6	14.7
14	6295	30.9	118.6	87.7	1.3
15	9891	31.7	128.8	97.1	0
16	605	32.9	120.9	88.0	0
17	52621	26.7	114.7	88.0	0.3
18	603	27.7	111.8	84.1	0.3
19	5391	71.4	130.6	59.2	14.7
20	6071	37.9	132.3	94.4	0.3
21	5762	62.8	149.3	86.5	1.00
	F. pro	<0.001	0.002	<.001	<0.001
	CV%	16.7	9.4	13.7	51.7
	LSD	11.4	19.9	19.7	3.3
	Grand mean	41.4	128.6	87.2	3.8

Note. *Fruiting zone length was the height between the first basal and the terminal capsules.

There was no significant difference ($p = 0.29$) among the varieties in the wet biomass yield (mean = 243 g/plant) although variety (999004) which is highly branched (17/plant) had a wet biomass yield of 454 g/plant. All other forage or fresh vegetable uses of the sesame varieties were significantly different (Table 3). The fresh capsule weight of the variety (9891), 2 g, was above the collection average of 1.4 g.

The 21 varieties had a mean harvest index/plant of 0.48 which was significant ($p < 0.001$). Variety (999004) also combined a low harvest index (0.29: Table 3) and a high wet biomass with good yield of 23 g/plant. Capsule numbers per plant of these varieties was the best index to predict yield (73% correlation: Table 4).

Table 3. Forage-use indices of the tested cultivars

No	Accession	Fresh capsule weight (g)	Fresh capsule weight/plant (g)	Wet Biomass yield/plant (g)	Harvest index/plant
1	965	1.380	155.4	317	0.5067
2	565	1.170	89.3	179	0.4800
3	506	1.403	110.8	182	0.5333
4	5951	1.253	88.1	182	0.4867
5	999004	1.177	126.6	454	0.2900
6	987	1.560	87.8	158	0.5667
7	5041	1.270	95.7	180	0.5167
8	958	1.367	86.4	168	0.5133
9	6197	1.423	73.7	121	0.6767
10	5183	1.077	82.6	173	0.4867
11	Sudan white	1.320	105.8	310	0.3600
12	Sudan white (sesamin rich)	1.347	87.2	264	0.3433
13	5261	1.057	84.0	768	0.3567
14	6295	1.287	53.0	112	0.4733
15	9891	2.007	98.2	166	0.5400
16	605	1.490	72.3	142	0.5233
17	52621	1.287	59.3	115	0.5233
18	603	1.197	88.5	146	0.5167
19	5391	1.503	203.8	475	0.4467
20	6071	1.183	77.0	149	0.5233
21	5762	1.287	123.0	335	0.3700
	F. pro	<0.001	0.004	0.29	<0.001
	CV%	9.90	36.5	102.9	12.80
	LSD	0.2185	58.7	412	0.1008
	Grand mean	1.335	97.5	243	0.4778

*Wet biomass yield was the fresh weight of leaves, stem and capsules on a plant. Harvest Index was taken as the ratio of seed yield to biomass yield.

The yield of any crop is a very complex quantitative character resulting from different factors, the more important of them being the yield per plant and number of plants per unit area (Menon, 1967). Although sesame has numerous varieties and ecotypes adapted to various ecological conditions (Nzikou et al., 2009; Adebowale et al., 2011) in the present study, we investigated the adaptability of temperate selected varieties in a tropical region. The tested sesame varieties had all the commercial seed colours between white and black. Seed colour is a major trait that affects consumer's acceptability. Paltak and Dixit (1992) reported that the consumer's preference for seed colour differs from region to region for which we need to confirm in Ghana. White sesame seeds, have higher oil, protein and moisture ratios as compared to black seeded sesame (Kanu, 2011). The physical appearance is a key marketing indicator and acceptability of sesame type varies greatly with cultural preferences. The Sudanese favour white seeded sesame whilst the Japanese prefer black seeded ones (Hossain et al., 2010). In particular, a larger seed size, coupled with a light-coloured seed coat like white, often command price premiums in a market-dependent manner (Cassells & Caddick, 2000; Graham et al., 2001). In parts of Asia for example, white seeded sesame sells at least 30% higher than dark coloured varieties (Chakraborty et al., 1984).

Table 4. Seed and capsule yield of sesame cultivars

No.	Accession	Number of capsules per plant	Number of seeds per capsule	Seed yield per plant (g)	1000-seed weight (g)
1	965	134.3	67.8	28.85	3.400
2	565	97.7	94.0	22.86	2.833
3	506	96.3	50.3	17.53	3.000
4	5951	111.7	53.3	17.80	3.467
5	999004	161.3	56.0	22.63	4.233
6	987	66.7	55.3	12.58	3.267
7	5041	134.3	83.7	17.69	2.533
8	958	78.0	80.7	13.62	2.800
9	6197	56.7	54.0	13.36	3.200
10	5183	81.0	81.0	19.21	3.167
11	Sudan white	60.0	60.0	19.02	3.833
12	Sudan white (sesamin rich)	60.3	60.3	15.71	3.700
13	5261	52.7	52.7	18.89	3.200
14	6295	62.7	62.7	10.55	2.733
15	9891	64.0	64.0	14.64	3.667
16	605	57.7	57.7	13.10	2.967
17	52621	82.0	43.0	11.76	3.133
18	603	82.0	66.0	12.29	2.767
19	5391	162.7	58.0	29.75	3.167
20	6071	87.7	51.0	15.97	2.833
21	5762	96.0	72.7	22.10	3.400
	F. pro	0.007	0.020	<0.001	<0.001
	CV%	34.4	23.1	29.40	9.80
	LSD	55.0	9.21	8.534	0.5204
	Grand mean	97.0	62.70	17.62	3.205

Plant density is a crucial index in sesame and an optimum of 400,000 plants/ha has been recommended (Katung, 2003). In earlier experiments in Ghana, Sintim and Yeboah Badu (2010) at 250,000 plants/ha obtained yields that were above quoted yield values in Africa. Seed yield is a complex trait governed by polygenes and therefore is influenced more by environmental factors (Basu et al., 2009). Another important factor that influences the seed yield of sesame is varietal differences hence the need to select varieties of good quality that can be adjusted to the local climatic conditions (Pham et al., 2010). In other sesame growing areas in the tropics, seed yield of 37 g per plant has been reported in controlled pollinated fields. The tested varieties with greater that 20g seed yield per plant are very promising and will be introduced into the local farming/cropping system. Although characters such as plant height of determinate varieties (Uzun & Cagirgan, 2001) and early maturation (Day, 2000; Day et al., 2002) make a great contribution to seed yield these attributes are variety specific. The most crucial index should rather relate to internode length and capsules per node. In the varieties tested, plant height and fruiting zone length were not a critical determinant of yield. Plants with tall stems should combine with short length of basal capsule from the ground, short internode and number of capsules per internode to ensure high yield. Early maturation is very important where sesame is grown as a second crop in areas with short seasons. Currently we have been able to grow sesame in a three season cycle in a year if we utilize the first rain in early March for our 12 week maturity sesame varieties. The selection of varieties should however not sacrifice plant features related to earliness for only seed yield. There have been instances where seed yield was negatively correlated with plant height (Ngala et al., 2013) which is similar to results from this experiment. In our case it was attributed to long internodes and low capsule number per node for some varieties.

Sesame has shown in other places to be highly sensitive to day length (Narayanan & Reddy, 1982; Suddihiyam et al., 1992) and has significant interactions with temperature (Suddihiyam et al., 1992). Oil content has been found to be influenced negatively by temperature (Kamal-Eldin & Appelqvist, 1994). It flowers in about 45 days under 10-hour day length. The short day length requirement was not considered to be crucial for these varieties, since summer day length is even longer in temperate regions from where the seeds were collected. Sesame is also reported to adapt to different photoperiod requirements (Ashri, 2007) as well as the existence of day-neutral cultivars. These new temperate selected varieties can fit into our cropping pattern as most of the land is normally fallow after harvesting a main crop due to the long term moisture requirements for native crops. Sesame is also known to be a good candidate when planted as a primary crop or a secondary crop after wheat (Langham et al., 2008). The performance of these adaptable varieties such as: those with more than 100 capsules per plant, 90 seeds per capsule, 4.23 g 1000-seed weight are comparable to the high yield value of 3.59 g 1000-seed weight reported elsewhere for high yielding varieties by Anwar et al. (2012) and Hassan (2012).

4. Conclusions

For other attributes not tested in this experiment: sesame can increase yields in subsequent cash crops by successfully improving soil moisture retention and soil texture. It is notable for its ability to grow under droughty conditions and in extreme heat and under conditions where few other crops can survive as it requires very few inputs. These attributes make sesame an excellent candidate for low-input sustainable food systems as required in Ghana.

References

Abhilash, P. C., & Singh, N. (2010). Effect of growing *Sesamum indicum* L. on enhanced dissipation of lindane (1,2,3,4,5, 6-hexachlorocyclohexane) from soil. *International Journal of Phytoremediation, 12*, 440-453. http://dx.doi.org/10.1080/15226510903213944

Adebowale, A. A., Sanni, S. A., & Falore, O. A. (2011). Varietal differences in the physical properties and proximate composition of elite sesame seeds. *World Journal of Agricultural Sciences, 7*, 42-46.

Anwar, M., Hasan, E.-U., Mahmood, T., Iqbal, J., & Hussain, M. (2012). TS-5: A new high yielding sesame cultivar. *Journal of Agricultural Research, 50*, 477-484.

Ashri, A. (1998). Sesame breeding. *Plant Breeding Reviews, 16*, 179-228. http://dx.doi.org/10.1002/9780470650110

Ashri, A. (2007). Sesame (*Sesamum indicum* L.). In R. J. Singh (Eds.), *Genetic Resources, Chromosome Engineering, and Crop Improvement* (Vol. 4, pp. 231-289, Oilseed Crops). Boca Raton, FL, USA: CRC Press.

Basu, S. K., Acharya, S. N., Bandara, M. S., Friebel, D., & Thomas, J. E. (2009). Effects of genotype and environment on seed and forage yield in fenugreek (*Trigonella foenum*-graecum L.) grown in Western Canada. *Australian Journal of Crop Science, 3*, 305-314. Retrieved from http://www.cropj.com/saikat_3_6_2009_305_314.pdf

Bedigian, D., & Harlan, J. (1986). Evidence for cultivation of sesame in the ancient world. *Economic Botany, 40*, 137-154.

Beech, D. (1985). Sesame research possibilities for yield improvement. In A. Ashri (Eds.), *Sesame and safflower: Status and potential* (pp. 121-126). FAO plant production and protection, Rome. Retrieved from http://safflower.wsu.edu/newsletter1985.pdf

Boureima, S., Eyletters, M., Diouf, M., Diop, T. A., & Van Damme, P. (2011). Sensitivity of seed germination and seedling radicle growth to drought stress in sesame (*Sesamum indicum* L.). *Research Journal of Environmental Sciences, 5*, 557-564. http://dx.doi.org/10.3923/rjes.2011.557.564

Cassells, J., & Caddick, L. (2000). Harvest timing critical for chickpea quality. *Farming Ahead, 107*, 41-42.

Chakraborty, P. K., Maiti, S., & Chatterjee, B. N. (1984). Growth analysis and agronomic appraisal of *Sesamum indicum* L. *Indian Journal of Agricultural Science, 54*, 291-295.

Day, J. S. (2000). Development and maturation of sesame seeds and capsules. *Field Crops Research, 67*, 1-9. http://dx.doi.org/10.1016/S0378-4290(00)00073-3

Day, J. S., Langham, D. R., & Wongyai, W. (2002). Potential selection criteria for the development of high-yielding determinate sesame varieties. *Sesame Safflower Newsletter, 17*, 29-35.

Dorothea, B. (2003). Evolution of sesame revisited: domestication, diversity and prospects. *Genetic Resources and Crop Evolution, 50*, 779-787. http://dx.doi.org/10.1023/A:1025029903549

FAO. (2012). *FAOSTAT Databases*. Retrieved from http://www.faostat.fao.org

Francis, E. (2013). *Sesame: For the Cooperative Extension Service University of Arkansas System Division of Agriculture*.

Genstat® Release 9.2. (2007). *Lawes Agricultural Trust*. Rothamted Experimental Station.

Graham, J., Matassa, V., Panozzo, J., & Starick, N. (2001). *Genotype and environment interaction for whole grain colour in chickpea* (pp. 372-373). 4th European Conference on Grain Legumes.

Hassan, M. A. M. (2012). Studies on Egyptian sesame seeds (*Sesamum indicum* L.) and its products 1-physicochemical analysis and phenolic acids of roasted Egyptian sesame seeds (*Sesamum indicum* L.). *World Journal of Dairy & Food Sciences, 7*, 195-201.

Hossain, S., Ford, R., McNeil, D., Pittock, C., & Pannozo, J. F. (2010). Inheritance of seed size in chickpea (*Cicer arietinum* L.) and identification of QTL based on 100-seed weight and seed size index. *Australian Journal of Crop Science, 4*, 126-135.

Jefferson Agricultural Institute. (2003). *Strategies for Commercializing New Crops* (Report of a Workshop). St. Louis, Missouri Thomas Jefferson Agricultural Institute. Retrieved from http://www.jerfersoninstitute.org

Kamal-Eldin, A., & Appelqvist, L. A. (1994). Variation in fatty acid composition of the different acyl lipids in seed oils four Sesamum species. *Journal of American Oil Chemists' Society, 71*, 135-139. http://dx.doi.org/10.1007/BF02541547

Kanu, P. J. (2011). Biochemical analysis of black and white Sesame seeds from China. *American Journal of Biochemistry and Molecular Biology, 1*, 145-157. http://dx.doi.org/10.3923/ajbmb.2011.145.157

Katung, P. D. (2003). The response of two varieties of sesame (*Sesamum indicum,* L.) to different plant populations and sowing dates at Samaru, Northern Guinea savanna. *Samaru Journal of Agricultural Research, 19*, 17-27.

Langham, D. R., Riney, J., Smith, G., & Wiemers, T. (2008). *Sesame grower guide*. Sesaco sesame coordinators, Lubbock, TX. Retrieved Decemeber 18, 2014, from http://www.sesaco.net

Menon, E. F. (1967). Effects of varying spacing on yield of sesame. *Indian Journal of Agronomy, 12*, 274-276.

Moazzami, A. A., Haese, S. L., & Kamal-Eldin, A. (2007). Lignan contents in sesame seeds and products. *European Journal of Lipid Science and Technology, 109*, 1022-1027. http://dx.doi.org/10.1002/ejlt.200700057

Morris, J. B. (2002). Food, industrial, nutraceutical, and pharmaceutical uses of sesame genetic resources. In J. Janick & A. Whipkey (Eds.), *Trends in New Crops and New Uses* (pp. 153-156). ASHS Press, Alexandria, VA.

Myers, L. R. (2002). *Sesame: A high value oilseed.* Alternative crop guide Jefferson Institute, Columbia, MO. Retrieved from http://www.jeffersoninstitute.org

Myers, R. (2002). *Alternative crop guide: Sesame.* Jefferson Institute, Columbia, MO. Retrieved July 15, 2014, from http://www.jeffersoninstitute.org

Narayanan, A., & Reddy, K. B. (1982). Growth, Development and yield of sesame (*Sesamum indicum* L.) cultivars. *Field Crop Research, 5*, 217-224. http://dx.doi.org/10.1016/0378-4290(82)90024-7

Nath, R., Chakraborty, P. K., & Chakraborty, A. (2001). Effect of climatic variation on yield of sesame (*Sesamum indicum* L.) at different dates of sowing. *Journal of Agronomy and Crop Science, 186*, 97-102. http://dx.doi.org/10.1046/j.1439-037X.2001.00456.x

Ngala, A. L., Dugje, I. Y., & Yakubu, H. (2013). Effects of inter-row spacing and plant density on performance of sesame (*Sesamum indicum* L.) in a Nigerian Sudan savanna. *Science International (Lahore), 25*, 513-519. http://dx.doi.org/10.13140/2.1.2230.3368

Nzikou, J. M., Matos, L., Bouanga-Kalou, G., Ndangui, C. B., Pambou-Tobi, N. P. G., Kimbonguila, A., & Desobry, S. (2009). Chemical composition on the seeds and oil of sesame (*Sesamum indicum* L.) grown in Congo-Brazzaville. *Advance Journal of Food Science and Technology, 1*, 6-11.

Obiajunwa, E. I., Adebiyi, F. M., & Omode, P. E. (2005). Determination of essential minerals and trace elements in Nigerian sesame seeds using TXRF technique. *Pakistan Journal of Nutrition, 4*, 393-395. http://dx.doi.org/10.3923/pjn.2005.393.395

Oplinger, E. S., Punta, D. H., Kaminski, A. R., Hanson, C. V., Oelke, E. A., Schutte, E. E. & Doll, J. D. (1990). *Sesame: Alternative field crop manual*. Retrieved from http://www.hort.purdue.edu/newcrop.afcm/sesame.html

Paltak, H. C., & Dixit, S. K. (1992). Genetic variability and inter-relationship studies in black seeded sesame (*Sesamum indicum* L.). *Madras Agricultural Journal, 79*, 4-100.

Pham, T. D., Nguyen, T. T., Carlsson, A. S., & Bui, T. M. (2010). Morphological evaluation of sesame (*Sesamum indicum* L.) varieties from different origins. *Australian Journal of Crop Science, 4*, 498-504. http://dx.doi.org/dn=536481490080115

Phillip, T. A. (1977). *Sesame: An agricultural notebook* (pp. 61-63). New edition, Longman Group Ltd., London.

Raw Materials Research and Development Council-RMRDC. (2004). Survey report of ten selected agro raw materials in Nigeria. *Beniseed* (1st ed., p. 108).

Saha, S. N., & Bhargava, S. C. (1980). Physiological analysis of growth, development and yield of oilseed sesame. *Journal of Agricultural Science, 958*, 733-736. http://dx.doi.org/10.1017/S0021859600088158

Schilling, R., & Cattan, P. (1991). Sesame cultivation in tropical Africa. *Oleagineux, 46*, 129-131.

Sheahan, C. M. (2014). *Plant guide for sesame (Sesamum orientale)*. USDA-Natural Resources Conservation Service, Cape May Plant Materials Center, Cape May, NJ.

Sintim, H. O., & Yeboah Badu, V. I. (2010). Evaluation of Sesame (*Sesamum indicum*) production in Ghana. *Journal of Animal and Plant Sciences, 6*, 653- 662. Retrieved from http://www.biosciences.elewa.org/JAPS

Sintim, H. O., Yeboah Badu, V. I., Akotsen Mensah, C., & Ansah Duodu, K. (2014). A faunistic description of endemic arthropods associated with sesame. *Advances in Natural Science, 7*, 1-7. http://dx.doi.org/10.3968%2F4788

Suddihiyam, P., Steer, B. T., & Turner, D. W. (1992). The flowering of sesame (*Sesamum indicum* L.) in response to temperature and photoperiod. *Australian Journal of Agricultural Research, 43*, 1101-1116. http://dx.doi.org/10.1071/AR9921101

Tiwari, S., Kumar, S., & Gontia, I. (2011). Biotechnological approaches for sesame (*Sesamum indicum* L.) and niger (*Guizotia abyssinica* L. f. Cass.). *Asia-Pacific Journal of Molecular Biology and Biotechnology, 19*, 2-9.

Uzun, B., & Cagirgan, M. I. (2001). Path-coefficient analysis for seed yield and related characters in a population of determinate and indeterminate types of sesame (*Sesamum indicum* L.). *Turkish Journal of Field Crops, 6*, 76-80. Retrieved from http://www.field-crops.org/assets/pdf/product51b2ec340cef9.pdf

Uzun, B., & Cagirgan, M. I. (2006). Comparison of determinate and indeterminate lines of sesame for agronomic traits. *Field Crops Research, 96*, 13-18. http://dx.doi.org/10.1016/j.fcr.2005.04.017

Uzun, B., Lee, D., Donini, P., & Cagirgan, M. I. (2003). Identification of an AFLP marker linked to the closed capsule mutant trait in sesame. *Plant Breeding, 122*, 95-97. http://dx.doi.org/10.1046/j.1439-0523.2003.00787.x

Yoshida, H., & Kajimoto, M. (1994). Composition and quality characteristics of sesame seed (*Sesamum indicum*) oil roasted at different temperature in an electric oven. *Journal of Food Science, 65*, 331-336. http://dx.doi.org/10.1002/jsfa.2740650311

10

Effect of Inorganic and Organic Fertilizers on the Growth and Yield of Garlic Crop (*Allium sativum* L.) in Northern Ethiopia

Abraha Gebrekiros Assefa[1], Solomon Habtu Mesgina[1] & Yirga Weldu Abrha[1]

[1] Department of Land Resources Management and Environmental Protection, Mekelle University, Mekelle, Ethiopia

Correspondence: Abraha Gebrekiros Assefa, Department of Land Resources Management and Environmental Protection, Mekelle University, Mekelle, Ethiopia. E-mail: abrshmekelle@gmail.com

Abstract

Among others, low soil fertility status is a limiting factor for crop production in the study area in particular and northern Ethiopia in general. Increment of garlic yield with balanced fertilization is a key important factor identified. However, use of correct type and rate of fertilizer applications are major problems in the study area. The study was conducted to investigate the effect of N, P, S, and Zn fertilizers and compost on yield and growth parameters of garlic (*Allium sativum* L.) and specify optimum fertilizer dose at field level. The experiment was laid out in randomized complete block design with three replications and five experimental plots (treated with N-P-S-Zn, N-P-S, N-P, compost and control) with unit plot size of 3 m × 3.5 m = 10.5 m^2 and subjected to analysis of variance. The results showed that a combination of N, P, S, and Zn fertilizers gave a significant higher yield over the control plot ($P = 0.05$). The mean bulb yield obtained from plots fertilized with N-P-S-Zn, N-P-S, N-P fertilizers and compost was 4760, 4388, 4240 and 3451 kg he^{-1}, respectively while 2996.5 kg he^{-1} was the lowest average yield obtained from control plot. Similarly, the yield and growth parameters were increased progressively with combined application of N, P, S, and Zn. It can therefore be concluded that for increased garlic yield in the study area, application of 130 kg N, 20 kg P, 21 kg S and 15 kg Zn fertilizers per hectare could be needed.

Keywords: bulb yield, combined fertilizer, compost, garlic cloves

1. Introduction

Garlic (*Allium sativum* L.) was originated from central Asia about 3000 years and later spread to the Mediterranean regions (Tindal, 1986). Garlic is the main spices crop grown in Ethiopia. The annual production of garlic in Ethiopia is about 22,254.793 ton during 2012/2013 growing season and it is the second most important spice crops next to onion (FDRECSA, 2013). It is being consumed for curing many human diseases (Durak, Kavutch, & Aytac, 2004; Pitman, 2008)0. Balanced fertilizer application is essential for the vegetative growth and, thus, for producing crops with top quality and high yields especially on soils that are cultivated continuously (Chintala, McDonald, & Bryan, 2012a, 2012b). Garlic growers in the central high lands of Ethiopia tend to rely on fertilizer sources that contain only nitrogen (N) and phosphorus (P), resulting in steady decline in other nutrients in the soil. Smaller potassium (K) and sulfur (S) uptake relative to N uptake can predispose the crop to serious disease and insect damage (ENAIA, 2003). All the plants are characterized by a shallow root system which explains why fertilizers should be banded at 8-10 cm below the seed row (SOPIB, 2001). Due to the shallow root system, it is preferable to split fertilizer application (before sowing or planting, at the fully expanded leaf stage, and just before bulb formation). High nutrient availability is important during bulb formation. The bulb comprises small bulb called cloves (Wadjito, Rubiatai, & Abidin, 1988). Availability of Nitrogen and Phosphorous is of prime importance for growing plants as it is a major and indispensable source of protein and nucleic acid molecules. It is also an integral part of chlorophyll molecules, which are responsible for photosynthesis. Sulfur also an essential plant nutrient with role in balanced fertilization and consequently in crop production is being increasingly appreciated. Sulfur performs many physiological functions like synthesis of sulfur and methionine (Baird, 1991). Sulfur application in garlic also enhances the uptake of N, P, K, and Ca by the crop (Hossain, 1997). Its pungency is due to the presence of a volatile oil (allyl propyl disulphide) (Malik, 1994). It is well known that the use of fertilizer helps in production and is a somewhat quick method for achieving maximum yield (Naruka & Dhaka, 2001).

In Gumselasa irrigated fields of northern Ethiopia (the study area), farmers grow two to three crops in a same piece of land in a year. It is expected that such cropping practice leads to nutrient imbalance or depletion in the soil. Hence, evaluating the initial soil fertility status of the area is very important to recommend a balanced fertilizer dosage for a specific crop. However, it is not common practice to assess how much fertilizer remains in the field for the next crop and how much would be applied to achieve higher yield in respect of benefits in the study area. In the absence of initial soil fertility status assessment practice and location-specific fertilizer application rate, farmers of the study area apply fertilizers simply based on their local practice called blanket fertilizer recommendation. Consequently, such field level soil fertility management makes the soil heterogeneous and fertility depletes in a long run that could be accompanied by low production. Garlic being a cash crop, increment of yield with balanced fertilization is one of the identified researchable issues in the area. This study was, thus, conducted to find out optimum dosage of fertilizer for higher production in the area by assessing the deficiency or sufficiency levels of N, P, S, and Zn nutrients in the field and evaluating the effect of combined N, P, S, and Zn fertilizations and compost treatments on the yield and yield components of garlic crop.

2. Materials and Methods

2.1 Area Description

The study area is located in the northern highlands of Ethiopia at Gumselassa irrigation scheme (13°14′N and 39°32′E) at an altitude of 1960 m.a.s.l. The scheme is situated about 4 km east of Adigudom town in the Tigray region. The area belongs to the "Weynadega" (characterized by an altitude of 1400-2600 m.a.s.l based on traditional classification) agro ecological zone. The rainfall in the study area is highly variable with space and time. It is in the range of 300-900 mm per year and uni-modal, with more than 85% of the rain falling within a period of four months from June to September. Drought periods of several weeks during the rainy season are also frequent (Mintesinot, Tigabu, & Fassil, 1999). The maximum and minimum average temperatures are 28 °C and 9 °C, respectively (SAERT, 1994). The largest part of the area is dominated by arable lands having a gentle slope and encircled by a smaller area with steep sloping hills. About 61% of the area is covered with black clayey soils (Mintesinot, Tigabu, & Fassil, 1999; Table 2) having pH value ranging between 7.72 and 8.11 (Table 3).

2.2 Methods

The experiment was conducted in farmer's field for two consecutive irrigation seasons in 2012/2013 and 2013/2014 from February to May. The experimental plot was laid out in randomized complete block design with three replications and five treatment plots. A local garlic cultivar called 'Rie' was used as planting material. The unit plot size was 3 m × 3.5 m = 10.5 m^2. The treatments consisted of three different combinations of N, P, S, and Zn fertilizers (N-P, N-P-S and N-P-S-Zn) assigned to three plots, fourth plot fertilized with compost and fifth plot was a control (Table 1).

Table 1. Fertilizer application rates

Treatments	Nutrient Doses (Kg he^{-1})				
	N	P	S	Zn	Compost
T_0	0	0	0	0	0
T_1	130	20	0	0	0
T_2	130	20	21	0	0
T_3	130	20	21	15	0
T_4	0	0	0	0	12,000

Note. T_0 is control; T_1 is nitrogen and phosphorous; T_2 is nitrogen, phosphorous and sulphur; T_3 is nitrogen, phosphorous sulphur and Zinc and T_4 is compost.

Two soil samples were randomly picked from each replication sites at a depth of 0-20 cm using auger to form a composite sample. The soil samples were air dried, crushed in a mortar and sieved through a 2 mm sieve. Chemical and physical soil analyses were done on the separates pass through 2mm mesh only. The pH of the soils was determined in 1:2.5 soil water suspensions using a glass electrode and the electrical conductivity of the soils was measured in saturated extract (Van Reeuwijk, 1992). Soil bulk densities were determined by analyzing the collected undisturbed soil samples following the core sampling method (Black, 1965). The Walkley and Black method was used to determine organic carbon content of the soils (Walkley & Black, 1934) and the organic matter content was obtained by multiplying the percent organic carbon by conversion factor of 1.724. Total nitrogen was analyzed using the Kjeldahll method (Black, 1965). Available phosphorus was determined by Olsen method (Olsen & Dean, 1965). Available sulfur was determined using the KCl-40 test, where 0.25 M KCl was added to the soil and extracted at 40 °C temperature for three hours (Blair, Chinoim, Lefroy, Anderson, & Crocker, 1991). Available zinc was determined after air-dried soils were extracted in 1% EDTA, the filtrate was aspirated into an air-acetylene flame of an atomic absorption spectrophotometer and the absorbance was recorded (Baker, 1974). The soil particle size distribution was determined using hydrometer method (Van Reeuwijk, 1992) and classified in to its soil class on the bases of particle size (Table 2).

Fertilizer doses were calculated according to initial soil nutrient status of the experimental plots. The soil texture was determined as clay soil (Table 2). Soil pH was fairly alkaline, the total N status, available phosphorus, sulfur, and zinc were either at average or below the critical level and organic matter content of the soil was low (Table 3).

Table 2. Soil textural class of the study area according to USDA

Sample Code	Soil Separate Value			Textural Class
	%Sand	%Silt	%Clay	
Replication -1	17	21	62	clay
Replication -2	17	19	64	clay
Replication -3	21	21	58	clay

Note. USDA - United States Department of Agriculture.

The entire quantity of P, S, Zn with one half of N were applied as basal dose at the time of final land preparation and the remaining half N was applied as top-dress at the fully expanded leaf stage, and just before bulb formation with side-dressing (Table 1). The sources of N, P, S, and Zn were urea, DAP, zinc-oxide and ammonium-sulfate, respectively. Sprouted garlic bulbs with diameter of 5-8 inch were planted within the first week of February in both seasons at a spacing of 20cm x 8cm between rows and plants, respectively. All the management practices except the experimental variables were the same to all plots. Land was ploughed three times to prepare a favorable seedbed. The crop was harvested on 16 May 2013 and 15 May 2014 at maturity stage when most of the leaves dried up and bended over. Care was taken to avoid any kind of bulb injury during lifting. Data on plant parameters (plant height, total bulb weight, bulb diameter, fresh and dry matter weight of garlic) were collected properly. The collected data at various growth stages in response to the applied combinations of fertilizer and compost were recorded and analyzed. The results were subjected to analysis of variance (ANOVA) using SAS statistical software version 9.0 and treatment effects were compared using the Fisher's Least Significant Differences test at 5% level of significance.

Table 3. Nutrient status of the initial soil

Soil parameters	Values			
	Replication -1	Replication -2	Replication -3	Average
pH (1:2.5 Soil-water)	8.11	8	7.72	7.94
Bulk density (g cm^{-3})	1.2	1.34	1.5	1.35
Organic matter (%)	2.07	2.6	2.49	2.39
Total N (%)	0.113	0.109	0.108	0.11
Available P (mg kg^{-1})	2.34	2.1	3.84	2.76
Available S (mg kg^{-1})	18.2	17.6	18.4	18.07
Available Zn (mg kg^{-1})	0.98	1.04	1.1	1.04

All data pertaining to growth yield and yield components were collected from 5 plants in each of the 15 experimental plots of garlic. A random sampling was done from three central rows of the plot for determination of plant height. After harvesting and curing, bulb yield and yield attributes (bulb weight per plot, number of cloves per bulb, bulb diameter, bulb fresh weight and dry matter weight) were recorded.

3. Results

3.1 Bulb Yield

During the irrigation seasons of 2012/2013 and 2013/2014 in the study area, garlic showed a significant variation in yield due to application of different combinations of fertilizer treatments at 5% probability. On an average of the two irrigation seasons, combined application of N, P, S, and Zn nutrients significantly influenced garlic yield ($P = 0.05$) over others with average yield of 4760 kg he^{-1} (Table 4). Plots fertilized with N, P, and S alone yielded up to 4388 kg he^{-1} and plots fertilized with N and P alone yielded on an average up to 4240 kg he^{-1}. Plots fertilized with compost also yielded 3451 kg he^{-1} on an average. The control treatment plots receiving no fertilizer, however, had the lowest yield (2996.5 kg he^{-1}) on average.

Table 4. Yield as affected by different combinations of fertilizer and compost

Treatments	Bulb Yield (kg he^{-1})		Mean Yield (kg he^{-1})
	2012/2013	2013/2014	
T_0	2183a	3810c	2996.5
T_1	3717a	4762b	4240.0
T_2	3700a	5076b	4388.0
T_3	3967a	5552a	4760.0
T_4	2933a	3969c	3451.0
CV%	38.53	4.5	

Note. he is hectare, T_0 is control; T_1 is nitrogen and phosphorous; T_2 is nitrogen, phosphorous and sulphur; T_3 is nitrogen, phosphorous sulphur and Zinc and T_4 is compost.

3.2 Plant Height

Combined application of N, P, S, and Zn showed a significant variation in plant height at different combinations including compost application (Table 5). Plant height increased gradually with increasing the combinations of nutrient types N, P, S, and Zn within the recommended doses. The tallest plants (49.47 cm in 2012/2013 and 68.67 cm in 2013/2014) were recorded in plots fertilized with combined N, P, S, and Zn fertilizers. Conversely, the shortest plants (43.73 cm in 2012/2013 and 46.33 cm in 2013/2014) were found in control plots. Plant height recorded in 2012/2013 across the treatments followed the pattern similar to those observed in 2013/2014.

3.3 Bulb Diameter

Application of different combinations of N, P, S, and Zn has created significant impact on bulb diameter of garlic (Table 5). The highest bulb diameter (3.28 cm in 2012/2013 and 5.37 cm in 2013/2014) was recorded in plots that received N, P, S, and Zn, while the lowest (2.99 cm in 2012/2013 and 3.49 cm in 2013/2014) was observed in control treatment.

3.4 Bulb Fresh Weight

A significant variation in fresh weight of bulb at harvest was observed due to application of different combinations of nutrients (Table 5). The maximum weight of fresh bulb (130.91 g in 2012/2013 and 157.31 g in 2013/2014) was obtained in plots fertilized with N, P, S, and Zn nutrients while the minimum weight of bulb (100.34 g in 2012/2013 and 83.26 g in 2013/2014) was observed in the control treatment.

3.5 Bulb Dry Weight

Dry bulb weight at harvest was significantly affected by different combinations of nutrients N, P, S, and Zn application (Table 5). The maximum bulb dry weight (42.83 g in 2012/2013 and 61.93 g in 2013/2014) was recorded in plots that received combined N, P, S, and Zn nutrients. Minimum dry weight of bulb (32.49 g in 2012/2013 and 36.58 g in 2013/2014) was, however, recorded in the control plots.

3.6 Number of Cloves Per Bulb

Application of combined fertilizers and compost significantly increased number of cloves per bulb of garlic over control in both seasons. The maximum numbers of cloves per bulb (8.69 in 2012/2013 and 11.0 in 2013/2014) were recorded in plots fertilized with N, P, S, and Zn nutrients while the minimum values (4.48 in 2012/2013 and 4.60 in 2013/2014) were recorded in plants where no fertilizers was applied (Table 5).

Table 5. Yield parameters of garlic as affected by different fertilizer combinations and compost

Treatments	Plant height (cm)		Bulb diameter (cm)		Fresh weight (gm.)		Dry weight (gm.)		Number of cloves per bulb	
	2012/2013	2013/2014	2012/2013	2013/2014	2012/2013	2013/2014	2012/2013	2013/2014	2012/2013	2013/2014
T_0	43.733b	46.33e	2.99a	3.49d	100.34a	83.260d	32.49a	36.58e	4.48a	4.60d
T_1	46.86ab	55.47c	3.10a	4.44b	120.14a	136.08b	40.54a	51.23c	5.23a	5.47c
T_2	47.67ab	64.67b	3.25a	4.59b	120.11a	154.88a	41.06a	53.11b	6.54a	8.33b
T_3	49.470a	68.67a	3.28a	5.37a	130.91a	157.31a	42.83a	61.93a	8.69a	11.0a
T_4	46.53ab	53.33d	3.01a	3.88c	120.95a	112.85c	40.73a	48.00d	6.05a	7.93b
CV%	15	2.495	131	7.081	15.86	4.265	42.09	1.314	57.74	15.32

Note. T_0 is control; T_1 is nitrogen and phosphorous; T_2 is nitrogen, phosphorous and sulfur; T_3 is nitrogen, phosphorous sulfur and Zinc and T_4 is compost; cm is centimeter and gm is gram.

4. Discussion

Nitrogen supply to the plant increases the rate of metabolism where more carbohydrate is synthesized. This in turn increases the bulb weight and total yield as confirmed by the results of Amin, Rahim, and Hashem (1995) who reported that 100 kg N he^{-1} resulted in maximum yield. Conversely, minimum yield was recorded in the control plot (Table 4); which could be due to low fertility status of the original soil indicating a need for balanced fertilization. Accordingly, higher yield of garlic was observed in plots fertilized with N, P, S, and Zn fertilizers over the others (Table 4). The possible justification for this higher result could be due to combined effect of the contributions of nitrogen to chlorophyll, enzymes, and proteins synthesis; as phosphorus is essential for root growth, phospho-proteins and phospho-lipids. The diameter and weight of bulbs were significantly improved with the application of S up to 24 kg/ha. It also might be due to adequate nutrient supply which favored the enlargement of bulb and its weight. The result was in conformity with the findings of Nasiruddin, Farooque, and Baten (1993) that reported the application of both potassium and sulfur either individually or in combined increased plant height, number of leaves, bulb diameter, bulb weight and bulb yield. Balasubramonium, Raman, and Moorty (1979) studied that the added S had positive effect in increasing the yield of onion but a reduction in yield with very high dose of S was also observed.

As shown in Table 5, there was a significant difference ($P = 0.05$) of plant height among the treatments. The observed increase in plant height particularly in the treatments fertilized with Zinc is due to the fact that zinc involves in auxin metabolism and other enzymatic reactions that increase leaf length as reported in Tisdale, Nelson, and Beaton (1985). It may be also due to the collective effect of both nitrogen and zinc that stimulate plant growth and thus increases leaf length. Significant difference ($P = 0.05$) was also observed on number of cloves per bulb among the treatments. The control plots significantly gave lower mean values on both plant height and number of cloves per bulb. Generally, significant difference of vegetative growth as affected by different combinations of nutrient elements and compost was observed among treatments.

5. Conclusion

The soils of Gumselassa irrigation scheme are low in organic matter and deficient in major nutrients. Consequently, yield of garlic crop including other crops growing in this area is very low. One of the major limitations responsible for the low yield in this area is indeed low soil fertility and incongruence between the capacity of the soils to provide nutrients and the amounts of nutrients added from fertilizers needed to supplement the soils. From the results obtained during the two consecutive irrigation seasons, the yield of garlic crop in terms of bulb and growth parameters increased with combined application of N, P, S and Zn. Based on the obtained results, combined application of nitrogen, phosphorous, sulfur, and zinc with the rate of 130 kg, 20 kg, 21 kg and 15 kg per hectare, respectively, is the most suitable and recommendable dose for maximum growth and yield of garlic in the study area.

Acknowledgements

The authors express their appreciation to the European Union and African Union cooperative research to increase food production in irrigated farming systems in Africa (EAU4FOOd) project for funding the research work and Mekelle University for facilitating the research work. The scientific legacy of the late Dr. Kiros Habtergebreal embedded in this piece of research work deserves a special praise. We would also like to thank the farmers in the study area as well as the staff members and laboratory technicians at the Mekelle Agricultural Research Center in Tigray who contributed technical and materials support to this research.

References

Amin, M. M. U., Rahim, M. A., & Hashem, M. A. (1995). Influences of planting time and nitrogen on the growth and yield of onion. *Bangladesh J. Scientific and Indust. Res., 30*, 2-3.

Aujla, T. S., & Madan, S. P. S. (1992). Response of onion (*Allliuncepa*) to irrigation and row spacing on deep sandy-loam soil in subtropical monsoon region. *Indian J. Agril. Sciences, 62*(2), 129-134.

Baird, J. (1991). *Soil facts/AG-439-15*. North Carolina cooperative extension service.

Baker, D. E. (1974). Determination of the relative availability H, Ca. Mg, K, Na, P, S, Fe, Mn, Zn, Cu, Al and other metals by a small exchangeable approach to soil testing. *Hand Book on Reference Methods for Soil Testing* (pp. 92-102). Council on Soil Testing and Plant Analysis Athens, Ga.

Balasubramonium, A. S., Raman, G. V. G., & Moorty, K. K. K. (1979). Effect of sulfur application in the yield and quality of onion (*Allium cepa* L.). *Agril. Res. J. Kerala, 17*(1), 138-140.

Black, C. A. (1965). *Method of soil analysis* (Part I). American Society of Agronomy. Madison, Wisconsin, USA.

Blair, G. J., Chinoim, N., Lefroy, R. D. B., Anderson, G. C., & Crocker, G. J. (1991). A soil sulphur test for pastures and crops. *Australian Journal of Soil Research, 29*, 619-626. http://dx.doi.org/10.1071/SR9910619

Chintala, R., McDonald, L. M., & Bryan, W. B. (2012a). Effect of soil water and nutrients on productivity of Kentucky bluegrass systems in acidic soils. *Journal of Plant Nutrition, 35*, 288-303. http://dx.doi.org/10.1080/01904167.2012.636131

Chintala, R., McDonald, L. M., & Bryan, W. B. (2012b). Optimization of water potential and nutrient levels for Kentucky bluegrass-white clover mixture on acidic soils. *Biotechnology, Agronomy, Society and Environment, 16*, 167-177.

Durak, I., Kavutch, M., & Aytac, B. (2004). Effect of garlic extract consumptions on blood lipid and oxidant/ antioxidant parameters in human with high blood cholesterol. *J. Nutr. Biochem, 15*(6), 373-377. http://dx.doi.org/10.1016/j.jnutbio.2004.01.005

Ethiopian National Agricultural Input Authority (ENAIA). (2003). *Crop Variety Register, 5*, 101-102.

FDRECSA. (2013). Federal Democratic Republic of Ethiopia, Central Statistical Agency, agricultural sample survey of 2012/2013 (2005EC), Volume I, Report on area and production of major crops. *Addis Ababa Ethiopia, statistical bulletin 532*.

Hossain, M. M. (1997). *Effect of different levels of nitrogen and potash on the growth and yield of garlic* (M.S. Thesis, p. 65). Dept. Hort. Bangladesh Agril. Univ., Mymensingh.

Malik, M. N. (1994). Bulb crops, Onion. *Horticulture* (pp. 500-501). National Book Foundation Islamabad Pakistan.

Mintesinot, B., Tigabu, L., & Fassil, K. (1999). An expletory study of farming systems under rain-fed and irrigated agriculture at Gumselassa Micro-Dam site in southern Tigray. *Project MUC-RUG*. Mekelle, Ethiopia.

Naruka, I. S., & Dhaka, R. S. (2001). Effect of row spacing and nitrogen fertilization ongrowth, yield and composition of bulb in garlic (*Allium sativum* L.) cultivars. *Journal of Spices and Aromatic Crops, 10*(2), 111-117.

Nasiruddin, K. M., Farooque, A. M., & Baten, M. A. (1993). Effect of potassium and sulphur on growth and yield of onion. *Bangladesh J. Agril. Sci., 20*(1), 35-40.

Olsen, S. R., & Dean, L. A. (1965). Phosphorus. In C. A. Blach (Ed.), *Methods of soil analysis* (Agronomy No. 9, pp. 1044-1046). American society of agronomy, Madison.

Pitman, D. (2008). Medina Research Station and John Burt, Horticulture Adviser, South Perth.

SAERT. (1994). *Sustainable Agriculture and Environmental Rehabilitation in Tigray*. Master book of Tigray, Mekelle, Ethiopia.

SOPIB. (2001). *Sulphate of Potash information board and Bulb Crops (Onions, Shallots, Garlic, Leeks)*. Retrieved from http://www.sopib .com

Tindal, H. D. (1986). *Vegetables in the Tropics* (p. 533). Macmillan Education Limited Hound Mills Basing Toke, Hamshire, England.

Tisdale, S. L., Nelson, W. L., & Beaton, J. D. (1985). Micronutrients and other beneficial elements in soils and fertilizers. *Soil fertility and fertilizers, Zinc* (pp. 387-388). Macmillan Publishing Company, New York.

Van Reeuwijk, L. P. (1992). *Procedure for soil analysis* (2nd ed.). Int. Soil Reference and Information Center (ISRIC), the Netherlands.

Wadjito, A., Rubiatai, A., & Abidin, Z. (1988). The effect of bulb size and spacing on the production of garlic (*Allium sativum* L.) cultivar Gombloh. *Bulletin Penelitian Horticulture, 16*(3), 63-68.

Walkley, A., & Black, C. A. (1934). Examination of different methods for determining organic matter and the proposed modification by the chromic acid titration method. *Soil Sci, 34*, 29-38. http://dx.doi.org/10.1097/00010694-193401000-00003

11

The Impact of Mineral Nitrogen Fertilization on the Occurrence of Native Diazotrophic Bacteria in Kohlrabi (*Brassica oleracea)* Shoots and Roots

Lempie K. Ekandjo[1] & Silke Ruppel[1]

[1] Leibniz-Institute of Vegetable- and Ornamental Crops, Großbeeren/Erfurt e.V., Theodor-Echtermeyer Weg 1, Grossbeeren, Germany

Correspondence: Lempie K. Ekandjo, Leibniz-Institute of Vegetable- and Ornamental Crops, Großbeeren/Erfurt e.V., Theodor-Echtermeyer Weg 1, D-14979 Grossbeeren, Germany.

Abstract

Biological Nitrogen Fixation (BNF) is a process of great importance in crop production systems, as it provides additional natural sources of mineral nitrogen. BNF is catalyzed by diazotrophs that are identified by the *nif* operon presence comprising the *nif*H gene that encodes for enzyme nitrogenase synthesis. Thoroughly understanding of factors that influence diazotrophic abundance is crucial for their utilization to enhance sustainability and prevent land degradation in modern agriculture. In this study the impacts of nitrogen fertilization on diazotrophic abundance in *Brassica oleracea* roots and leaves was investigated in greenhouse experiments by real-time qPCR. One way ANOVA was used to compare means and bivariate Pearson correlation tested for relationships between variables. Increased nitrogen fertilization significantly increased the nitrogen content in leaves but not in roots. No significant changes in *nif*H gene copy numbers nor in proportion of *nif*H gene copy numbers were detectable. This indicates no effect of mineral N fertilization on the abundance of total native diazotrophic bacterial numbers in *Brassica oleracea* plants.

Keywords: diazotrophs, NH_4NO_3 fertilization, nitrogen fixation, *nif*H gene

1. Introduction

Biological nitrogen fixation (BNF) is a process through which the atmospheric inert gas N_2 is converted to nutrient ammonia (NH_3). The process is carried out by bacteria. In both symbiotic and free-living bacteria, BNF is genetically controlled by a group of *nif*, *anf* or *vnf* genes (Joerger et al., 1991; Waugh et al., 1995; Coelho et al., 2007a; Sugitha & Kumar, 2009). The *nif*H gene codes for dinitrogenase reductase and is part of the *nif* cluster that produces the nitrogenase catalytic enzyme complex and its regulatory proteins that are responsible for reducing atmospheric N_2 to ammonia (Sugitha & Kumar, 2009; Rosado et al., 1998; Poly et al., 2001). This functional gene has been a center of focus for researchers studying detection, dynamics and diversity of bacterial nitrogen fixing communities (Coelho et al., 2007b) and thus a potential indicator of microorganisms' ability to carry out BNF. The *nifH* gene is often used to quantify diazotrophs using real-time qPCR (Juraeva et al., 2006) and for their molecular identification. Among all the nitrogenase genes, *nifH* is the most extensively studied, with well-documented sequences from both cultured and uncultured diazotrophs (Coelho et al., 2007a; Tan et al., 2003; Deslippe & Egger, 2006). Consequently, *nifH* is utilized in phylogenetic, diversity, and abundance studies of diazotrophic microorganisms (Gaby & Buckley, 2012; Sugitha & Kumar, 2009).

The N_2 fixing enzyme nitrogenase is highly influenced by the availability of the mineral form (such as ammonium nitrate) in which nitrogen is present (Lovell, 2000; Fritzsche & Niemann, 1990; You et al., 2005). The existence of adequate availability of mineral nitrogen in any form in the plant growth substrate and eventually in plant tissues should inhibit N_2 fixation by bacteria, as the process is energetically expensive than mineral nitrogen uptake (Fritzsche & Niemann, 1990; Hartmann, 1988). The impact of nitrogen fertilization on plant growth, photosynthetic capacities, *nif*H expression and isolated diazotrophs has been extensively studied (Vos & van der Putten, 1998; Zhao et al., 2005; Cechin & Fumiss, 2004; Pati & Chandra, 1992), but surprisingly no study has yet focused on the quantification of diazotrophs or diazotrophs abundance dynamics in the phyllosphere and in rhizosphere in relation to nitrogen fertilization and the relationship between shoots and roots. More knowledge is still needed on the relationships between mineral nitrogen availability, N_2 fixing activity, and the colonization and abundance of diazotrophic organisms around the plant. In order to characterize the interconnections between plant mineral N availability and native diazotrophs abundance we quantified diazotrophs (*nif*H-gene real-time qPCR) in leaves and roots of a vegetable plant kohlrabi (*Brassica oleracea*) grown at increasing nitrogen fertilization. *Brassica oleracea* is one of the non-leguminous traditional German vegetable of the cabbage family, with edible tuber and leaves and it can be eaten raw or cooked. Kohlrabi have high content of amino acids and glucosinolates which of nutritional importance to humans (Coi et al., 2010). We hypothesized that rising nitrogen content within the plant tissues by NH_4NO_3 fertilization should negatively affect the abundance of diazotrophic bacteria in both roots and leaves. This was based on the assumption that provision of plant species' optimum or above-optimum mineral nitrogen nutrients decreases the need to get additional mineral nitrogen via BNF and thereafter down-regulate the growth and settlement of diazotrophic bacteria around and within plant tissues.

2. Materials and Methods

2.1 Greenhouse Experiment and Harvesting

The study was conducted and repeated at the Leibniz-Institute of Vegetable and Ornamental Crops in Großbeeren (Brandenburg, Germany). Kohlrabi (*Brassica oleracea*) plants were grown at 5 different nitrogen fertilization levels including a negative control. Each nitrogen level consisted of 24 plants. All were grown for 8 weeks in the greenhouse in pots (33 cm diameter) containing 1kg of non-sterilized quartz sand (0.5-1 mm particle size). Three times a week essential nutrients were provided as 100 ml half strength nitrogen-free Hoagland solution (Hoagland & Arnon, 1950). Plants were provided with reverse-osmosis purified water according to demand (about 150 ml per day); the water that leaked through the plant pots was collected in plates and reused. The different N fertilization treatments were applied by injecting 10 ml NH_4NO_3 solution around the root zone with a pipette once a week. The total nitrogen administered per plant was applied during the growth period and administered at weekly intervals to prevent any negative impact of high salinity due to high nitrogen ion concentrations. The N levels applied as total amount of NH_4NO_3 per plant for the entire experiment were: 0g for N level 0 (N0 negative control), 0.5 g for N level 1 (N1), 1 g for N level 2 (N2); 1.5 g for N level 3 (N3) and 2 g for N level 4 (N4). Other essential macronutrients were provided at a uniform rate at all nitrogen levels in this experiment (Table 1). After 8 weeks plants were harvested and separate samples of shoots and roots collected (three plants pooled per replicate). For DNA extraction and molecular analyses, samples were stored at -20 °C overnight before lyophilization. The remaining roots and leaves were dried at 60 °C for one week, weighed and milled with a high speed plant grinder (Fritsch Pulverisette; Oberstein, Germany) for nitrogen content analysis. This experiment was repeated once.

Table 1. NPK fertilization rates (grams per plant of nitrogen (N), phosphorus (P) and potassium (K))

N level	N (g)	P (g)	K (g)
N0	0	2.8	5
N1	10	2.8	5
N2	20	2.8	5
N3	30	2.8	5
N4	40	2.8	5

2.2 Nucleic Acid Extraction and qPCR

DNA was extracted from 50 mg lyophilized plant material using DNeasy plant mini kits (Qiagen, Hilden GmbH, Germany) according to the manufacturer's instructions. Lysis of bacterial cells was ensured by adding five 5 mm sterile metal beads and using a Retsch MM200 mechanical disrupter (Haan, Germany) at 30 rpm for 5 minutes. Quantitative real-time PCR (qPCR) was run using an Advanced™ Universal SYBR® green I dye Supermix system (Bio-Rad Laboratories, Hercules, CA, USA) as an indicator of resultant PCR amplicons. Diazotroph amplification was done with *nif*H specific primer pair 19F (5′-GCIWTYTAY GCIAARGGIGG-3′) and 388R (5′AAICCRCCRCAIACI ACRTC-3′) (Juraeva et al., 2006) by the following protocol: Cycle 1, 95 °C/5 min; cycle 2, 94 °C/30 sec, 50 °C/60 sec, 72 °C 75 sec (recording fluorescence data) 40 repeats; cycle 3, 72 °C/10 min;

cycle 4, melt curve 55 °C increasing temperature every 10 sec by 0.5 °C (recording fluorescence data), 85 repeats; and cycle 5, 4 °C hold (Juraeva et al. 2006). Total bacterial copy numbers were quantified using the primers 519f (5′-CAGCMGCCGCGGTAANWC-3′) and 907r (5′-CCGTCAATTCMTTTRAGTT-3′) specific for the bacterial domain (Lane, 1991) with the PCR protocol: Cycle 1, 95 °C/5 min; cycle 2, 95 °C/15 sec, 53 °C/30sec, 72 °C/42 sec (recording fluorescence data) 40 repeats; cycle 3, 72 °C/10 min; cycle 4, melt curve 55 °C increasing temperature every 10 sec by 0.5 °C (recording fluorescence data), 85 repeats; and cycle 5, 4 °C hold (Ruppel et al., 2006). Fluorescence was recorded at 520 nm after excitation at 490 nm with a Bio-Rad CFX96 real-time qPCR detection system (BIO-RAD, Foster City, California, United States of America). Total bacterial and *nif*H gene copy numbers were quantified according to a standard curve generated from PCR products specific for bacteria (*Escherichia coli*) and *nif*H (*Kosakonia radicincitans*) genes of known copy numbers. The results were log10 transformed and calculated as gene copy numbers per µg DNA.

2.3 Nitrogen Content Analysis

Dried root and leaf samples were milled to a fine powder with a high speed plant grinder (Fritsch Pulverisette, Oberstein, Germany) and total nitrogen content was analyzed using a CN-Vario EL analyzer (Elementar, Hanau, Germany) (Berger et al., 2013; Chintala et al., 2013).

2.4 Statistical Analysis

Comparison of mean values of 8 replicates was performed using one-way ANOVA analysis and Bonferroni-test at a P-level of ≤ 5% (IBM® SPSS® 20). Statistical differences between means are indicated by uppercase letters in roots and lowercase letters in leaves. Bars indicate the standard errors. Pearson-type correlations (r) were calculated at a P-level of ≤ 5% and at ≤ 1% (IBM® SPSS® 20).

3. Results

3.1 NH_4NO_3 Fertilization Impact on Dry Weight and Plant Nitrogen Content

To demonstrate the N-fertilization impact on plant growth and nitrogen content, leaves and roots were sampled and analyzed separately after a growth period of 8 weeks. For both leaves and roots, provision of nitrogen fertilization resulted in a substantially increased plant dry matter between N0-N1, however further increase in fertilization did not result in continuous increase in plant dry matter (Figure 1). That is to say, increased shoot to root growth ratio with increasing N fertilization could only be seen from N0 to N1. As a result, there were significant differences ($p < 0.05$) in dry matter between N0 and fertilized N levels in both roots and leaves. The mean nitrogen content was significantly higher in leaves than in roots at all levels ($p < 0.05$). There was a continuous and significant increase ($p < 0.05$) in leaves' mean nitrogen content with the lowest mean recorded at N0 (0.94 ± 0.09 g, n = 8) and the highest recorded at N4 (7.18 ± 0.61 g, n = 8) (Figure 2). Roots did not exhibit the same trend; the lowest mean was also recorded at N0 (0.62 ± 0.79 g, n = 8) but the highest mean was at N1 (1.69 ± 0.22 g, n = 8) and values remained essentially constant from N1 to N4 (Figure 2). However, plant root dry weight was significantly low ($p < 0.05$) at N0 than at fertilized N levels.

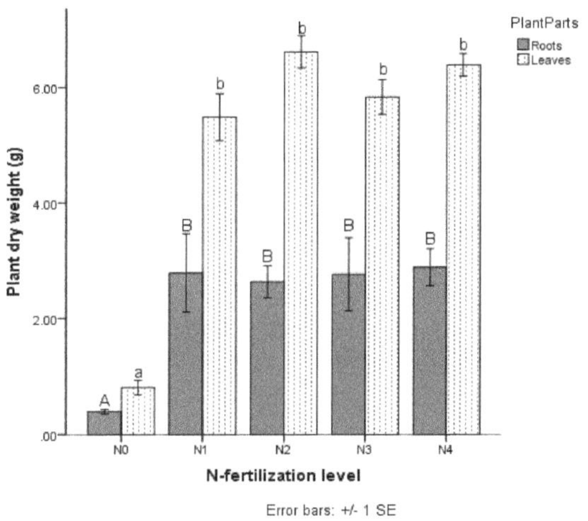

Figure 1. Dry weight (g plant^{-1}) of kohlrabi leaves and roots (both n = 8) at increasing levels of NH_4NO_3 fertilization. Statistical differences between means are indicated by different uppercase letters in roots and different lowercase letters in leaves

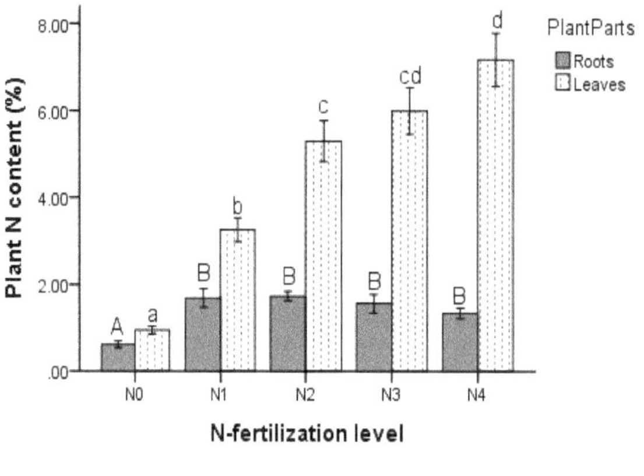

Figure 2. Plant nitrogen content (%, dry weight basis) of kohlrabi leaves and roots (both n = 8) at increasing levels of NH_4NO_3 fertilization. Statistical differences between means are indicated by different uppercase letters in roots and different lowercase letters in leaves

3.2 Diazotrophic Bacterial (nifH) Quantification

Quantification of total diazotrophic organisms using functional *nif*H gene copy numbers in real-time qPCR analysis revealed significantly ($p < 0.05$) higher *nif*H gene copy numbers in leaves than in roots (Figure 3). Nonetheless, there was no detectable influence ($p > 0.05$) of increased N-fertilization on mean *nif*H gene in either roots or leaves (Figure 3). To further evaluate the N-fertilization impact on the plant microbiome community structure at each N-fertilization level, *nif*H gene was calculated as a percentage of the total bacteria (16S rDNA gene copy numbers) to reveal the extent to which the total bacteria population within and around the plant was accounted for by diazotrophs. This additional calculation was done to validate the *nif*H quantity results in Figure 3. Similar to *nif*H copy numbers results in Figure 3 diazotrophs proportion (*nif*H gene copy numbers of total bacterial copy numbers) also did not reveal a decreasing or an increasing trend with increasing N-fertilization on the plants (Figure 4). In addition these results also indicated a high variation in *nif*H proportion at different N-fertilization levels. Thus there were no significant difference ($p > 0.05$) in the mean *nif*H proportion in leaves and roots. Similar to *nif*H quantification, diazotrophs proportion at the total bacterial counts was higher in leaves than in roots when molecular methods were applied.

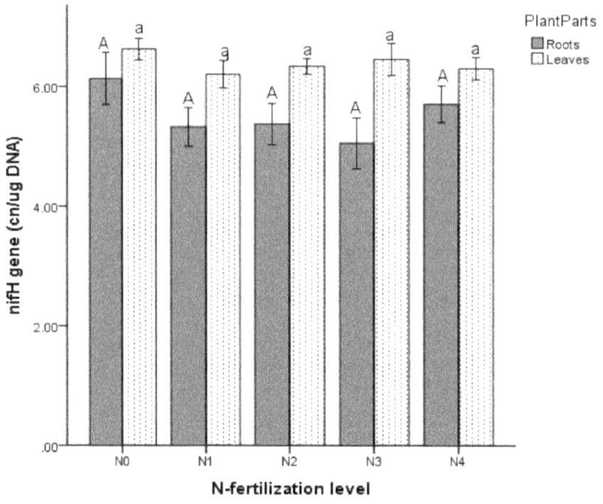

Figure 3. Total *nif*H gene copy numbers (cn) (*nif*H gene cn μg^{-1} DNA, n = 8) of kohlrabi leaves and roots (both n = 8) at increasing levels of NH_4NO_3 fertilization. Statistical differences between means are indicated by different uppercase letters in roots and different lowercase letters in leaves

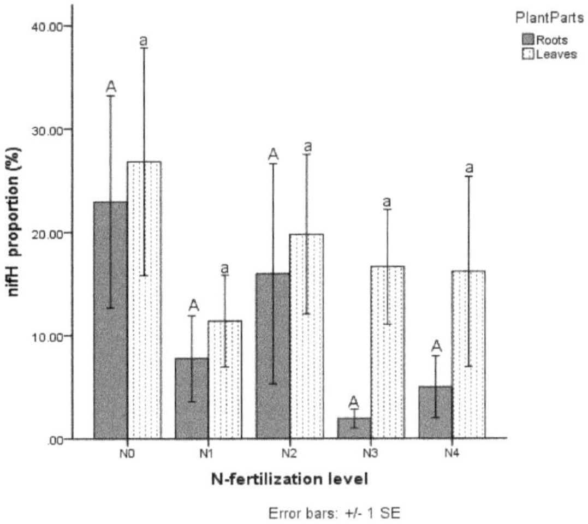

Figure 4. Proportion of *nif*H gene copy numbers in total bacterial copy numbers (%) of kohlrabi leaves and roots (both n = 8) at different levels of NH_4NO_3 fertilization. Statistical differences between means are indicated by uppercase letters in roots and lowercase letters in leaves

3.3 Correlation

As expected, increasing NH_4NO_3 fertilization was significantly positively correlated to kohlrabi leaf dry matter (r = 0.72**) and leaf nitrogen content (r = 0.86**) (Table 2). Root dry matter also showed significant correlation to NH_4NO_3 fertilization (r = 0.46**) while, N-content showed no significant correlation to NH_4NO_3 fertilization (r = 0.31). Nevertheless, increased NH_4NO_3 fertilization did not significantly impact *nif*H gene copy numbers in leaves (r = 0.10) or in roots (r = -0.15). Likewise the *nif*H gene copy number proportion of total bacterial copy numbers was not significantly correlated to increased N-fertilization in either leaves (r = -0.11) or roots (r = -0.30) (Table 2). Furthermore to test for interactions between leaves and roots correlation test was done for different variables. In this test only *nif*H quantity roots was positively correlated (0.512**) to the *nif*H quantity in leaves among other variables.

Table 2. The impact of N fertilization other studied variables

Variables	Leaves (r values)	Roots (r values)
Dry weight	+ (0.72**)	+ (0.46**)
Nitrogen content	+ (0.86**)	0 (0.31)
*nif*H quantity	0 (-0.10)	0 (-0.15)
*nif*H percentage	0 (-0.11)	0 (-0.30)

Keys: + Positive correlations; - Negative correlations; 0 insignificant correlations; * Correlation is significant at the 0.05 level; ** Correlation is significant at the 0.01 level.

4. Discussion

The unexpected insignificant differences between means and correlation coefficients found between *nif*H, *nif*H proportion in relation to N-fertilization could be explained by the fact that most diazotrophic bacteria have multiple plant growth promoting functions in associated plants, and BNF is one of those (Yuan et al., 2012; Fernando & Linderman, 1995; Xu et al., 2004). Other functions around the plants include phosphate transformer genes, siderophore genes as wells as phytohormone production genes and of course these functions vary between strains (Witzel et al., 2012). These other functions can contribute to a selective advantage unrelated to nitrogen fixation for plants colonized by diazotrophic bacteria. In this model, the absence of the need for the BNF does not necessarily lead to the loss of diazotrophs from the plant as they still provide other benefits to the plant. It is worth noting that presence of *nif*H gene-containing microorganisms does not invariably indicate ongoing BNF, especially when there is mineral nitrogen available to the plant (Fritzsche & Niemann, 1990). Fritzsche and Niemann (1990) also documented a decrease in BNF activity with provision of increased NH_4Cl but did not report the impact on the nitrogen fixing bacterial population.

The regulation of diazotrophic activities seems very complex. In this study, provision of sufficient ammonium nitrate was ensured by adding the compound above the documented optimum requirements in the literature (Carmen et al., 2007). From the Fritzsche and Niemann (1990) findings, it appears that BNF in the rhizosphere does not solely depend on total absence of mineral nitrogen sources. Similar findings recorded by Lovell (2000) indicate that there was still BNF activity in open marine water that had abundant ammonium availability. In addition some findings indicated that high level of mineral nitrogen supply might inhibit N_2 fixation in some diazotrophic species but not in all (Rosado, 1998; Lovell, 2000; Ruppel & Merbach, 1995). Lovell (2000) even reported an increase in nitrogenase expression with increase in mineral N availability and concluded that short term manipulation in nitrogen supply does not have significant impact on the diazotrophic communities. Similarly Juvaera et al. (2006) reported that although there were physiological changes in plants due to N fertilization these did not cause significant changes to *nif*H gene abundance or to the diazotrophic culturable organisms around the plants.

5. Conclusion

These results revealed that increased NH_4NO_3 fertilization had no significant impact on diazotrophs abundance in rhizosphere and in the rhizosphere of the studied kohlrabi plants (*Brassica oleracea)*, despite the increased nitrogen content observed within plant tissues particularly in the phylosphere. However, these results highlight the potential importance of the phyllosphere in BNF, since diazotrophs abundance (*nif*H gene copy numbers) were higher in leaves than in roots. Finally there is a need for more extensive investigations in both phyllosphere and rhizosphere to determine if the *nif*H genes were being expressed or not in the diazotrophic communities under adequate supply of mineral nitrogen.

Acknowledgements

I am gratefully to DAAD and IGZ.e.V institutes for funding my research and to the entire plant nutrition team at IGZ for technical support.

References

Berger, B., Brock, A., & Ruppel, S. (2013). Nitrogen supply influences plant growth and transcriptional responses induced *Enterobacter radicincitans* in *Solanum lycopersicum*. *Plant and Soil, 370*(2), 641-652. http://dx.doi.org/10.1007/s11104-013-1633-0

Burgmann, H., Wider, F., Sigler, W. V., & Zeyer, J. (2003). mRNA extraction and reverse transcriptase-PCR protocol for detection of *nif*H gene expression by *Azotobacter vineandii* in soil. *Appl. Environ. Microbiol., 69*, 1928-1935. http://dx.doi.org/10.1128/AEM.69.4.1928-1935.2003

Carmen, F., Fink, M., Laber, H., Maync, A., Paschold, P., Acharpf, H., et al. (2007). *Gartenbauliche Berichte* (4th ed.). Großbeeren: IGZ.

Cechin, I., & Fumis, T. F. (2004). Effect of Nitrogen supply on growth and photosynthesis of sunflower plants grown in the greenhouse. *Plant Science, 166*, 1379-1385. http://dx.doi.org/10.1016/j.plantsci.2004.01.020

Chintala, R., Clay, D. E., Schumacher, T. E., Malo, D. D., & Julson, J. L. (2013). Optimization of Oxygen parameters for analyzing carbon and nitrogen in biochar materials. *Analytical Letters, 46*, 532-538. http://dx.doi.org/10.1080/00032719.2012.721103

Choi, S., Ryu, D., Park, S., Ahn, K., Lim, Y., & An, G. (2010). Composition Analysis between Kohlrabi (*Brassica oleracea* var. *gongylodes*) and Radish (*Raphanus sativus*). *Korean Journal of Horticultural Science Technology, 28*(3), 469-475.

Coelho, M. R. R., de Vos, M., Carneiro, N., Marriel, I. E., Paiva, E., & Seldin, L. (2007a). Diversity of *nif*H pools in two cultivars of sorghum (*Sorghum bicolor*) treated with contrasting levels of nitrogen fertilizer. *FEMS Microbiology Letter, 279*, 15-22. http://dx.doi.org/10.1111/j.1574-6968.2007.00975.x

Coelho, M. R. R., Mota, F. F., Carneiro, N. P., Marriel, I. E., Paiva, E., Rosado, A. S., & Seldin, L. (2007b). Diversity of *Penibacillus* spp. In the rhizosphere of four sorghum (*Sorghum bicolor*) cultivars sown with two contrasting levels of nitrogen fertilizer accessed by rpoB based PCR-DGGE and sequencing analysis. *Journal of Microbiology Biotechnology, 54*, 753-760.

Deslippe, J., & Egger, K. (2006). Molecular diversity of *nif*H genes from bacteria associated with high artic dwarf shrubs. *Microbial Ecology, 51*, 516-525. http://dx.doi.org/10.1007/s00248-006-9070-8

Farnedlid, H. (2013). *Distribution and activity of nitrogen-fixing bacteria in marine and estuarine waters.*

Linneaus University Dissertations: Sweden: Linnaeus University Press.

Fernando, W. G. D., & Linderman, R. G. (1995). Inhibition of *Phytophthora vignae* and root rot of cowpea by soil bacteria. *Biological Agricultural Horticulture, 12*, 1-4. http://dx.doi.org/10.1080/01448765.1995.9754719

Fritzsche, C., & Niemann, E. G. (1990). Nitrogen fixation in activities continuous culture with NH_4Cl containing media. *Applied Environmental Biology, 56*, 1160-1123.

Fuentes-Ramirez, L. E., Caballero-Mellado, J., Sepulveda, J., & Martinez-Romero, E. (1999). Colonization of sugarcane by *Acetobacter diazotrophicus* is inhibited by high N fertilization. *FEMS Microbiol. Ecol., 29*, 117-128. http://dx.doi.org/10.1111/j.1574-6941.1999.tb00603.x

Gaby, J. C., & Buckley, D. H. (2012). A comprehensive evaluation of PCR primers to amplify the *nif*H of Nitrogenase. *Plos One, 7*, e42149. http://dx.doi.org/10.1371/journal.pone.0042149

Hartmann, A. (1988). Eco-physiological aspects of growth and nitrogen fixation in *Azosprillum* spp. *Plant Soil, 110*, 225-238.

Hoagland, R., & Arnon, I. (1950). *The water-culture methods for growing plants without soil*. United States of America: California agricultural experiment station.

Joerger, R. D., Wolfinger, E. D., & Bishop, P. E. (1991). The encoding dinitrogenase reductase 2 is required for expression of the second alternative nitrogenase from *Azotobacter vinelandii*. *Journal of Bactriology, 173*, 4440-4446.

Juraeva, D., George, E., Davranov, K., & Ruppel, S. (2006). Detection and quantification of the *nif*H gene in the shoot and root of cucumber plants. *Canadian Journal of Microbiology, 52*, 731-739. http://dx.doi.org/10.1139/W06-025

Lane, D. J. (1991). 16S/23S rRNA sequencing. In E. Stackebrandt & M. Goodfellow (Eds.), *Nucleic acid techniques in bacterial systematics* (pp. 205-248). Chichester, England: Wiley.

Limmer, C., & Drake, H. R. (1998). Effects of carbon, nitrogen, and electron acceptor availability in anaerobic N2-fixation in a beech forest soil. *Soil Biol Biochem, 30*, 153-158. http://dx.doi.org/10.1016/S0038-0717(97)00099-0

Lovell, C. R., Piceno, Y. M., Quarttro, J. M., & Bagwell, C. E. (2000). Molecular analysis of diazotroph diversity in the rhizosphere of the smooth Cordgrass, *Spartina alterniflora*. *Applied and Environmental Microbiology, 66*(9), 3814-3822. http://dx.doi.org/10.1128/AEM.66.9.3814-3822.2000

Pati, B. R., & Chandra, A. K. (1992). Nitrogen fixing potentialities of the phyllospheric bacteria in relation to the concentration of sucrose in the media. *Zentralblatt für mikrobiologie, 147*(7), 435-440.

Poly, F., Monrozier, L. J., & Bally, R. (2001). Improvement in the RFLP procedures for studying the diversity of *nif*H genes in communities of nitrogen fixers in soil. *Res. Microbiol., 152*, 95-103. http://dx.doi.org/10.1016/S0923-2508(00)01172-4

Qiagen. (2006). *DNeasy plant handbook*. Germany, Hilden: Qiagen.

Rosado, A. S., Duarte, G. F., Seldin, L., & Elsas, J. D. (1998). Genetic diversity of *nif*H gene sequences in *Paenibacillus azotofixans* strains and soil samples analyzed by Denaturing Gradient Gel Electrophoresis of PCR-Amplified gene fragments. *Applied and Environmental Microbiology, 64*(8), 2770-2779.

Ruppel, S., & Merbach, W. (1995). Effects of different nitrogen sources on nitrogen fixation and bacterial growth of *Pantoea agglomerans* and *Azospirillum* spp. in bacterial pure culture: An investigation using $^{15}N_2$ and acetylene incubation. *Microbiol. Res., 150*, 409-418. http://dx.doi.org/10.1016/S0944-5013(11)80023-6

Ruppel, S., & Merbach, W. (1997). Effect of ammonium and nitrate on specific $^{15}N_2$-fixation of *Azospirillum* spp. and *Pantoae agglomerans* in association with wheat plants. *Microbiol. Res., 152*, 377-383. http://dx.doi.org/10.1016/S0944-5013(97)80055-9

Ruppel, S., Rühlmann, J., & Merbach, W. (2006). Quantification and localization of bacteria in plant tissues using quantitative real-time PCR and online emission fingerprinting. *Plant and Soil, 286*(1-2), 21-35. http://dx.doi.org/10.1007/s11104-006-9023-5

Sugitha, T. C., & Kumar, K. (2009). Identification of *nif*H genes of heterophic and endophytic diazotrophs associated with rice (*Oryza sativa* L.) by targeted DNA finger printing. *African Journal of Biotechnology, 8*(22), 6249-6252.

Sutton, S. (2010). The most probable number and its use in quality control microbiology. *Journal of GXP Compliance, 14*(4), 28-33.

Tan, Z., Suga, Y., & Reinhold-Hurek, B. (2003). Effect of N fertilization, plant genotype and environmental conditions on *nif*H gene pools in roots of rice. *Environmental Biology, 5*, 1009-1015. http://dx.doi.org/10.1046/j.1462-2920.2003.00491.x

Vos, J., & van der Putten, P. E. L. (1998). Effects of nitrogen supply on leaf growth, leaf nitrogen economy and photosynthetic capacity in potato. *Field Crops Research, 59*, 63-72. http://dx.doi.org/10.1016/S0378-4290(98)00107-5

Waugh, S. I., Paulsen, D. M., Mylona, P. V., Maynard, R. H., Premakumar, R., & Bishop, P. E. (1995). The gene encoding the delta subunits of dinitrogenase 2 and 3 are required for Mo-independent diazotrophic growth by *Azotobacter vinelandii*. *Journal of Bacteriology, 177*, 1505-1510.

Witzel, K., Gwinn-Giglio, M., Nadendla, S., Shefchek, K., & Ruppel, R. (2012). Genome sequences of Enterobacter radicincitans DSM16656, a plant growth promoting endophyte. *Journals of Bacteriology, 194*(19), 5468-5471. http://dx.doi.org/10.1128/JB.01193-12

Xu, C. K., Mo, M. H., Zhang, L. M., & Zhang, K. Q. (2004). Soil volatile fungistasis and volatile fungistatic compounds. *Soil Biology Biochemistry, 36*, 1997-2004. http://dx.doi.org/10.1016/j.soilbio.2004.07.020

You, M., Nishigushi, M., Saito, A., Isawa, T., Mitsui, H., et al. (2005). Expression of the *nif*H gene of a *Herbaspirillum* endophyte in wild rice species: daily rhythm during the light dark cycle. *Applied and Environmental Microbiology, 71*(12), 8183-8190. http://dx.doi.org/10.1128/AEM.71.12.8183-8190.2005

Yuan, J., Raza, W., Shen, Q., & Huang, Q. (2012). Antifungal activity of *Bacillus amyloliquefaciens* NJN-6 volatile compounds against *Fusarium oxysporum* f. sp. Cubanse. *Applied Environmental Microbiology, 78*(16), 5942-5994. http://dx.doi.org/10.1128/AEM.01357-12

Zhao, D., Reddy, R. K., Kakani, G. V., & Reddy, V. R. (2005). Nitrogen deficiency effects on plant growth, leaf photosynthesis, and hyperspectral reflectance properties of Sorghum. *European Journal of Agronomy, 22*, 391-403. http://dx.doi.org/10.1016/j.eja.2004.06.005

12

Study on Nutritional Requirements of Nematophagous Fungi in Terms of Carbon and Nitrogen Sources

Anamika[1]

[1] Department of Agriculture, CAFF, Fiji National University, Fiji Island

Correspondence: Anamika, Department of Agriculture, CAFF, Fiji National University, Fiji Island. E-mail: anamika_bhu01@rediffmail.com

Abstract

Nematophagous fungi act as natural enemies for biological control of plant parasitic nematodes. *Dactylaria eudermata* and *Arthropotrys oligospora* are the nematohogous fungi producing one to three dimensional hyphal bails and adhesive nets as a capturing device were selected and it was observed that growth and trapping effectiveness is influenced by temperature and C:N ratio. For the assessment of favorable temperature radial growth of both the fungi was studied at different temperature and 25 °C was recorded as an optimum temperature. Carbon and nitrogen content in media affect their predacity against plant parasitic nematode so their growth and sporulation were studied by providing different carbon and nitrogen sources. Among carbon sources, Glucose was the best source for both the fungi whereas Potassium nitrate for *D. eudermata* and Sodium nitrate for *A. oligospora* were recorded as the best Nitrogen sources.

Keywords: *Dactylaria eudermata*, *Arthropotrys oligospora*, temperature, carbon and nitrogen sources

1. Introduction

Nematode trapping fungi are responsible for keeping the nematode population in check and are an important part of the subsoil ecosystem. Nematophagous fungi predominantly present in the soil as saprophytes but act as parasites in presence of nematodes so as to fulfill the requirements of additional nutrient source in terms of nutritional advantage. The fungi converts its saprophytic stage to their parasitic phase and change their morphological structure into traps or mature spores. The development of infection structures is a prerequisite for the trapping of nematodes. The nematodes releases nemin like chemical substances which evokes the fungi to form trapping devices and these devices attracts nematodes towards itself and finally nematodes are trapped. If nematophagous fungi can be weaponized against nematode disease, then we will be able to see an even more noticeable difference. These fungi comprise more than 200 species found in all major taxonomic groups including lower (oomycetes, chytridiomycetes, zygomycetes) and higher fungi (ascomycetes, basidiomycetes and deuteromycetes). Most of them, including both nematode trapping and endoparasitic species, are asexual fungi (deuteromycetes), that all share the ability to attack living nematodes and use them as nutrients (Nordbring-Hertz et al., 2002). The nematophagous fungi, which infect nematodes with their conidia, are mostly more dependent on nematodes as a nutrient and are considered as obligate parasites (Jansson & Nordbring-Hertz, 1979; Jansson, 1982). These fungi are widely distributed in all types of agricultural soils but they are abundant in dung or surface litter and decaying plant debris. These fungi act on the principle of attraction, adhesion, penetration and killing live nematodes and digest it by producing fascinating trapping devices. The ability to capture nematodes is connected with the development of structures on the fungal mycelia. The nematophagous fungi have formed hyphal structures, such as adhesive hyphal nets, knobs, branches and constricting or non-constricting rings, in which nematodes are captured by adhesion or mechanically. Fungal hyphae then penetrate the nematode and utilize its content for food. Many nematode trapping deuteromycetes are indeed also good saprophytes and have evolved among cellulolytic or lignolytic fungi, they can use cellulose and other polysaccharides as carbon sources (Barron, 1992). The nitrogen content is less in plant debris and due to extremely low nitrogen levels, nematophagous fungi have evolved parasitic behavior to satisfy their nitrogen requirements as nutritional demands (Thorn & Barron, 1984; Barron, 2003). Also in the soil environments with a high carbon: nitrogen ratio, nematodes might serve as an important source of nitrogen. Part of the problem is that fungus population density, and especially trapping, is difficult to quantify in soil. Moreover, trapping usually cannot be inferred simply from suppression of nematodes following organic amendment because such amendments can suppress nematodes in many different ways (Wang et al., 2002; Widmer et al., 2002). With organic matter of a C:N ratio greater than 20:1, N will temporarily be immobilized in microbial tissue, creating a nitrogen deficiency (Akhtar & Malik, 2000). The nematode management potential of an organic soil amendment is directly related to N-content or inversely related to the C:N ratio (Mian & Rodriguez-Kabana, 1982a, 1982b). Therefore it is necessary to study about the different sources of carbon and nitrogen with respect to the temperature as changes in climatic condition also affect the growth and development of the nematophagous fungi. So the objectives of the study were taken to determine the suitable temperature and appropriate carbon nitrogen sources for the radial growth of *D. eudermata* and *A. oligospora in vitro* condition.

2. Materials and Methods

Isolation of nematophagous fungi (*Dactylaria eudermata* and *Arthrobotrys oligospora*) was done from the soil sample collected from different villages of Allahabad, India by the method described by Duddington (1955). Single spore isolation technique (Tuite, 1969) was used for obtaining the pure culture of both *D. eudermata* and *A. oligospora* and were maintained on maize meal agar and used for further experiment.

2.1 Temperature Effect on Radial Growth of D. eudermata and A. oligospora

Maize meal agar medium was selected as an ideal medium to check the temperature effect on radial growth of both the fungi. 5 mm fungal disc of both the fungus were cut from the periphery of seven days old culture with the help of sterilized cork borer and inoculated into Petri dishes containing cooled and solidified maize meal agar medium and were incubated at 15, 20, 25, 30 and 35 °C.

Three petriplates were used for both fungi at varied temperature. Radial growths were measured on 6th days of inoculation and experiment was repeated three times in Complete Randomize Design and pooled data was subjected to statistical analysis.

2.2 Effect of Different Carbon Sources on Radial Growth of D. eudermata and A. oligospora

The radial growths of both the fungus (*D. eudermata* and *A. oligospora*) were studied on Yeast extract peptone soluble starch (YPSS) medium. The carbon source of this medium soluble starch was replaced by five other carbon sources, viz. Dextrose, Sucrose, Glucose, Mannitol and Fructose. Medium with soluble starch served as check. All the culture media were prepared, sterilized and poured separately into several sterilized Petri dishes. 5 mm fungal disc, duly cut with a sterilized cork borer, were taken from the periphery of seven days old culture and inoculated into Petri dishes containing different media. Three replications were maintained for each treatment and were incubated at 25 ±1 °C.

Radial growth and sporulation of both the fungus was measured after six days. The experiment was repeated three times in Complete Randomize Design and pooled data was subjected to statistical analysis.

2.3 Effect of Nitrogen Sources on Radial Growth of D. eudermata and A. oligospora

The radial growth of both the fungus (*D. eudermata* and *A. oligospora*) was studied on Czapek's medium. The nitrogen source of this medium sodium nitrate was replaced by five other nitrogen sources, viz. Calcium nitrate, Ammonium nitrate, Ammonium chloride, Potassium nitrate and Ammonium sulphate. Medium with Sodium nitrate served as check. All the culture media were prepared, sterilized and poured separately into several sterilized Petri dishes. 5 mm fungal disc were taken from the periphery of seven days old culture with a sterilized cork borer and inoculated into Petri dishes containing different media. Three replications for each treatment were maintained and incubated at 25 ±1 °C.

Radial growth and sporulation of both the fungus was measured after six days. The experiment was repeated three times in Complete Randomize Design and pooled data was subjected to statistical analysis.

3. Results and Discussion

3.1 Temperature Effect on the Radial Growth of Nematophagous Fungi

The effect of temperature on the radial growth of the both nematophagous fungi are presented in Table 1 (Figure 1). Radial growth of the test fungi was significantly influenced by temperature. For the radial growth of the fungus at different temperature, 25 °C was recorded as an optimum temperature. From the observations recorded, it is quite clear that at this optimum temperature *A. oligospora* grows faster than *D. eudermata*. In general, radial growth of both the fungus showed decrease in growth with increase or decrease in temperature beyond the optimum temperature. At 30 °C the growth was appreciable while above 30 °C, the growth rate decreased significantly. At temperature below 20 °C, the growth of both the fungus was also very slow.

Table 1. Effect of temperature on growth of nematophagous fungi in MMA medium

Temperature (°C)	**Radial Growth (mm) After 6 Days**		**Mean**
	Nematophagous fungi		
	D. eudermata	***A. oligospora***	
T_1 (15)	15.2	16.2	15.7
T_2 (20)	34.8	40.5	37.6
T_3 (25)	69.1	78.2	73.6
T_4 (30)	65.5	71.5	68.5
T_5 (35)	58.2	60.2	59.2
Mean	48.5	53.3	50.9
F – Test	S	S	
SEm±	0.49	0.98	
CD (P = 0.05)	1.09	1.088	

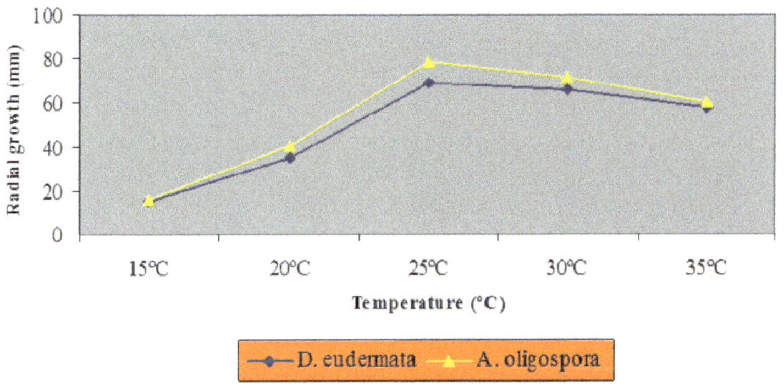

Figure 1. Effect of temperature on growth of nematophagous fungi in MMA medium

Similar observations on different temperature effect for different fungi have been reported by earlier workers. Pandey (1973) reported that 25 °C as optimum temperature for *Arthrobotrys robusta* while Gueye et al. (1997) found optimal growth of *A. musiformis* at 25-30 °C.

3.2 Effect of Carbon Sources on the Radial Growth of D. eudermata and A. oligospora

Data on the radial growth of *D. eudermata* and *A. oligospora* on different carbon sources are presented in Table 2 (Figure 2). Among the carbon sources, glucose supported maximum radial growth of *D. eudermata* followed by fructose. Mannitol, however, supported minimum growth in comparison to other carbon sources. Starch, sucrose and Dextrose also supported moderate growth of this fungus. The radial growth of *D. eudermata* on different carbon sources was recorded in the following order:

Glucose > Fructose > Starch > Sucrose > Dextrose > Mannitol

The test fungus *D. eudermata* sporulates well in all the test carbon sources. However, excellent sporulation of the fungus was observed on glucose followed by dextrose.

Glucose also supported maximum growth of *A. oligospora* followed by starch. Mannitol however, supported minimum rate of growth in comparison to glucose and Starch. Fructose, sucrose and dextrose showed appreciable growth of this fungus. The radial growth of *A. oligospora* on different carbon sources was recorded in the following order:

Glucose > Starch > Fructose > Sucrose > Dextrose > Mannitol

The test fungus *A. oligospora* sporulates in all the tested carbon sources. However, excellent sporulation of the fungus was observed in glucose and starch followed by fructose. Similar observation on glucose as best carbon sources for different fungi have been reported by earlier workers (Ajello, 1948; Colderone & Barnett, 1972;

Srivastava, 1981; Prasad, 1985; Singh, 2007).

Table 2. Effect of carbon sources on the radial growth of nematophagous fungi

Treatments	Nematophagous fungi			
	D. eudermata		***A. oligospora***	
	Radial Growth (mm)	***Sporulation***	**Radial Growth (mm)**	***Sporulation***
T_0 (Starch)	73.1	+++	86.3	+++++
T_1 (Fructose)	75.1	+++	83.2	++++
T_2 (Dextrose)	65.9	++++	76.9	++
T_3 (Sucrose)	70.1	++	81.2	+++
T_4 (Mannitol)	54.5	+	64.9	+
T_5 (Glucose)	77.9	+++++	89.1	+++++
Mean	69.4		80.2	
F – Test	S		S	
SEm±	0.64		0.68	
CD (P = 0.05)	1.39		1.47	

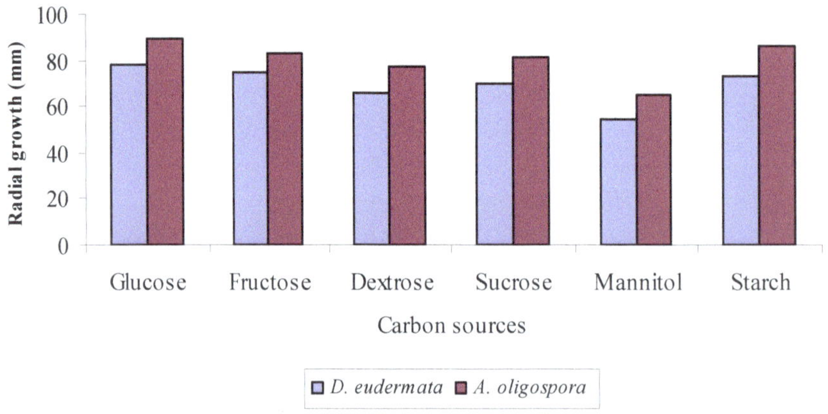

Figure 2. Effect of carbon source on the radial growth of nematophagous fungi

3.3 Effect of Nitrogen Sources on the Growth of D. eudermata and A. oligospora

Data recorded in Table 3 (Figure 3) shows that the radial growth of *D. eudermata* and *A. oligspora* differ significantly on different nitrogen sources. In case of *D. eudermata,* among six nitrogen sources, potassium nitrate supported maximum radial growth followed by sodium nitrate and calcium nitrate. Ammonium chloride was found to be the poorest source of nitrogen. The radial growth of *D. eudermata* on different nitrogen sources was recorded in the following order:

Potassium nitrate > Sodium nitrate > Calcium nitrate > Ammonium nitrate > Ammonium sulphate > Ammonium chloride

Among six nitrogen sources, sodium nitrate was found to be the best for the radial growth of *A. oligospora* followed by potassium nitrate and calcium nitrate while ammonium sulphate was found to be the poorest source of nitrogen. The radial growth of *A. oligospora* on different nitrogen sources were recorded in the following order:

Sodium nitrate > Potassium nitrate > Calcium nitrate > Ammonium nitrate > Ammonium chloride > Ammonium sulphate

Both the fungus sporulate in all the six nitrogen sources tested except *D. eudermata* does not sporulate on ammonium chloride, while *A. oligospora* sporulated comparatively lesser in ammonium chloride. Excellent sporulation of both the fungus was observed in sodium nitrate and potassium nitrate. Similar observations on best nitrogen sources for different fungi have been reported by earlier workers (Lilly & Barnett, 1951; Srivastava, 1981; Prasad, 1985; Subramanian & Tyagi, 1968). The capacity of *D. eudermata and A. oligospora* to use nitrate indicate that the fungi produces reductase enzyme or other associated enzymes like hypernitrite reductase and hydroxylamine reductase, involved in the metabolism of the nitrate (Walker & Nicholas, 1962; Medina & Nicholas, 1957; Singh, 2007). The fungal growth in carbon and nitrogen deficient media may be accredited to the presence of these sources in form of impurities in other chemicals used during study.

Table 3. Effect of nitrogen sources on the radial growth of nematophagous fungi

Treatments (Nitrogen Sources)	Nematophagous fungi			
	D. eudermata		*A. oligospora*	
	Radial Growth (mm)	*Sporulation*	Radial Growth (mm)	*Sporulation*
T_0 (Sodium Nitrate)	75.3	+++++	88.5	+++++
T_1 (Potassium Nitrate)	78.6	++++	84.9	+++++
T_2 (Calcium Nitrate)	72.1	+++	82.1	+++
T_3 (Ammonium Nitrate)	69.5	++	74.5	++
T_4 (Ammonium Chloride)	61.2	-	68.2	+
T_5 (Ammonium Sulphate)	64.5	+	64.5	++
Mean	70.2		77.1	
F – Test	S		S	
SEm±	0.44		0.46	
CD (P = 0.05)	0.95		1.00	

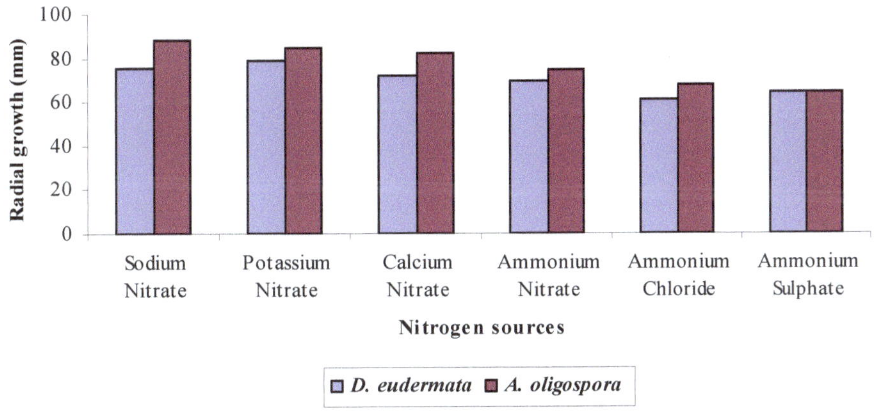

Figure 3. Effect of nitrogen sources on the radial growth of nematophagous fungi

Acknowledgements

Author is thankful to Professor Sobita Simon for constant help and moral support in conducting the experiments.

References

Ajello, L. (1948). A Cytological and Nutritional Study of Polychytrium aggregatum. Part II. Nutrition. *American J. Bot., 35*, 135-140. http://dx.doi.org/10.2307/2438236

Akhtar, M., & Malik, A. (2000). Roles of organic soil amendments and soil organisms in the biological control of plant-parasitic nematodes. *A Review. Bioresour. Technol., 74*, 35-47. http://dx.doi.org/10.1016/S0960-8524(99)00154-6

Barron, G. L. (1992). *Lignolytic* and *Cellulolytic* fungi as predators and parasites. In G. C. Carroll & D. T. Wicklow (Eds.), *The Fungal Community, its Organization and Role in the Ecosystems* (pp. 311-326).

Marcel Dekker, New York.

Barron, G. L. (2003). Predatory fungi, wood decay, and the carbon cycle. *Biodiversity, 4*, 3-9. http://dx.doi.org/10.1080/14888386.2003.9712621

Coldrone, R. A., & Barnett, H. L. (1972). Axenic growth and nutrition of gonatobotryum fuscum. *Mycologia, 64*, 153-164. http://dx.doi.org/10.2307/3758023

Duddington, C. L. (1955). Notes on the technique of handling predacious fungi. *Trans. Brit. Mycol. Soc., 38*(2), 97-103. http://dx.doi.org/10.1016/S0007-1536(55)80021-6

Gueye, M., Duponnois, R., Samb, P. I., & Mateille, T. (1997). Study on 3 strains of *Arthrobotrys oligospora*: Biological characterization and effects on *Meloidogyne mayaguensis* parasitic on tomato in Senegal. *Tropicultura, 15*(3), 109-115.

Jansson, H. B. (1982). Attraction of nematodes to endoparasitic nematophagous fungi. *Transactions of the British Mycological Society, 79*, 25-29. http://dx.doi.org/10.1016/S0007-1536(82)80187-3

Jansson, H. B., & Norbring-Hertz, B. (1979). Attraction of nematodes to living mycelium of nematophagous fungi. *Journal of General Microbiology, 112*, 89-93. http://dx.doi.org/10.1099/00221287-112-1-89

Lilly, V. G., & Barnett, H. L. (1951). *Physiology of fungi*. Mc Graw Hill, New York.

Medina, A., & Nicholas, D. J. D. (1957). Hyponitrite Reductase in *Neurospora*. *Nature, 179*, 533-534. http://dx.doi.org/10.1038/179533a0

Mian, I. H., & Rodriguez-Kabana, R. (1982b). Soil amendments with oilcakes and chicken litter for control of *Meloidogyne arenaria*. *Nematropica, 12*, 205-220.

Mian, I. H., & Rodriguez-Kabana, R. (1982a). Organic amendments with high tannin and phenolic contents for control of *Meloidogyne arenaria* in infested soil. *Nematropica, 12*, 221-234.

Nordbring-Hertz, B., Jansson, H. B., & Tunlid, A. (2002). Nematophagous fungi. *Encyclopedia of Life Science, 12*, 681-690.

Pandey, V. S. (1973). Predatory activity of nematode trapping fungi against the larvae of Trichostrongylus axei and Ostertagia ostertagi: A possible method of biological control. *Journal of Helminthology, XLVII*(1), 35-48. http://dx.doi.org/10.1017/S0022149X00023725

Prasad, G. (1985). *Studies on nematophagous fungi* (Ph.D. Thesis). Mycology and Plant Pathology, I.A.S., Bananas Hindu University.

Singh, U. B. (2007). *Ocuurance Characterization and Performance of some Predacious Fungi* (M.Sc Thesis). Banaras Hindu University.

Srivastava, S. S. (1981). *Studies on some predaceous fungi* (Ph.D. Thesis). Banaras Hindu University.

Subramanian, C. V., & Tyagi, P. D. (1968). *Proc. Indian Acad. Sci., 58*, 111-129.

Tuite John. (1969). *Plant Pathological Methods* (pp. 1-81). Burges Publishing Company, USA.

Walker, C. G., & Nicholas, D. J. D. (1962). An Iron Requirement for a Dissimilatory Nitrate Reductase in *Neurospora crassa*. *Nature, 189*, 141-142. http://dx.doi.org/10.1038/189141a0

Wang, K. H., Sipes, B. S., & Schmitt, D. P. (2002). Crotalaria as a cover crop for nematode management: A review. *Nematropica, 32*, 35-57.

Widmer, T. L., Mitkowski, N. A., & Abawi, G. S. (2002). Soil organic matter and management of plant-parasitic nematodes. *Journal of Nematology, 34*, 289-295.

13

Total Coliforms and *Escherichia coli* in Surface and Subsurface Water from a Sugarcane Agroecosystem in Veracruz, Mexico

Juan Valente Megchún-García[1], Cesáreo Landeros-Sánchez[1], Alejandra Soto-Estrada[1], María del Refugio Castañeda-Chávez[3], Juan Pablo Martínez-Dávila[1], Iouri Nikolskii-Gavrilov[2], Itzel Galaviz-Villa[3] & Fabiola Lango-Reynoso[3]

[1] Colegio de Postgraduados, Campus Veracruz, Manlio Fabio Altamirano, Veracruz, México

[2] Colegio de Postgraduados, Campus Montecillo, Montecillo, Texcoco, México

[3] Instituto Tecnológico de Boca del Río, Carretera Veracruz, Boca del Río, Veracruz, México

Correspondence: Cesáreo Landeros-Sánchez, Colegio de Postgraduados, Campus Veracruz Apartado postal 421, Veracruz, Ver., México. E-mail: clandero@colpos.mx

Abstract

Water contamination is a phenomenon of global concern resulting from human activities. Coliform bacteria reduce water quality and negatively affect public health. The pollution of surface and groundwater by coliform bacteria, including *Escherichia coli*, originate, in general, from point sources of pollution derived from human settlements, such as those located in Module I-1, Irrigation District 035, La Antigua, Veracruz, Mexico. The objective of this study was to assess the level of contamination of surface and groundwater by coliform bacteria and *E. coli*, as well as to identify point sources of water contamination by these bacteria in the sugarcane agroecosystem of Irrigation Module I-1, La Antigua. Sampling sites included deep wells, irrigation canals and natural streams near point sources of pollution. The determination of total coliform bacteria and *E. coli* were made in accordance with Mexican Standard NMX-AA-042-1987. Total coliform results revealed differences between groundwater (198.6 MPN/100 mL) and surface water concentrations (52,419.2 MPN/100 mL) ($p < 0.05$), and between irrigation water (76,501.1 MPN/100 mL) and concentrations in natural streams (28,337.3 MPN/100 mL). The highest concentration of *E. coli* was found in groundwater and surface water samples from the municipality of La Antigua. The primary sources of contamination are the discharges from drains and septic tanks. Total coliform values exceeded permissible limits established by NOM-127-SSA1-1994 that regulates the permissible water quality limits for human use and consumption. The presence of *E. coli* represents a significant public health risk.

Keywords: water contamination, irrigation canals, sugarcane, enteric bacteria

1. Introduction

Water contamination is a global concern for humanity, and is defined as an addition of energy or substances which alter the natural state of the environment, and are caused by anthropogenic activities on the surface of the earth (Krupa, 1999; Pérez y Landeros, 2009; Rebolledo et al., 2011). On the other hand, water pollution is the contamination of natural water bodies by chemical, physical, radioactive or pathogenic microbial substances directly or indirectly discharged into water bodies without adequate treatment to remove harmful compounds (Hogan, 2014). Contamination can be atmospheric, aquatic, edaphic, genetic, chemical and biological. As well, pollution and contamination are classified as being from point (entering the environment from definable points; e.g. a drainage pipe) or non-point sources (entering the environment diffusely; e.g. surface runoff). Microbiological contamination can be due to viruses, bacteria, protozoa and fungi that come mainly from sewage discharge and other human activities, as well as physicochemical and ecophysical processes occurring in aquatic systems due to increased human population levels and industrial development (Wong & Barrera, 1996; Martinez, Fonseca, Ortega, & Garcia, 2009; Lango et al., 2012). Bacterial contamination from total coliforms in surface water and groundwater originates from point sources such as latrines, septic tanks, trails, cattle pens or stables, discharges of untreated sewage and runoff of rainwater containing fecal material (Sasson, 1993; Herrero et al., 2000; Perdomo, Casanova, & Ciganda, 2001; Castañeda, Pardio, Orrantia, & Lango, 2005; Camargo & Alonso, 2007; Rebolledo et al., 2011). Non-point sources of microbiological and physicochemical contamination are associated with high production levels in agroecosystems. It is estimated that a person can defecate between 100 to 500 million coliform bacteria per gram of fecal matter and generate a number of infectious diseases such as typhoid, paratyphoid, cholera, infant diarrhea and bacillary dysentery (caused by *Shigella* bacteria) (Orozco & Gutiérrez, 1983; Landeros-Sánchez et al., 2012). Gastrointestinal diseases related to bacterial contamination are influenced by economic, health and cultural factors (Camargo & Alonso, 2007). Of particular importance is the bacterium *Escherichia coli* which is an indicator of fecal contamination, and is well-studied genetically, biochemically and physiologically (Ramirez, Contreras, & Gómez, 2005).

Ecosystems have biological, ecological and aesthetic value, but agroecosystems have social values in terms of food production and other forms of use or consumption. The agroecosystem concept is presented as an option for efficient economic, social and environmental management, and as a conceptual model of farming with a minimum level of human cybernetic control. It is the optimal unit for studying agriculture and its transformation because it is integrated into regional agricultural and rural systems through human production-consumption chains, and is affected by politics and the culture of public and private institutions. The agroecosystem is an open system built from the social modification of a natural system, has dynamic feedback and control processes, and is internally and externally regulated in response to internal and external variations in its environment. The spatial area, biodiversity and objectives of an agroecosystem depend on the type of controller that regulates the resources it manages and its complex interactions with the environment (Martínez, Gallardo, Bustillo, & Pérez, 2011; Del Angel, Diaz, Guajardo, & Linares, 2011; Galaviz-Villa, Landeros-Sánchez, Castañeda-Chávez, & Lango-Reynoso, 2012). In the present study, the focus on the agroecosystem was from the structural-functionalist perspective, and did not consider the involvement of a driver, as is done when only one process occurring in the agroecosystem is studied, such as bacteriological contamination. Here, the sugarcane agroecosystem is highlighted by the cultivation area of Irrigation Module I-1, La Antigua, Veracruz, Mexico, within which the aquatic sampling points and small cities and rural villages are located. The surface area covered by sugarcane in the study area is 13,793 hectares, and includes portions of the municipalities of La Antigua, Paso de Ovejas, Puente Nacional and Úrsulo Galván. Irrigation Module I-1 was legally established as a Civil Association (A.C.) in 1993, and is called Asociación de Usuarios de Riego del Modulo de Riego I-1, La Antigua, A.C. Subsequent references to the Association will be made as Irrigation Module I-1, La Antigua.

Extreme weather can cause spills, runoff and infiltration of wastewater containing fecal matter into streams, swamps, reservoirs and coastal waters that increase the volume of contaminated surface water and groundwater (Perdomo et al., 2001; Bautista, Aguilar, & Batllori, 2011). As well, human settlements within an area contribute to pollution and contamination (Juarez, Martinez, Diaz, Perez, & Brust, 1990; Perdomo et al., 2001; Galaviz-Villa et al., 2012; Landeros-Sánchez et al., 2012). Contamination from total coliforms and *E. coli* in surface and groundwater is also related to political and technological actions, such as the lack of effective enforcement of the National Water Act and Mexican Official Standards (Moreno et al., 2002). Chemical or biological contamination of aquifers is considered an irreversible phenomenon that must be addressed by relevant authorities (De las Heras & Moreno, 2005). Based on this, the hypothesis of this work consider that point and non-point sources of pollution located in sugarcane Irrigation Module I-1, La Antigua, Veracruz, Mexico, are the main contaminants of surface and groundwater by coliform bacteria, including *E. coli*. Thus, this study was conducted to assess total coliform bacteria and *E. coli* contamination in surface and groundwater in sugarcane agroecosystem, Irrigation Module I-1, La Antigua, Veracruz, Mexico.

2. Method

2.1 Characterization of the Study Area

The research was conducted in sugarcane Irrigation Module I-1, La Antigua, Veracruz, Mexico, located in Irrigation District 035 (Figure 1). The study area is located in the central portion of the state of Veracruz between 19°09′ and 19°25′N and 96°17′ and 96°25′W, and covers 13,793 hectares and supplies 247,179 kilometers of drainage for the municipalities of La Antigua, Manlio Fabio Altamirano, Paso de Ovejas, Puente Nacional and Úrsulo Galván, whose populations are 25 855, 23 807, 34 141, 22 510, and 30 260 people, respectively.

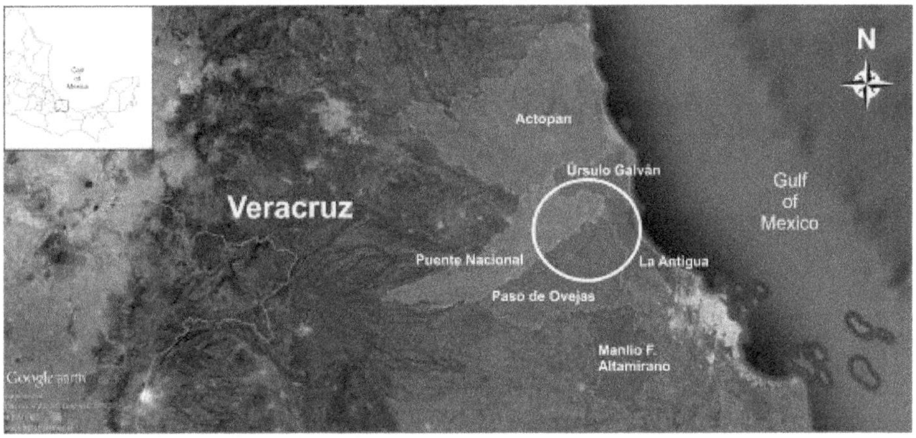

Figure 1. Location of the study area (Irrigation Module I-1, La Antigua, Veracruz, Mexico)

2.2 Diagnostic Phase of the Study Area

To determine the sampling points for bacterial contamination, data were provided by technicians from the Irrigation Module I-1, La Antigua A.C., which consisted of identifying mapped locations and distributions of groundwater and surface water supplies in the field during October to December, 2012. The point sources of bacterial contamination were identified visually with help of users of the irrigation Module. Sampling points (32) were within the municipalities of La Antigua, Paso de Ovejas and Úrsulo Galván, and included groundwater sources from deep wells (for drinking water and irrigation), domestic wells, natural surface waters (Río La Antigua, Río Chico, and Río Tolome) and irrigation canals (main distribution frame, main channel, side channel, Drain 1).

2.3 Sampling Groundwater and Surface Water

Water samples were collected from wells, waterwheels, irrigation canals and streams according to Mexican standard protocols NOM-230-SSA1-2002, NMX-AA-003-1980 and NOM-014-SSA1-1993 in February 2013. Samples were transported to Instituto Tecnologico de Boca del Rio, Laboratorio de Investigación en Recursos Acuáticos (LIRA).

2.4 Analysis of Total Coliforms and E. coli

Total coliforms and *E. coli* were determined for each sample according to Mexican standard operating protocol NMX-AA-042. The most probable number (MPN) was determined for total coliforms while *E. coli* was assessed using presence/absence. For this, the Colilert method was used and consisted in utilizing Quanti-Trays of 48 holes. In a sterile beaker, 90 mL of sterile distilled water was mixed with 10 mL of each sampling water and the Colilert reagent; the mixture was poured on the Quanti-Trays, which were incubated at 35 ± 1 °C for 24 and 48 h to determine total coliforms and *E. Coli*, respectively. Three replicates per sample of water were used.

2.5 *Statistical Analysis*

For this research work three replicates were used in each water sampling site (32). The results analytic process was carried out by means of graphic parametric variance analyses (F test) using "Statistica" Version 11 Software. In Figure 2 and 3 can be observed that the populations which are statistical different do not overlap. The value of $Z = 1.645$ is the point where the confidence probability of a standard normal distribution is equal to 95 % and if the boxes of the mean values of the populations do not overlap then it is interpreted that these populations are statistical different. Therefore, there are statistical differences between the ground and surface populations (Figure 2) and between La Antigua and the municipalities of Paso de Ovejas and Ursulo Galván (Figure 3).

3. Results and Discussion

Concentrations of total coliforms were higher in surface waters (52,419.2 MPN/100 mL) than in groundwater (198.6 MPN/100 mL) ($P < 0.01$) (Figure 2); this might indicate the existence of point sources of bacterial contamination within sugarcane Irrigation Module I-1, La Antigua. In Mexico, it is reported that homes, small factories, roads and hospitals discharge thousands of cubic meters of waste per year into channels, drains, main reservoirs and streams (Gordillo et al., 2010). Méndez, San Pedro, Castillo, and Vazquez (2010) and Camargo and Alonso (2007) reported that coliform concentrations in groundwater and surface water imply the presence of bacterial genera *Citrobacter*, *Enterobacter*, *Escherichia*, *Hafnia*, *Klebsiella*, *Serratia* and *Yersinia*; so the presence of these bacteria could pose a risk to human health in the communities surrounding sugarcane Irrigation Module I-1, La Antigua, Veracruz.

Total coliform concentrations varied among municipalities, with La Antigua having the highest concentration (38,000MNP/100 mL) followed by Paso de Ovejas (11,000 MPN/100 mL) and Úrsulo Galván (4,500 MPN/100 mL) (Figure 3). Thus, it is likely that there are point sources of bacterial contamination affecting water irrigation channels, streams, shallow water and deep wells. Even though the water samples were equally distributed between regions, the differences in the coliform concentrations appeared to be so because of the presence of the main cities, some rural villages, hospitals, and a sugarcane factory, which likely contributed through their various waste discharges to the referred high bacteria concentrations.

The concentration of total coliforms was higher in water from irrigation canals; followed by streams, shallow water and deep wells ($P < 0.01$, Table 1). No statistically significant differences between shallow and deep wells were found when using Fisher LSD Test. Total coliform levels in shallow and deep wells, streams and irrigation canals exceeded permissible limits established by Mexican Official Standard NOM-127-SSA1-1994 that regulates water quality for human use and consumption. However, total coliform concentrations from deep and shallow wells did not exceed the permissible limits for total coliforms established in Mexican Official Standard NOM-033-ECOL-1997, which regulates pollutants in treated wastewater to be reused in public services.

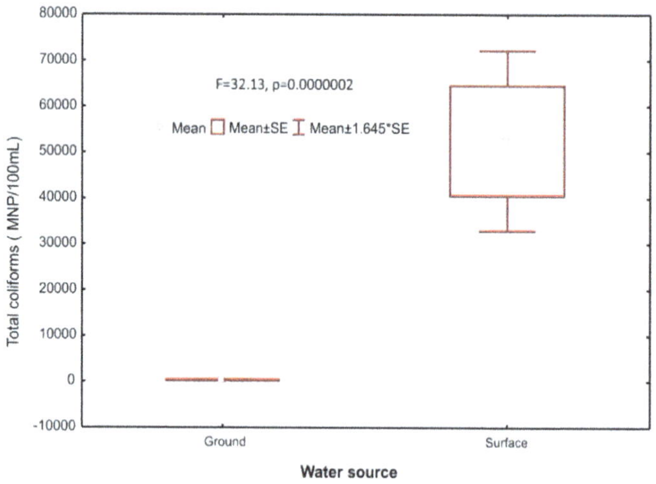

Figure 2. Total coliform bacteria in ground and surface water samples from Irrigation Module I-1, La Antigua, Veracruz

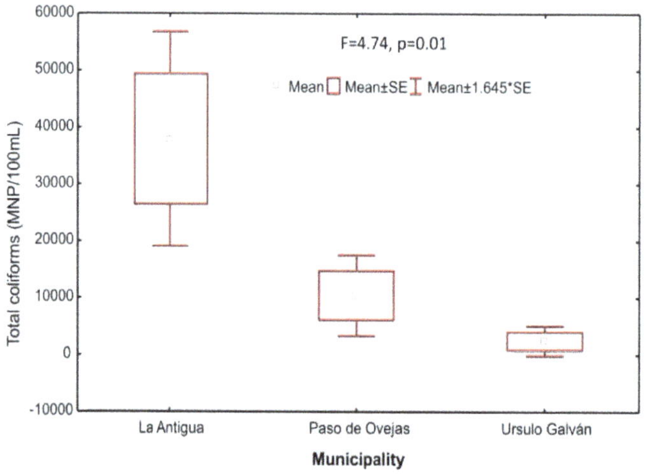

Figure 3. Total coliform concentrations found in ground and surface water from three municipalities in Irrigation Module I-1, La Antigua samples (sample data were combined in each municipality)

Total coliform values found in this study (Table 1) exceeded the maximum concentrations of total coliforms reported for shallow water from 120 agricultural wells in the valley of San Luis Potosi, Mexico (2,400 MPN/100 mL); values which also exceeded limits set for ecological criteria (CE-CCA-001/89) established by the Diario Oficial de la Federación Mexicana and by the Ley Federal de Derechos de Agua, 2009 (Sarabia, Cisneros, Aceves De, Durán, & Castro, 2011). These criteria indicate that the organisms in fresh or sea water should not exceed 200 NMP/100 mL.

The average total coliform concentration among the sampling points (28,337.2 MPN/100 mL) exceeded 300 MPN/100 mL reported by Chagas et al. (2006) from surface waters in Argentina. As well, Lango et al. (2012) and Landeros-Sánchez et al. (2012) reported that total coliforms in lagoon systems were affecting public health.

The total coliform concentrations in samples from the irrigation canals in Irrigation Module I-1 (Table 2) were higher in water leaving the canals. This occurs because some canal routes cross communities and are used by people to discharge sewage, domestic waste and that from animal husbandry activities. The irrigation water eventually mixes with streams and aquifers, resulting in the deterioration of water quality (Pacheco, Cabrera, & Pérez, 2004). Coliform bacteria can survive in slow moving and highly polluted groundwater. The increased concentrations of coliform bacteria in irrigation canals probably indicate that the water also contains other pathogens (Perdomo et al., 2001).

Table 2 shows statistically significant differences among the total coliform concentrations in samples from three irrigation canals and Drain 1, with the latter having the highest mean value (129,332.5 MPN/100 mL). Drain 1 receives discharges from the Instituto Mexicano del Seguro Social (IMSS) clinic in the city of Cardel, Veracruz, wastewater runoff from the urban and rural zones in Irrigation Module I-1, La Antigua, and agrochemicals from sugarcane cultivation (Galaviz et al., 2012). The rate with which contamination occurs depends on chemical, physical and biological soil characteristics (Pacheco et al., 2004).

Table 1. Comparison of mean values for concentrations of total coliforms in samples of groundwater and surface water from Irrigation Module I-1, La Antigua, Veracruz, and relevant official Mexican standards

Water source	Total coliforms (MPN/100 mL)
Deep wells	87.3c
Shallow water wells	309.9c
Streams	28337.2b
Irrigation canals	76501.1a
NOM-127-SSA1-1994	2
NOM-033-ECOL-1997 (Fecal coliforms)	240-1000

Note. Mean values comparison by using Fisher LSD Test at a significance probability $p < 0.01$. The presence of different letters means that there exists a statistical difference. NOM = Official Mexican standards, MPN=Most probable number.

Table 2. Mean concentrations of total coliforms in water samples entering and exiting irrigation canals in Irrigation Module I-1, La Antigua, Veracruz

Irrigation canals	Total coliforms (MPN/100 mL)	Irrigation canals	Total coliforms (MPN/100 mL)
Water entering	52,951.5b	Area sampled	5,559.3c
Water exiting	64,580.0a	Principal canal	69,100.0b
		Lateral canal	31,071.1c
		Drain 1	129,332.5a

Note. Mean values comparison by using Fisher LSD Test at a significance probability $p < 0.01$. The presence of different letters means that there exists a statistical difference. MPN = Most probable number.

All water samples showed presence of *E. coli* (Figure 4). This bacterium was found in all water samples from streams (100%), followed by the samples from irrigation canals (83%), deep wells (30%) and shallow wells (22%). Of the total water samples obtained from 32 sampling points, 63% were from groundwater and 37% from surface water. Of the groundwater samples analyzed, 27% were contaminated with *E. coli* and 73% were negative. Among surface water samples, 92% were positive for *E. coli*, and 8% were negative (Figure 5). It has been found that among groundwater samples from deep wells, *E. coli* and other coliforms increased the total assimilation rate of NO_{3-} and O_2 to synthesize amino acids and proteins (Perdomo et al., 2001). The biophysical and environmental conditions, nitrogen content, resistance to pH conditions and soil factors allow *E. coli* to survive in groundwater (Cifuentes & Ruiz, 1994; Juarez, Silva, Uribe, & Cifuentes, 2003).

Contamination of surface and groundwater by *E. coli* is a risk to public health since this bacterium causes human illnesses such as diarrhea in children and adults, as well as inflammatory diarrhea. The latter is characterized primarily by bacteria attacking intestinal epithelia (Landeros-Sánchez et al., 2012). In Figures 3 and 4, values for *E. coli* exceeded the limits set by the official standard NOM-127-SSA1 1994; water for potable use and consumption should not contain *E. coli*. Several groups of these bacteria can severely affect children 1 to 8 years of age and cause death, especially if Hemolytic Uremic Syndrome (HUS) is involved, which primarily occurs when water contaminated with this bacteria is ingested (Michanie, 2003; Perez, Rodriguez, & Rulli, 2009). Therefore, it is recommended that water be tested for the presence of *Escherichia coli*, particularly enteropathogenic *E. coli* (EPEC), which causes diarrhea in children less than two years of age in developing countries. This pathogen causes histological damage to the mucosa of the small intestine and can cause death in children (Vidal, Canizalez, Gutierrez, & Navarro, 2007). Thus, health authorities in Mexico (federal, state and local, and public health institutions) must monitor water bodies vigilantly to ensure that official standards are not being surpassed and that the tests are effective and timely.

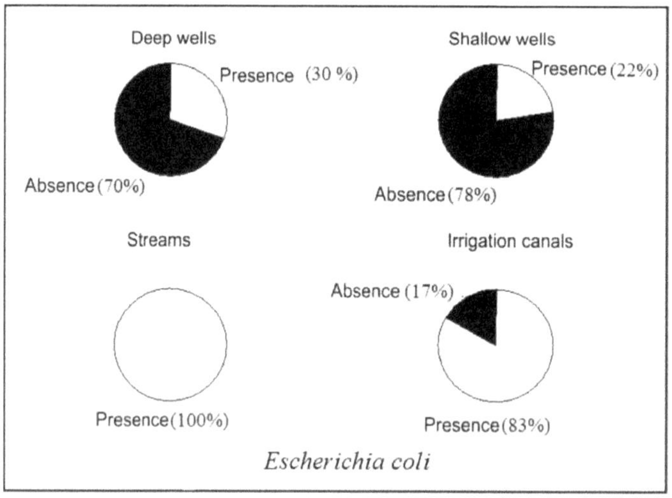

Figure 4. Percentage of water samples from deep and shallow wells, streams and irrigation canals showing presence or absence of *Escherichia coli*

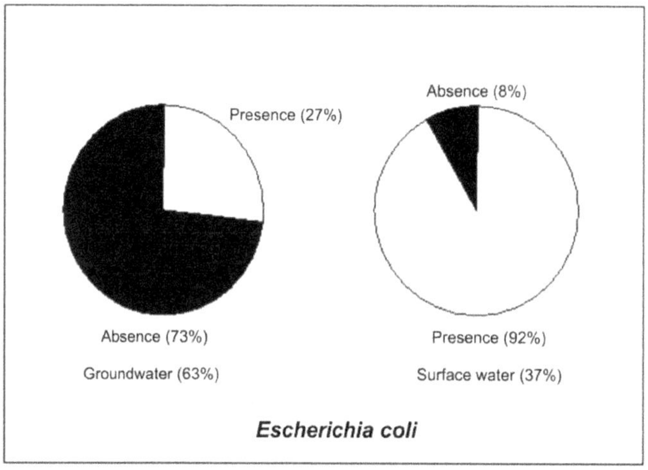

Figure 5. Percentage of samples of ground and surface water from Irrigation Module I-1, La Antigua, showing the presence or absence of *Escherichia coli*

Surface waters transport high concentrations of coliform bacteria and *E. coli* into groundwater, revealing that bacterial contamination of groundwater occurs due to runoff and infiltration processes (Ferguson, De Rodahusman, Altavilla, Deere, & Ashbolt, 2003; Guber, Sheldon, & Pachepsky, 2005). The approach used in the present study (Figure 6) facilitated understanding of the phenomenon and identification of point sources of bacterial contamination occurring within sugarcane Irrigation Module I-1, La Antigua. Therefore, point sources of bacterial contamination were identified as septic tanks, latrines, and sewage discharges to streams, irrigation canals and drains.

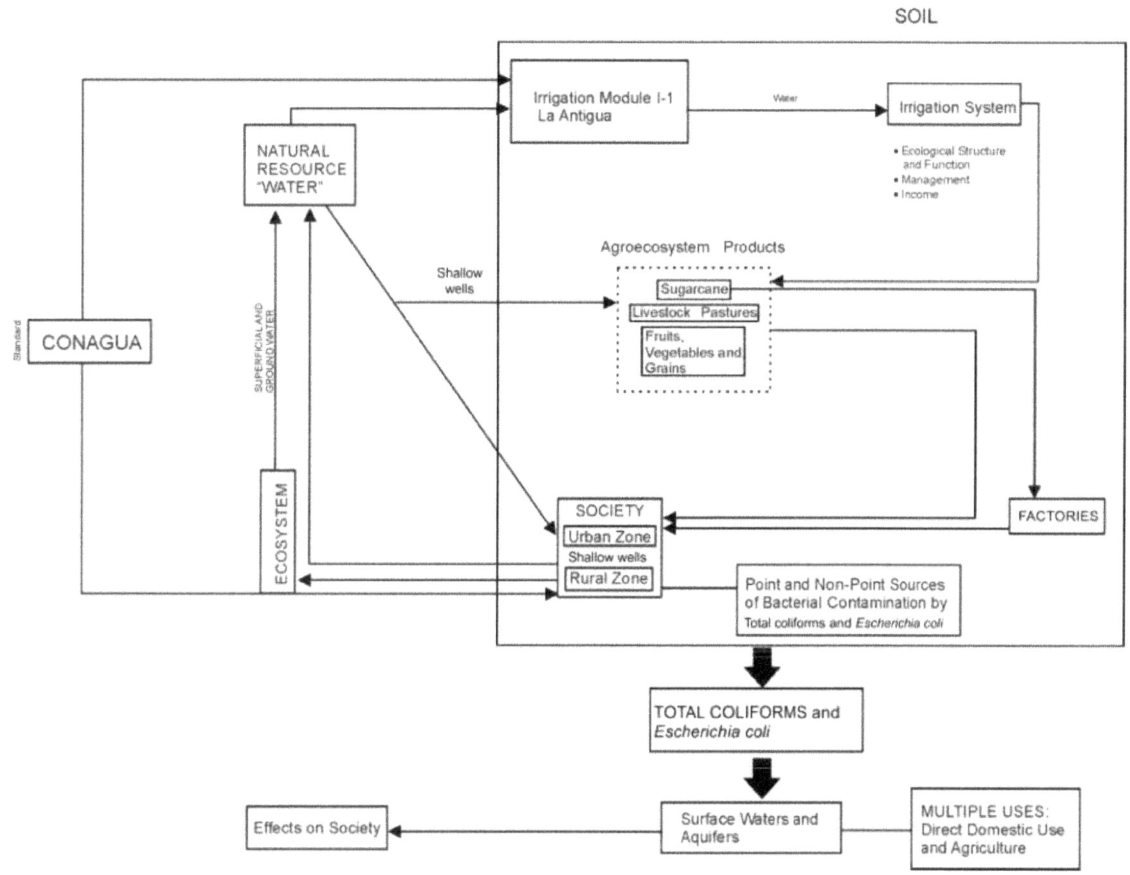

Figure 6. Conceptual model for studying contamination from total coliforms and *Escherichia coli* in sugarcane Irrigation Module I-1, La Antigua

4. Conclusions

The concentration of total coliforms and *E. coli* in both surface and groundwater from deep and shallow wells, streams and irrigation canals exceeded permissible limits (NOM-127-SSA1-1994, Mexican Official Standard for human use and consumption). The concentration of total coliforms from deep and shallow wells did not exceed the maximum permissible limits for contaminants (NOM-033-ECOL-1997, Wastewater). Contamination from total coliform bacteria and *E. coli* was provided by point sources such as septic tanks, sewage and hospital waste which are discharged into irrigation canals and streams. In addition, some portion of the total coliforms and *E. coli* that enter the water conveyance system used to irrigate sugarcane comes from the La Antigua River, which supplies Irrigation Module I-1, La Antigua. Therefore, contamination and pollution from total coliform bacteria and *E. coli* are risks to the health of the residents of the urban and rural areas contained therein. Thus, it is necessary to establish a culture of environmental stewardship, particularly water quality, to ensure that Mexican Official Standards be applied efficiently in order to prevent further negative impacts to the environment and public health.

Acknowledgements

To Colegio de Postgraduados Campus Veracruz (LPI-8 Impacto y Mitigación del Cambio Climático) and to the Instituto Tecnológico de Boca del Río (ITBOCA) for financial support. Thanks also to Ing. Juan Manuel

Hernández Pérez for assistance in the preparation of tables and figures.

References

Bautista, F., Aguilar, D. Y., & Batllori, E. (2011). Amenazas, vulnerabilidad y riesgo de contaminación de las aguas subterráneas en la península de Yucatán. *Teoría y Praxis, 9*, 9-31.

Camargo, J. A., & Alonso, A. (2007). Contaminación por nitrógeno inorgánico en los ecosistemas acuáticos: problemas medioambientales, criterios de calidad del agua, e implicaciones del cambio climático. *Ecosistema, 16*, 98-110.

Castañeda, M., Pardio, S., Orrantia, B., & Lango, F. (2005). Influence of water temperature and salinity on seasonal occurrences of Vibrio cholerae and enteric bacteria in oyster–producing areas of Veracruz, México. *Marine Pollution Bulletin, 50*, 1641-1648. http://dx.doi.org/10.1016/j.marpolbul.2005.06.036

Chagas, C. I., Morettón, J., Santanatoglia, O. J., Paz, M., Muzio, H., De Siervi, M., & Castiglioni, M. (2006). Indicadores de contaminación biológica asociados a la erosión hídrica en una cuenca de pampa ondulada Argentina. *CI. Suelo, 24*, 1-10.

Cifuentes, E., & Ruiz, P. G. (1994). Escenarios epidemiológicos del uso agrícola del agua residual: el valle del Mezquital, México. *Salud Pública de México, 36*, 3-9.

De la Heras, I. J., & Moreno, A. J. L. (2005). La calidad del agua contaminación de las aguas producida por la agricultura. Limitaciones para su uso en la agricultura. In O. M. F. Martín de Santa, F. P. López & B. A. Calera (Eds.), *Agua y Agronomía* (pp. 295-318). Grupo Mundi Prensa. Madrid, España.

Del Ángel, P. A. L., Díaz, P. G., Guajardo, P. R., & Linares, B. B. C. (2011). Landscapes and fragile economy: ecosystems and agroecosystems in the Coatepec - La Antigua basin, Veracruz, Mexico. *Tropical and Subtropical Agroecosystems, 14*, 629-642.

Ferguson, C., De Roda Husman, A. M., Altavilla, N., Deere, D., & Ashbolt, N. (2003). Fate and transport of surface water pathogens in watersheds. *Critical Reviews in Environmental Science and Technology, 33*, 299-361. http://dx.doi.org/10.1080/10643380390814497

Galaviz-Villa, I., Landeros, S., Castañeda-Chávez, M. del R., & Lango, R. (2012). Theoretical conceptual assembly for the analysis of sugarcane agroecosystems in the central Gulf of Mexico: an eclectic model. *Journal of Agricultural Science, 4*, 214-220. http://dx.doi.org/10.5539/jas.v4n12p214

Gordillo, M. A. J., Cabrera, C. R. B. E., Hernández, M. M., Galindo, E., Otazo, E., & Prieto, F. (2010). Evaluación regional del impacto antropogénico sobre el aire, agua y suelo, caso: Huasteca Hidalguense, México. *Revista Internacional de Contaminación Ambiental, 26*, 229-251.

Guber, A. K., Shelton, D. R., & Pachepsky, Y. A. (2005). Effect of manure on *Escherichia coli* attachment to soil. *Journal of Environmental Quality, 34*, 2086-2090. http://dx.doi.org/10.2134/jeq2005.0039

Herrero, M. A., Maldonado, M. V., Sardi, G., Flores, M., Orlando, A., & Carbo, L. (2000). Distribución de la calidad del agua subterránea en sistemas de producción agropecuarios bonaerenses. 2. Condiciones de manejo y grado de contaminación. *Revista Argentina Producción Animal, 20*(3-4), 237-245.

Hogan, M. C. (2014). Water pollution. In C. Cleveland (Ed.), *Encyclopedia of Earth*. Washington, DC: Mark McGinley.

Juárez, F. L. A., Silva, S. J., Uribe, S. F. J., & Cifuentes, G. E. (2003). Microbiological indicators of water quality in the Xochimilco Canals, Mexico City. *Salud Pública de México, 45*, 389-395.

Juárez, M. J., Martínez, R. G., Díaz, S. J., Pérez, G. M. L., & Brust, C. H. (1990). Evaluación del sistema integral para tratamiento del agua en un hospital general de la Ciudad de México. *Salud Pública, 32*, 566-574.

Krupa, S. V. (1999). *Polución, población y plantas*. Editorial Mexicana. Edición Rogelio Dromundo Salazar. *Registro Número 306* (p. 184).

Landeros-Sánchez, C., Lango-Reynoso, F., Castañeda-Chávez, M. del R., Galaviz, V. I., Nikolskii, G. I., Palomarez, G. M., ... Mínguez, R. M. M. (2012). Assessment of water pollution in different aquatic systems: aquifers, aquatic farms on the Jamapa River and coastal lagoons of Mexico. *Journal of Agricultural Science, 4*, 186-196. http://dx.doi.org/10.5539/jas.v4n7p186

Lango, F., Castañeda-Chávez, M., Zamora-Castro, J. E., Hernández-Zárate, G., Ramírez-Barragán, M. A., & Solís-Morán, E. (2012). La acuariofilia de especies ornamentales marinas: un mercado de retos y oportunidades. *Latin American Journal of Aquatic Research, 40*, 12-21. http://dx.doi.org/10.3856/vol40-issue1-fulltext-2

Martínez, D. J. P., Gallardo, L. F., Bustillo, G. L. C., & Pérez, V. A. (2011). El agroecosistema, unidad de estudio y transformación de la diversidad agrícola. In Comisión Nacional para el Conocimiento y Uso de la Biodiversidad (CONABIO), *La Biodiversidad en Veracruz: Estudio de Estado* (Volumen I, pp. 453-462). Gobierno del Estado de Veracruz, Universidad Veracruzana, Instituto de Ecología, A.C. México.

Martínez, R. A., Fonseca, G. K., Ortega, S. J. L., & García, L. C. (2009). Monitoreo de la calidad microbiológica del agua en la cuenca hidrológica del Río Nazas, México. *Revista Química Viva, 8*, 35-47.

Méndez, N. R. I., San Pedro, C. L., Castillo, B. E. R., & Vázquez, B. E. (2010). Modelación del tiempo de conservación de muestras biológicas de agua. *Revista Internacional de Contaminación Ambiental, 26*, 327-335.

Michanie, S. (2003). *Escherichia coli* O157:H7: la bacteria que disparó el HACCP en la industria de la carne. *Énfasis Alimentos, 9*, 1-7.

Moreno, C. P., Rojas, G. J. L., Zárate, L. D., Ortiz, P. A., Lara, D. A. L., & Saavedra, V. T. (2002). Diagnóstico de los manglares de Veracruz: distribución, vínculo con los recursos pesqueros y su problemático. *Madera y Bosques, 8*, 61-88.

NMX-AA-003. (1980). *Norma Oficial Mexicana.* Aguas residuales. Muestreo. Retrieved from http://www.conagua.gob.mx/CONAGUA07/Noticias/NMX-AA-003-1980.pdf

NMX-AA-042. (1987). Norma Oficial Mexicana. Calidad del agua determinación del número más probable (NMP) de coliformes totales, coliformes fecales (termotolerantes) y *Escherichia coli. Presuntiva*, 1-21. Retrieved from http://www.conagua.gob.mx/CONAGUA07/Noticias/NMX-AA-042-1987.pdf

NOM-014-SSA1. (1993). *Norma Oficial Mexicana.* Procedimientos sanitarios para el muestreo de agua para uso y consumo humano en sistemas de abastecimiento de agua publicos y privados. Retrieved from http://www.salud.gob.mx/unidades/cdi/nom/014ssa13.html

NOM-033-ECOL. (1997). *Norma Oficial Mexicana.* Límites máximos permisibles de contaminantes para las aguas residuales tratadas que se reúsan en servicios al público. Retrieved from http://www.elaw.org/system/files/NOM-003-ECOL-1997.pdf

NOM-127-SSA1. (1994). *Norma Oficial Mexicana.* Salud ambiental, agua para uso y consumo humano-límites permisibles de calidad y tratamientos a que debe someterse el agua para su potabilización. Retrieved from http://www.salud.gob.mx/unidades/cdi/nom/127ssa14.html

NOM-230-SSA1. (2002). *Norma Oficial Mexicana.* Salud ambiental. Agua para uso y consumo humano, requisitos sanitarios que se deben cumplir en los sistemas de abastecimiento públicos y privados durante el manejo del agua. Procedimientos sanitarios para el muestreo. Retrieved from http://www.salud.gob.mx/unidades/cdi/nom/230ssa102. html

Orozco, B. V., & Gutiérrez, G. E. A. (1983). Contaminación fecal contera en la zona del Puerto de Ensenada, Baja California. *Ciencias Marinas, 9*, 27-34.

Pacheco, A. J., Cabrera, S. A., & Pérez, C. R. (2004). Diagnóstico de la calidad de agua subterránea en los sistemas municipales de abastecimiento en el Estado de Yucatán, México. *Ingeniería, 8*, 165-179.

Perdomo, C. H., Casanova, O. N., & Ciganda, V. S. (2001). Contaminación de aguas subterráneas con nitratos y coliformes en el litoral sudoeste del Uruguay. *Agrociencia, 5*(1), 10-22.

Pérez, V. A., & Landeros, S. C. (2009). Agricultura y deterioro ambiental. *Elementos: Ciencia, Arte y Cultura, 16*(73), 19-25.

Pérez, C., Rodríguez, D., & Rulli, F. (2009). Innovación docente. Infecciones infantiles provocadas por *Escherichia coli*: síndrome urémico hemolítico y otras. *Revista Iberoamericana de Educación, 50*, 1-9.

Ramírez, S. J., Contreras, F. G., & Gómez, E. M. C. (2005). La fase estacionaria en la bacteria *Escherichia coli. Revista Latinoamericana de Microbiología, 43*, 92-101.

Rebolledo, M. A., Del Ángel, P. A. L., Megchún, G. J. V., Adame, G. J., Nataren, V. J., & Capetillo, B. A. (2011). Coberteras vivas para el manejo de malezas en mango (*Mangifera indica* L.) cv. Manila. *Tropical and Subtropical Agroecosistemas, 13*, 327-338.

Sarabia, M. I. F., Cisneros, A. R., Aceves De, A. J., Durán, G. H. M., & Castro, L. J. (2011). Calidad del agua de riego en suelos agrícolas y cultivos del Valle de San Luis Potosí, México. *Revista Internacional de Contaminación Ambiental, 27*, 103-113.

Sasson, A. (1993). *La alimentación del hombre del mañana* (p. 807). UNESCO.

Vidal, J. E., Canizález, R. A., Gutiérrez, J. J., & Navarro, G. F. (2007). Patogénisis molecular, epidemiología y diagnóstico de *Escherichia coli* enteropatógena. *Salud Pública de México, 49*, 376-386. http://dx.doi.org/10.1590/S0036-36342007000500008

Wong, I., & Barrera, G. (1996). Niveles de contaminación microbiológica en el Golfo de México. In A. V. Botello, G. J. L. Rojas, J. Benítez & L. D. Zárate (Eds.), *Golfo de México, Contaminación e Impacto Ambiental: Diagnóstico y Tendencias* (pp. 383-397). EPOMEX Serie Científica 5. Universidad Autónoma de Campeche México.

14

Evaluation of Four Sunflower Hybrids (*Helianthus annuus*) under Three Irrigation Regimes and Two Doses of Fertilization on Flower Production

E. Carrillo-Ávila[1], C. García-Acedo[2], J. Arreola-Enríquez[1], C. Landeros-Sánchez[3], M. L. Osnaya-González[1] & C. Castillo Aguilar[1]

[1] Colegio de Postgraduados Campus Campeche, Carretera Haltunchén-Edzná, Sihochac, Champotón, Campeche, México

[2] Secretaria de Desarrollo Rural, Calle Ricardo Castillo Oliver Lote 12, Barrio de San Francisco, Campeche, México

[3] Colegio de Postgraduados, Campus Veracruz, Carretera Federal Xalapa-Veracruz, vía Paso de Ovejas entre Puente Jula y Paso San Juan, Tepetates, Veracruz, México

Correspondence: C. Landeros-Sánchez, Colegio de Postgraduados, Campus Veracruz, Km. 88.5 Carretera Federal Xalapa-Veracruz, vía Paso de Ovejas entre Puente Jula y Paso San Juan, Tepetates, Veracruz, C.P. 91690, México. E-mail: clandero@colpos.mx

Abstract

Flower production is an income source for farming families. However, their productivity has been limited due to the lack of information on technical managements of crop species. In Mexico, 21,129 hectares are planted every year, with an annual production of 83,000 tons of flowers for decoration, and a total production value of 6,097 million Mexican pesos. In order to design agronomic management practices for sunflower production in the state of Campeche, three factors were evaluated: 1) sunflower hybrids, 2) time of irrigation and 3) fertilization doses, with four (hybrids: "Sunbright", "Prado Red Shade", "Full Sun" and "GH-382"), three (soil moisture tension at the start of irrigation of -10, -35 and -60 kPa), and two (Fertilization formulas: 60-50-0, and 30-25-0) levels, respectively using a sub split plot arrangement nested in a complete random block design with three replicates. Response variables evaluated were: plant height (PH), stem diameter (SD), leaf width (LW), leaf length (LL), distance between nodes (DN), number of inflorescences (NF), inner diameter (IDF), external diameter (EDF) and capitulum weight (CW). Irrigation affected PH, SD, LL, IDF, EDF, and CW. In all these variables, the highest values were found when irrigation was conducted at the lowest moisture tension (-10 kPa). Likewise, in all those variables significant effect of hybrids were found, which confirms that the morphology of flowers is defined by genetic factors. Specifically, hybrid "Full Sun" had significantly higher PH, SD, and DN compared to the rest of the hybrids. "Full Sun" and "Sunbright" had the highest values for PH and LL. "Full Sun" and "GH 382" had significantly higher IDF and EDF. "Prado Red Shade" had significantly lower PH, SD, LL, CW and higher NF. Fertilization only had significant effects on LW, DN and CW. In conclusion, irrigation improves morphological characteristics of sunflower plants. If the farmer's objective is to produce larger flowers, then hybrids "Fun Sun" and "GH 382" are recommended. However, if the objective is to produce small flowers, hybrids "Sunbright" and "Prado Red Shade" are recommended. The latter is also suitable to produce a larger number of capitula.

Keywords: flower production, moisture tension, irrigation, morphological characteristics

1. Introduction

Floriculture represents a source of income for farming families. Thanks to traditional knowledge, families can produce a wide number of flower species, such as sunflower, madonna lily, heliconia, spikenard, tulips, etc. However, farmer's productivity is somehow limited due to the lack of information regarding the technical management of each species. In Mexico, there is a planted surface of 21,129 hectares, which are devoted to the production of flowers for decoration, with a total production of 83,000 tons of flowers, and a total production value of 6,097 million Mexican pesos. 80% of the total flower production is sold locally, 90% of the rest is exported to the United States (De la Madrid, 2009; Musalem et al., 2006). The most important producers by state are the State of Mexico with 60.80%, Puebla with 15.24%, Morelos with 8.27% and Mexico City with 7.89%. These states make up a total national production of 92.99%. As far as the production of sunflower, in 2012, 91,722 tons were harvested in Mexico, with an average yield of 626.9 tons of flower per hectare. The most important producers were located in the states of Morelos and the State of Mexico (SIAP, 2014).

It is important to note the difference between the production of sunflower for decoration and the production of sunflower used for forage or processed to produce oil for human consumption. When sunflower is used for forage or oil production, farmers want to produce plants with large capitula, with a higher seed production per plant. Furthermore, when sunflower us used as forage, farmers want to produce robust, taller and heavier plants. However, when sunflower used for decoration moderate size of capitula is preferred. In general, capitula with a diameter between 10 and 15 cm are preferred for flower arrangements. Sunflowers having a diameter between 7 and 8 cm are preferred for bouquets-type arrangements (Melgares, 2002).

The production of sunflower can be conducted not only under greenhouse conditions, but also outdoors. However, the latter system is limited to some region of Mexico, spring and summer time are the two seasons when this system can be established (Altman et al., 1977; Melgares, 2002). Among the sunflower hybrids produced in Mexico, some of the most important are Sun Rich Lemon, Sun Rich Orange, Sun Deep, Moonbright, Sunbirght, Sunbeam, Full Sun, Sun Goddes, Sunwheel, Type 61, Type 556, Type 555, Sun King, Schnittgold, Sungold, Golden Globe, Eversun, Golden Yellow, Eversun Bright Yellow, Evening Sun, Prado Red Shade and Sunseed (Melgares, 2002; Musalem et al., 2006).

In the state of Campeche, although flowers are commonly produced, this activity has not been able to reach a social and economic impact as observed with other activities such as horticulture and fruit production. This indicates that flowers production represents a potential opportunity because the state of Campeche has the soil and weather condition needed for the majority of tropical flowers, particularly sunflower. Furthermore, information describing the amounts and total values of products generated are lacking. In spite of that, the income generated due to flower marketing is 850,000 Mexican pesos annually (Mausalem et al., 2006). In the last five years, floriculture has been among the most important activities in the state of Campeche. Nonetheless, it is still considered as a relatively new activity for farmers, who commonly use family members to perform small scale agricultural activities. Many of these farmers having traditional knowledge can produce flowers from different species such as Madonna lylies, heliconia, spikenards, tulips, sunflower, etc. However, their productivity is limited due to the lack of information regarding the technical management of each species, which further decreases the productivity. Therefore, these small producers can only sell the flowers in local markets.

Because of that, further research on flower production is important in order to create ways for improving productivity. Increasing the size of land for flower production is also important to provide a wider range of products to the market.

The objective of the present study was to develop a set of management practices for sunflower production, which is expected to be beneficial to producers improving the quality of flower produced, and consequently increase their income. To do so, the response of four sunflower hybrids were evaluated (Sunbright, Prado Red Shade, Full Sun and GH-382) when applying differential irrigation regimes, under two doses of fertilization on flower production. We hypothesize that the evaluated hybrids have the quality and productive characteristics of the flower that will respond to the application of different levels of moisture soil tension at the start of irrigation and to different doses of fertilization.

2. Materials and Methods

2.1 Experimental Location and Climate Conditions

This study was conducted under field condition at the Center for Research, Production and Teaching "Xamantun " of the Technological Institute of Chiná, which is located on the Campeche-Tixmucuy road, km. 15. The geographic location are north latitude 19°43′20.9″, west longitude 90°25′04″ and being at 16 m above sea level. The predominant climate corresponds to AW_0, according to the classification described by Köppen and modified by Garcia (1973), ambient characteristics are warm weather and sub humid with summer precipitations, distributed primarily from May to October, with an annual precipitation of 1,290 mm. When averaged by month, maximum ambient temperature ranges from 28.5 °C (January) to 36.0 °C (May) and minimum ambient temperatures ranges from 15.8 °C (January) to 22.5 °C (May).

This study was conducted in a total Surface area of 500 m^2 in a ferric Luvisol soil, also known as Kan-kab in the Mayan language. The infiltration of water in this type of soil is fast, and commonly has low levels of organic matter.

2.2 Land Preparation

Preparation of the land was conducted using heavy machinery two months before planting. The preparation consisted on weed elimination, land plowing, harrowing and land leveling.

2.3 Installation of the Irrigation System

After land preparation, a trickle irrigation system was installed by using PVC pipes having a diameter of 2", butterfly valves of 2" diameter, and connectors having a caliber of 0.04 and T-Tape with a caliber of 8000. The water output capacity was 5.28 l m^{-1} h^{-1}.

In order to control the irrigation, butterfly shut-off valves were installed. These valves allowed having a total control of the initiation and end of the irrigation in each experimental block.

2.4 Planting

Before planting, the land was irrigated continuously during 4 h in order to provide the appropriate conditions for seed germination. The distance between plants was 30 cm and the distance between furrows was 1 m. Two seeds were deposited at each planting site. Planting was conducted in April, 2007.

2.5 Experimental Design, Experimental Units and Treatments

The following factors were evaluated: 1) moisture tension at the time of irrigation, 2) hybrids and 3) fertilization doses in a sub Split plot arrangement, nested in a complete random block design with three replicates. The factor moisture tension at the time of irrigation was located in the larger plots (blocks), hybrids were located in the medium size plots and the soil fertilization doses were located in the small size plots.

1) *Experimental units*: The larger plots (blocks), where the moisture tension levels were nested had dimensions of 10 m length × 5 m width (six furrows and 1 m spacing among them). The medium size plot, where hybrids were found had dimensions of 2.5 m width × 5 m length (three furrows widthwise) and the smaller size plots where the fertilization doses were found had dimensions of 2.5 m width × 2.5 m length.

2) *Factor moisture tension at the time of irrigation*: The effect of irrigation was evaluated based on different values of the moisture tension in the soil at the time of irrigation. By doing so, we took into account that the water output from the soil to the plant root is determined primarily by this variable. This procedure has been widely accepted in similar studies. For example, Lu et al. (2000), Salaya et al. (2002), Kang et al. (2004), Orozco and Pérez (2006), Wang et al. (2007a, 2007b), Rivera-Hernández et al. (2009, 2010), Sudhir-Yadav et al. (2011), Mahajan et al. (2012), and Carli et al. (2014).

Three levels of moisture tension were tested, levels were -10, -35 and -60 kPa. The irrigation program applied to the larger plots was planned according to the moisture tension in the soil corresponding to each treatment, which was measured using the tensiometers of manometer installed at the center of each large plot, which had a porous capsule installed at 0.3 m deep immediately below the irrigation line.

The irrigation of each large plot was initiated when the reading of the manometer of the tensiometer reached the predetermined value and the irrigation continued until the manometer returned to cero (soil saturated with water). The irrigation time was recorded in each one of the larger plots within blocks. By doing so, the amount of water applied at each irrigation time for each treatment was calculated.

3) *Factor hybrid*: The levels (hybrids) were H1: "Sunbright". H2: "Prado Red Shade"; H3: "Full Sun", and H4: "GH-382".

4) *Factor fertilization*: The levels of fertilization of soil were F1: Fertilization formula 60-50-0; F2 and formula 30-25-0. These levels were defined based on results of the chemical composition analysis of the soil from the experimental plot. The physical and chemical analysis of the soil was conducted by taking samples 20 cm beneath the surface in each block, and then these samples were composited by block. The composited samples were shipped to the department of soil science of the National Institute of Forestry, Agriculture and Livestock Research (INIFAP, for its acronym in Spanish), in the state of Campeche for the analysis of chemical composition. The analytical methods and their interpretation were based on the Mexican Official Norm (NOM-021-RECNAT, 2000), and in the IFA (1992) standards. The physical and chemical analysis comprised: texture, pH, electric conductivity, organic matter content, and N, P, K, Ca, and Mg contents. Based on the results, the amount of fertilizer applied to each plot was calculated adjusting the levels of nitrogen and phosphorous, taking into account that the soils of the region have very high Ca content, and, in calcareous soils, P interacts with Ca forming low solubility calcium phosphates, reducing its availability for crops (Hedley & McLaughlin,

2005; Wang et al., 2005; Mc Beath et al., 2006; Sánchez, 2007). Potassium was not added because the soil already had high concentrations of this nutrient.

The chemical ingredients used in the fertilizers were urea and triple lime super phosphate; the calculations were conducted according to Fernandez (2007). The amount of fertilizer to be used was divided in three parts to avoid nutrient loss by lixiviation, given the high water conductivity of the soil. The first fertilization was carried out at the time of planting, and the second was carried out 15 days later, with the third fertilization being applied 40 days after the planting date.

5) *Treatments*: Each of the treatments evaluated in this study resulted from the combination of the levels of each factor as listed in Table 1.

Table 1. Treatments evaluated

Treatment	Moisture tension at the time of irrigation (kPa)	Hybrid	Fertilizer formula	Nomenclature
1	-10	Sunbright.	60-50-00	I1 H1 F1
2	-10	Sunbright.	30-25-00	I1 H1 F2
3	-10	Prado Red Shade.	60-50-00	I1 H2 F1
4	-10	Prado Red Shade.	30-25-00	I1 H2 F2
5	-10	Full Sun.	60-50-00	I1 H3 F1
6	-10	Full Sun.	30-25-00	I1 H3 F2
7	-10	GH 382.	60-50-00	I1 H4 F1
8	-10	GH 382.	30-25-00	I1 H4 F2
9	-35	Sunbright.	60-50-00	I2 H1 F1
10	-35	Sunbright.	30-25-00	I2 H1 F2
11	-35	Prado Red Shade.	60-50-00	I2 H2 F1
12	-35	Prado Red Shade.	30-25-00	I2 H2 F2
13	-35	Full Sun.	60-50-00	I2 H3 F1
14	-35	Full Sun.	30-25-00	I2 H3 F2
15	-35	GH 382.	60-50-00	I2 H4 F1
16	-35	GH 382.	30-25-00	I2 H4 F2
17	-60	Sunbright.	60-50-00	I3 H1 F1
18	-60	Sunbright.	30-25-00	I3 H1 F2
19	-60	Prado Red Shade.	60-50-00	I3 H2 F1
20	-60	Prado Red Shade.	30-25-00	I3 H2 F2
21	-60	Full Sun.	60-50-00	I3 H3 F1
22	-60	Full Sun.	30-25-00	I3 H3 F2
23	-60	GH 382.	60-50-00	I3 H4 F1
24	-60	GH 382.	30-25-00	I3 H4 F2

Note. H = Hybrid; I = irrigation; F = Fertilizer.

2.6 Variables Evaluated

1) *Vegetative development*: In order to quantify the response variables to each of the factor levels, the following variables were evaluated during the vegetative development: Plant height (PH), considered to be the distance between the base of the plant to the last pair of leaves; stem diameter (SD), measured using a vernier caliper at the base of the stem; width (LW) and length (LL) of the first pair of leaves measured from the tip of each leaf to

the base of the petiole; distance between nodes (DN), considered as the length from the first two lower nodes.

2) *Flowering*: During this stage, variable evaluated were the following: total number of inflorescences (NF), inner diameters (IDF), and external diameter of the capitulum (EDF), these measurements were taken using a vernier caliper, and weight of the capitulum at the time of cutting (CW).

3) *Quantification of variables*: Six plants were randomly collected from each experimental unit (small plot), which were used to determine the response variables.

2.7 Harvest

The harvest was conducted manually as the capitula from each of the hybrids reached the desired commercial size.

2.8 Statistical Analysis of Data

Data collected were subjected to analysis of variance as split plot arrangement nested in a complete randomized block design using SAS (2003). When differences were detected among variables from different treatments, the Tukey (1991) test was further conducted. Significance was declared at $\alpha = 0.10$.

3. Results and Discussion

3.1 Physical and Chemical Analysis of the Soil from Experimental Plot

Results of the physical and chemical analysis of soil in the experimental terrain are presented in Table 2. The type of soil is classified as clayey, with adequate content of organic matter, neutral pH, and is a non-saline soil. The soil nitrogen and phosphorous contents were low, and can limit the sunflower growth; Instead, the potassium, calcium and magnesium content were high and very high for the latter two (IFA, 1992; Table 2) because of the calcareous substrate found. Because of that, the elevated concentrations of calcium resulted within the normal range found in the state of Campeche, which generally causes phosphorous deficiency because there is a fast fixation phosphorous in this type of soil, therefore immobilizing this nutrient (Sharpley, 2000; Hedley & McLaughlin, 2005; Chintala et al., 2010, 2014). As a result of the soil chemical analysis, the following formulae were prepared: 60-50-0 and 30-25-0 using only nitrate and phosphate fertilizers, which are the nutrients that the soil was lacking.

Table 2. Results of the physical and chemical analysis of the experimental plot

Determination	Value	Observations
Sand (%)	22.61	
Mud (%)	12.25	Clay, according to the texture triangle of the USDA
Clay (%)	65.13	
pH	6.73	Neutral
EC (dS/m)	0.30	Non-saline soil
OM (%)	4.51	High
Nitrates (meq/l)	0.27	Low
Phosphates (mg/l)	2.28	Low
Potasium (meq/l)	12.24	High
Calcium (meq/l)	296.65	Very high
Magnesium (meq/l)	104.79	Very high

Note. EC = Electrical conductivity of the extract of saturation; MO = Content of organic matter. Analytical methods and interpretation based on the Mexican Official Norm (NOM-021-RECNAT, 2000). Observations also based on IFA (1992) standards.

3.2 Temporal Evolution of Moisture Tension

Figure 1 illustrates the temporal evolution of the soil moisture tension, which increased as the radical systems absorbed the water from soil. The way irrigation times were defined in each large plot is also included. The arrows indicate the time of the initiation of the irrigation according to each treatment. Irrigations were more frequent for treatment R1 where the soil was maintained wetter, and irrigations was less frequent for treatment R3 where soil was maintained drier, though irrigations in the latter were of greater magnitude because the irrigation was stopped until the manometer of the tensiometer returned to cero. In other words, this happened when the soil was saturated with water. Once the irrigation was applied, the moisture tension decreased as a response to water supply.

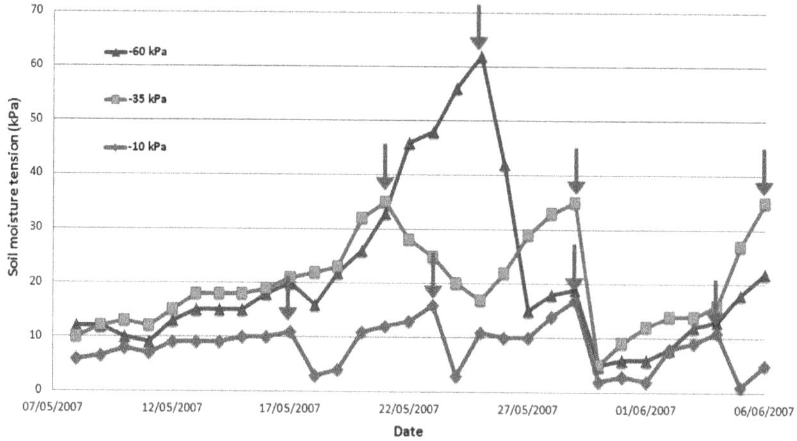

Figure 1. Evolution of the soil moisture tension

Note. Arrows indicate the time of the initiation of the irrigation for each treatment.

Table 3 lists the number of irrigations and the water sheet applied to each treatment at each moisture tension level in each block. It also includes the average water sheet and the total water sheet that each treatment received at each moisture tension level. In order to maintain a lower moisture tension in the soil, more irrigation is required, with the total amount needed being more than three times greater compared to the amount supplied to the drier treatment.

Table 3. Number of irrigations applied and water sheet applied to treatments at each moisture tension evaluated.

Irrigation treatment	Block one		Block two		Block three		Aiw (mm)	Aiw+pp (mm)
	IN	WL (mm)	IN	WL (mm)	IN	WL (mm)		
-10 kPa	13	325	12	300	14	350	325	405.6
-35 kPa	10	250	8	200	7	175	208.2	288.8
-60 kPa	5	125	4	100	3	75	100	180.6

Note. IN = irrigation number, WL = water sheet, Aiw = Average irrigation water applied, Aiw+pp = Average irrigation water plus precipitation.

3.3 Analysis of Variance of Response Variables Evaluated

3.3.1 Factor Moisture Tension

Table 4 lists the mean comparison found from the three irrigation treatments for variables studied. Irrigation applied in the wetter treatments (R1) yielded values significantly higher ($p \leq 0.10$) for variables PH, SD, LL, IDF, EDF and CW, and, even though no significant differences were found for the other variables, the greatest values were obtained when irrigation was applied at the lowest moisture tension (-10 kPa), except for variable NF. These results suggest that when the water stress is increased, blooming is stimulated, and a greater number of inflorescences are produced. Similar responses have been observed in other cultivated species (Southwick & Davenport, 1986; Cuevas et al., 2009; Uçan & Killi, 2010; Cuevas et al., 2012). It is important to note, however, that this also resulted in smaller size inflorescences (Table 4). Nonetheless, this effect was only observed in the hybrids "Prado Red Shape" and "GH 382", which are characterized for producing several inflorescences. The other hybrids only produce one flower per plant.

Table 4. Comparison of variables evaluated for irrigation treatments

Variable	Irrigation treatment			MSD
	-10 kPa	-35 kPa	-60 kPa	
PH (cm)	145.3 a	127.0 ab	115.0 b	29.8
SD (mm)	23.0 a	18.1 b	17.5 b	4.0
LW (cm)	13.4 a	11.0 a	11.2 a	3.4
LL (cm)	44.1 a	38.3 b	37.7 b	5.0
DN (cm)	10.6 a	8.9 a	8.8 a	5.3
NF	6.7 a	6.4 a	7.2 a	2.4
IDF (mm)	68.4 a	61.5 ab	50.5 b	17.9
EDF (mm)	76.2 a	66.7 ab	55.8 b	20.1
CW (g)	160.3 a	147.9 a	115.9 b	42.8

Note. PH = Plant height; SD = Stem diameter; LW = Leaf width; LL = Leaf length; DN = Distance between nodes; NF = Total number of inflorescences; IDF = Inner diameter of the flower; EDF – External diameter of the flower; CW = Weight of the flower; MSD = Minimum significant difference. Within rows, means with the same letter are no statistically different (Tukey, $p \leq 0.10$).

The positive effect of irrigation when applied at low (near cero) moisture tension has been widely reported in a range of plant species (Chintala et al., 2012a, 2012b). For example, Kang et al. (2004) and Wang et al. (2007b) obtained better yields for potato when irrigation was applied at -25 kPa; Orozco and Perez (2006) obtained higher yields for banana when irrigation was applied at -10 kPa. However, the best profit to cost ratio was obtained when irrigation was applied at -25 kPa. Rivera-Hernandez et al. (2009, 2010) tested different values of moisture tension at the time of irrigation on the growth and quality of corn, and they found that treatments that were wetter (-5 and -30 kPa) had the best results from growth and quality. Sudhir-Yadav et al. (2011) tested different irrigation treatments for rice, defined based on the moisture tension in the soil at 20 cm deep: when irrigation was applied at -20 kPa maximum values for crop and maximum water productivity were obtained.

For the production of sunflower, the effects of irrigation has been tested, response variables that have been tested are seed yield and the quality of the oil; significant effects of irrigation have been found for both variables (Flagella et al., 2002; Sezen et al., 2011a, 2011b; Jalilian et al., 2012). However, none of these studies has utilized soil moisture tension as an indicator of the time when irrigation is needed. As for the production of sunflower for decoration, there is limited published information regarding the evaluation of irrigation. Friedman et al. (2007) tested the effect of irrigation with drinkable water and with treated waste water on the quality of the flower, but they did not evaluate different levels of moisture of the soil.

3.3.2 Factor Hybrid

The hybrid "Full Sun" was, in general, of larger size not only for variables characteristics of the vegetative development, but also for those defining the distinctive traits of the inflorescences: it was significantly taller, with thicker stems and longer distance between nodes compared to the rest of the hybrids (Table 5). As far as the size of the flower, the hybrid "Full Sun" and "GH-382" had similar inner and external diameters of the flowers. Hybrid "Full Sun" had significantly lower number of flowers compared to "GH-382" and "Prado Red Shade", respectively.

On the other hand, the latter hybrid consistently had lower values for almost all the evaluated variables, being of lower height, with mid-size inflorescences having a size statistically similar to those found for "Sunbright". However, unlike the rest of the hybrids, "Sunbright" produced in average almost 22 inflorescences per plant, which contrasts the production of the others: "GH-382" barely produced a liter over 3 inflorescences. "Sunbright" and "Full Sun" only have one inflorescence per plant.

Table 5. Comparison mean of evaluated variables for the hybrids

Variable	Hybrids				MSD
	GH 382	Sunbright	Full Sun	Prado Red Shade	
PH (cm)	117.3 b	112.3 b	192.8 a	94.1 c	14.1
SD (mm)	19.3 b	18.8 b	25.0 a	15.1 c	2.5
LW (cm)	11.3 b	13.0 a	13.1 a	10.2 b	1.6
LL (cm)	38.4 bc	43.6 a	42.5 ab	35.7 c	4.7
DN (cm)	8.8 b	8.8 b	12.1 a	8.1 b	2.6
NF	3.2 b	1.0 b	1.0 b	21.7 a	2.8
IDF (mm)	68.7 a	50.7 b	76.6 a	44.7 b	11.2
EDF (mm)	76.4 a	54.0 b	84.0 a	50.5 b	12.5
CW (g)	253.4 a	129.7 b	119.7 b	62.6 c	56.8

Note. PH = Plant height; SD = Stem diameter; LW = Leaf width; LL = Leaf length; DN = Distance between nodes; NF = Total number of inflorescences; IDF = Inner diameter of the flower; EDF = External diameter of the flower; CW = Weight of the flower; MSD = Minimum significant difference. Within rows, means with the same letter are no statistically different (Tukey, $p \leq 0.10$).

Furthermore, hybrid "GH 382" produced capitula that were significantly heavier compared to the capitula of the other hybrids: in average their capitula weighed more than 250 g, maybe because this hybrid is primarily used for seed and for oil production (Diaz-Zorrita et al., 2003), whereas the capitula for hybrid "Prado Red Shape" barely exceeded 60 g. In spite of the fact that the hybrid "Full Sun" produced larger capitula, its average weight was 119 g, a value much lower than hybrid "GH 382". These latter two hybrids were the largest. It is important to note that morphological differences found are a consequence of genetic factors.

3.3.3 Factor Fertilization

In spite of the low level of the soil fertility, mainly in the macronutrients nitrogen and phosphorous (Table 2), the supply of the fertilization doses evaluated in this study did not have significant statistical differences in most of the variables. In addition, there were no edaphological problems observed in the plants due to nutrient deficiency. Fertilization level only had significant effects on the variables width of the first leaves, distance between nodes and weight of the inflorescences. Although, surprisingly, in the latter variable, the averaged maximum values were obtained when the lower fertilization level was applied (Table 6). For the rest of the variables, the higher values were obtained when the higher level of fertilizer was applied. Similar reports regarding the fertilization of sunflower have been conducted to evaluate the yield and quality of the seed (Ogunremi, 1986; Figueiredo & Grassi, 2007; Amanullah & Khan, 2011). However, there is limited information evaluating the fertilization of the crop on the quality of the flower. For example, Ahmad et al. (2012) evaluated five levels of nitrogen on the development and nutritional state of two hybrids of sunflower; they found that fertilization with nitrogen stimulated the growth of the stem; it also stimulated the concentration of nitrogen in the tissues. However, level of nitrogen did not affect the days needed for cutting the inflorescences. Pereira et al. (2009) evaluated the fertilization of the hybrid "Sunbright" with potassium silicate, similar to our findings; they did not find clear differences in the morphological characteristics of the flower as a response to the treatments.

Table 6. Comparison of the mean of variables evaluated for the fertilization treatments

Variable	Fertilization treatments		MSD
	30-25-00	60-50-00	
PH (cm)	127.9 a	130.4 a	2.8
SD (mm)	19.3 a	19.8 a	0.6
LW (cm)	11.5 b	12.2 a	0.3
LL (cm)	39.7 a	40.4 a	0.9
DN (cm)	9.0 b	9.8 a	0.5
NF	6.6 a	6.9 a	0.5
IDF (mm)	59.5 a	60.7 a	2.1
EDF (mm)	65.3 a	67.1 a	2.2
CW (g)	150.5 a	132.2 b	5.6

Note. PH = Plant height; SD = Stem diameter; LW = Leaf width; LL = Leaf length; DN = Distance between nodes; NF = Total number of inflorescences; IDF = Inner diameter of the flower; EDF = External diameter of the flower; CW = Weight of the flower; MSD = Minimum significant difference. Within rows, means with the same letter are no statistically different (Tukey, $p \leq 0.10$).

3.3.4 Effects of the Double and Triple Interactions

We only found significant effects in the double interaction hybrid × fertilization for the variable weight of the capitulum. For the rest of the double interactions of the other variables, no statistical significant effects were found. These findings indicate that practically none of the factors interacted with each other. However, significant effects were found for the triple interaction for the variable number of total inflorescences per plant. Furthermore, highly significant effects were found for the triple interaction for the rest of the variables. These interactions, both double and triple, were primarily a consequence of morphological differences attributed to genetic factors among hybrids. These data are presented in section 3.3.2, which were considered as a factor in the experimental design.

4. Conclusions

Irrigation significantly affected not only growth variables, but also the morphological characteristics of the inflorescences of sunflower. We found that when irrigation is applied when the soil moisture tension is -10 kPa, maximum values variables for vegetative growth and quality of the inflorescences were obtained. More specifically, plants were significantly taller, with thicker stems, larger leaves, and larger inner and external inflorescence diameters. In addition, the weight of the flowers were significantly greater compared to treatments where lower levels of water were applied.

We found clearly no significant effects of the factor irrigation and fertilization on the variable number of capitula. This finding indicates that that variable depend on genetic factors rather than on the stress that the plants were experiencing.

Significant effects were found in all variables among hybrids for variables attributed to morphological differences due to genetic factors. More specifically "Full Sun" was taller with a larger diameter of the stem, whereas "Prado Red Shade" was significantly smaller. The latter had significantly higher number of capitula. The capitula of hybrids "Full Sun" and "GH 382" had diameters significantly greater and the capitula of "GH 382" were significantly heavier.

We only found significant effects of fertilization factor on the variable weight of the capitula. Double interactions for the hybrid × fertilization were not statistically significant, except for the variable CW.

Triple interactions of all evaluated variables were significant. Double interactions and triple interaction were observed as a consequence of morphological differences existing among hybrids, which are attributed to genetic factors.

Acknowledgements

To Colegio de Postgraduados, Campus Campeche and Campus Veracruz (LPI-8 Impacto y Mitigación del Cambio Climático) for financial support. Thanks also to Ing. Juan Manuel Hernández Pérez for assistance in the preparation of tables and figures.

References

Ahmad, I., Doleb, J. M., & Nelson, P. (2012). Nitrogen application rate, leaf position and age affect leaf nutrient status of five specialty cut flowers. *Scientia Horticulturae, 142*(13), 14-22. http://dx.doi.org/10.1016/j.scienta.2012.04.009.

Amanullah, & Khan, M. W. (2011). Interactive Effect of Potassium and Phosphorus on Grain Quality and Profitability of Sunflower in Northwest Pakistan. *Pedosphere, 21*(4), 532-538. http://dx.doi.org/10.1016/S1002-0160(11)60155-0.

Carli, C., Yuldasheva, F., Khalikova, D., Condorib, B., Maresc, V., & Monneveuxc, P. (2014). Effect of different irrigation regimes on yield, water use efficiency and quality of potato (*Solanum tuberosum* L.) in the lowlands of Tashkent, Uzbekistan: A field and modeling perspective. *Field Crops Research, 163*, 90-99. http://dx.doi.org/10.1016/j.fcr.2014.03.021

Chintala, R., McDonald, L. M., & Bryan, W. B. (2010). Grouping soils by taxonomy order to improve lime recommendations. *Communications in Soil Science and Plant Analysis, 41*, 1594-1603. http://dx.doi.org/10.1080/00103624.2010.485239

Chintala, R., Schumacher, T. E., McDonald, L. M., Clay, D. E., Malo, D. D., Clay, S. A., ... Julson, J. L. (2014). Phosphorus sorption and availability in biochars and soil biochar mixtures. *CLEAN-Soil Air Water, 42*(5), 626-634. http://dx.doi.org/10.1002/clen.201300089

Chintala, R., McDonald, L. M., & Bryan, W. B. (2012a). Effect of soil water and nutrients on productivity of Kentucky bluegrass systems in acidic soils. *Journal of Plant Nutrition, 35*, 288-303. http://dx.doi.org/10.1080/01904167.2012.636131

Chintala, R., McDonald, L. M., & Bryan, W. B. (2012b). Optimization of water potential and nutrient levels for Kentucky bluegrass-white clover mixture on acidic soils. *Biotechnology, Agronomy, Society and Environment, 16*, 167-177.

Cuevas, J., Pinillos, V., Cañete, M. L., González, M., Alonso, F., Fernández, M. D., & Hueso, J. J. (2009). Optimal levels of postharvest deficit irrigation for promoting early flowering and harvest dates in loquat (Eriobotrya japonica Lindl.). *Agricultural Water Management, 96*, 831-838. http://dx.doi.org/10.1016/j.agwat.2008.11.002

Cuevas, J., Pinillos, V., Cañete, M. L., Parra, S., González, M., Alonso, F., ... Hueso, J. J. (2012). Optimal duration of irrigation withholding to promote early bloom and harvest in 'Algerie' loquat (Eriobotrya japonica Lindl.). *Agricultural Water Management, 111*, 79-86. http://dx.doi.org/10.1016/j.agwat.2012.05.006

De la Madrid, E. (2009). *La floricultura en México*. Revista El Mexicano. Retrieved from http://www.oem.com.mx

Díaz-Zorita, M., Duarte, G. A., & Plante, D. Z. (2003). *El cultivo del girasol*. Asociación Argentina de Girasol ASAGIR. Retrieved from http://www.asagir.org.ar/Publicaciones/Cuadernillo_web.pdf

Fernández, R. (2007). *El Cultivo de Flores en Venezuela*. Boletín Informativo.

Figueiredo, L. T., & Grassi, F. H. (2007). Level of sewage sludge for the productivity of sun flower. *J. Soil Sc. Plant Nutr., 7*(3), 16-25. http://dx.doi.org/10.4067/S0718-27912009000300007

Flagella, Z., Rotunno, T., Tarantino, E., Di Caterina, R., & De Caro, A. (2002). Changes in seed yield and oil fatty acid composition of high oleic sunflower (Helianthus annuus L.) hybrids in relation to the sowing date and the water regime. *European Journal of Agronomy, 17*(3), 221-230. http://dx.doi.org/10.1016/S1161-0301(02)00012-6

Friedman, H., Bernstein, N., Bruner, M., Rot, I., Ben-Noon, Z., Zuriel, A., ... Hagiladi, A. (2007). Application of secondary-treated effluents for cultivation of sunflower (*Helianthus annuus* L.) and celosia (*Celosia argentea* L.) as cut flowers. *Scientia Horticulturae, 115*, 62-69. http://dx.doi.org/10.1016/j.scienta.2007.07.003

García, E. (1973). *Modificaciones al sistema de clasificación climática de Köppen (para adaptarlo a las condiciones de la República Mexicana)*. Segunda edición corregida y aumentada. Instituto de Geografía. Universidad Nacional Autónoma de México. México.

Hedley, M. J., & McLaughlin, M. J. (2005). Reactions of phosphate fertilizers and byproducts in soils. In J. T. Sims & A. N. Sharpley (Eds.), *Phosphorus: Agricultura and the environment* (Agron. Monogr., pp. 181-252, No. 46). ASA, CSSA, and SSSA, Madison, WI. http://dx.doi.org/10.2134/agronmonogr46.c7

IFA (International Fertilizer Association). (1992). *World fertilizer use manual* (p. 427). Germany.

Jalilian, J., Modarres-Sanavy, S. A. M., Saberali, S. F., & Sadat-Asilan, K. (2012). Effects of the combination of beneficial microbes and nitrogen on sunflower seed yields and seed quality traits under different irrigation regimes. *Field Crops Research, 127*, 26-34. http://dx.doi.org/10.1016/j.fcr.2011.11.001

Kang, Y., Wang, F. X., Liu, H. J., & Yuan, B. Z. (2004). Potato evapotranspiration and yield under different drip irrigation regimes. *Irrig. Sci., 23*, 133-143. http://dx.doi.org/10.1007/s00271-004-0101-2

Lu, J., Ookawa, T., & Hirasawa, T. (2000). The effects of irrigation regimes on the water use, dry matter production and physiological responses of paddy rice. *Plant and Soil, 223*, 207-216. http://dx.doi.org/10.1023/A:1004898504550

Mahajan, G., Chauhan, B. S., Timsina, J., Singh, P. P., & Kuldeep, S. (2012). Crop performance and water- and nitrogen-use efficiencies in dry-seeded rice in response to irrigation and fertilizer amounts in northwest India. *Field Crops Research, 134*, 59-70. http://dx.doi.org/10.1016/j.fcr.2012.04.011

Mc Beath, T. M., Smernik, R. J., Lombi, E., & McLaughlin, M. J. (2006). Hydrolysis of pyrophosphate in a highly calcareous soil: a solid-state phosphorus-31 NMR study. *Soil Sci. Soc. Am. J., 70*, 856-862. http://dx.doi.org/10.2136/sssaj2005.0184

Melgares de, A., J. (2002). El cultivo del girasol para flor cortada. *Revista Flormarket Ed. Verdimedia, España, año II*(2), 55-61.

Musalem, L. O., Olvera, G. J., Ochoa, B. R., & Ortega, R. C. (2006). La floricultura Mexicana, el gigante que está despertando. *Revista Claridades Agropecuarias, 154*, 3-38. Retrieved from http://www.infoaserca.gob.mx/claridades/revistas/154/ca154.pdf

NOM-021-RECNAT. (2000). Especificaciones de fertilidad, salinidad y clasificación de suelos. *Estudios, muestreo y análisis* (p. 88). SEMARNAT, México.

Ogunremi, E. A. (1986). Effects of nitrogen fertilization and harvest time on sunflower yield and hollow seededness. *Field Crops Research, 13*, 44-53. http://dx.doi.org/10.1016/0378-4290(86)90006-7

Orozco, R. J., & Pérez, Z. O. (2006). Tensión de humedad del suelo y fertilización nitrogenada en plátano (Musa AAA Simmonds) cv. Gran Enano. *Agrociencia, 40*, 149-162.

Pereira, C. M., Zanão, J. L. A., Saraiva, G. J. A., & Barbosa, J. G. (2009). Silicon improves ornamental pot sunflower production and quality. *Ciência Rural, Santa Maria, 39*(8), 2394-2399. http://dx.doi.org/10.1590/S0103-84782009005000194

Rivera-Hernández, B., Carrillo-Ávila, E., Obrador-Olán, J. J., Juárez-López, J. F., Aceves-Navarro, L. A., & García-López, E. (2009). Soil moisture tension and phosphate fertilization on yield components of A-7573 sweet corn (*Zea mays* L.) hybrid, in Campeche, Mexico. *Agricultural Water Management, 96*, 1285-1292. http://dx.doi.org/10.1016/j.agwat.2009.03.020

Rivera-Hernández, B., Carrillo-Ávila, E., Obrador-Olán, J. J., Juárez-López, J. F., & Aceves-Navarro, L. A. (2010). Morphological quality of sweet corn (*Zea mays* L.) ears as response to soil moisture tension and phosphate fertilization in Campeche, Mexico. *Agricultural Water Management, 97*(9), 1365-1374. http://dx.doi.org/10.1016/j.agwat.2010.04.001

Salaya-Domínguez, J. M., Carrillo-Ávila, E., Palacios-Vélez, O. L., Aceves-Navarro, L. A., & Juárez-López, J. F. (2002). Respuesta del cultivo de sandía (Citrullus vulgaris Schrad) al potencial del agua en el suelo. *Revista Fitotecnia Mexicana, 25*(2), 127-133.

Sánchez, C. A. (2007). Phosphorus. In A. V. Broker & D. J. Pilbeam, (Eds.), *Handbook of Plant Nutrition* (pp. 51-82). Grup. Boca Raton. London, New York: Taylor Francis.

SAS (Statistical Analysis System). (2003). *Statistical Analysis System for Windows 9.1.3*. Service Pack 2. SAS Institute Inc., Cary, NC, USA.

Sezen, S. M., Yazar, A., Kapur, B., & Tekinb, S. (2011a). Comparison of drip and sprinkler irrigation strategies on sunflower seed and oil yield and quality under Mediterranean climatic conditions. *Agricultural Water Management, 98*, 1153-1161. http://dx.doi.org/10.1016/j.agwat.2011.02.005

Sezen, S. M., Yazar, A., & Tekinb, S. (2011b). Effects of partial root zone drying and deficit irrigation on yield and oil quality of sunflower in a Mediterranean environment. *Irrigation and Drainage, 60*(4), 499-508. http://dx.doi.org/10.1002/ird.607

Sharpley, A. (2000). Phosphorus availability. In M. E. Sumner (Ed.), *Handbook of Soil Science* (pp. D18-D37). CRC Press, New York.

SIAP. (2014). Servicio de Información Agroalimentaria y Pesquera (SIAP). Secretaría de Agricultura, Ganadería, Desarrollo Rural Pesca y Alimentación (SAGARPA), México. Retrieved from http://www.siap.gob.mx

Southwick, S. M., & Davenport, T. L. (1986). Characterization of water stress and low temperature effects on flower induction in citrus. *Plant Physiol., 81*, 26-29. http://dx.doi.org/10.1104/pp.81.1.26

Sudhir-Yadav, E., Humphreys, S. S., Kukal, G. G., & Rangarajan, R. (2011). Effect of water management on dry seeded and puddled transplanted rice Part 2: Water balance and water productivity. *Field Crops Research, 120*, 123-132. http://dx.doi.org/10.1016/j.fcr.2010.09.003

Tukey, J. W. (1991). The philosophy of multiple comparisons. *Statistical Science, 6*, 100-116. http://dx.doi.org/10.1214/ss/1177011945

Uçan, K., & Killi, F. (2010). Effects of different irrigation programs on flower and capsule numbers and shedding percentage of sesame. *Agricultural Water Management, 98*, 227-233. http://dx.doi.org/10.1016/j.agwat.2010.08.005

Wang, Q. R., Li, J., Li, Z., & Christie, P. (2005). Screening Chinese wheat germoplasm for phosphorus efficiency in calcareous soils. *J. Plant Nutr., 28*, 489-505. http://dx.doi.org/10.1081/PLN-200049186

Wang, D., Kang, Y., & Wan, S. (2007a). Effect of soil matric potential on tomato yield and water use under drip irrigation condition. *Agricultural Water Management, 87*, 180-186. http://dx.doi.org/10.1016/j.agwat.2006.06.021

Wang, F. X., Kang, Y., Liu, S. P., & Hou, X. Y. (2007b). Effects of soil matric potential on potato growth under drip irrigation in the North China Plain. *Agricultural Water Management, 88*, 34-42. http://dx.doi.org/10.1016/j.agwat.2006.08.006

15

Productive Efficiency of Crop Farms in Viet Nam: A DEA with a Smooth Bootstrap Application

Truong Tuan Linh[1,2], Teruaki Nanseki[3] & Yosuke Chomei[3]

[1] Laboratory of Agricultural and Farm Management, Department of Agricultural and Resource Economics, Kyushu University, Higashiku, Fukuoka, Japan

[2] Thai Nguyen University of Information and Communication Technology, Thai Nguyen, Viet Nam

[3] Faculty of Agriculture, Kyushu University, Higashiku, Fukuoka, Japan

Correspondence: Truong Tuan Linh, Graduate School of Bio-resource and Bio-environmental Science, Laboratory of Agricultural and Farm Management, Department of Agricultural and Resource Economics, Kyushu University, 6-10-1 Hakozaki, Higashiku, Fukuoka 812-8581, Japan.

Abstract

Rice and maize are the two most important annual crops in Viet Nam. Rice production is sufficient for self-consumption and exportation, whereas maize production is insufficient and must be imported because of the growing feed industry. In Son La, the largest area in the Northwest with the greatest production of maize, maize is cultivated continuously all year, whereas paddy rice is cultivated mostly in the spring. To our knowledge, this article is the first study to estimate farm-level technical and scale efficiencies and to identify the factors influencing them. This study first applied a standard and smooth bootstrap DEA method to estimate the technical and scale efficiencies, then used a Tobit regression method to identify factors influencing these efficiencies among farms. The results showed that the opportunity for both technical and scale inefficiencies of maize and rice crops is insignificant. Findings from the second stage indicated that the age of the head of household, the numbers of family, the national electricity source, the distance to the nearest market, and the access to credit, extension services and milling machines are the main factors affecting the technical and scale efficiencies of rice and maize crops. The findings suggest continuing improvement of management, co-operation in cultivation, crop diversity and optimal use rice plots. The results also recommend expanding the national electricity source, emphasizing policies for adjusting customs and knowledge, using credit in appropriate ways and continuing to enhance extension services.

Keywords: annual crops, rice crop, maize crop, DEA, smooth bootstrap, technical efficiency, scale efficiency, Vietnam

1. Introduction

Although the role of agriculture in the economy of Viet Nam has recently declined, it is still the most important economic sector. Agriculture contributed 20% of the GDP and comprised 47% of the total employment in Viet Nam in 2012 (World Bank, 2014). Decision No. 432/QD-TTg on approving the Viet Nam Sustainable Development Strategy for 2011-2020 emphasizes that "*To shift the structure of agriculture and rural areas towards industrialization, promote regional advantages; develop quality agricultural products; combine production with local and foreign market in order to lift the efficiency of using natural resources...*" (Prime Minister of Vietnam, 2012). This strategy extends the application of scientific and technological advances to increase the quantity and quality of agricultural production.

Rice and maize are the two most important annual crops. Their planted area and production are the highest compared with all types of annual crops, such as sugar cane, soya bean, and peanut. By 2012, the planted area of rice and maize was 7761.2 and 1172.5 thousand ha, respectively; production was 43737.8 and 5193.5 thousand tons, respectively (GSO, 2013). Rice is mostly used for food consumption and export, whereas maize production is insufficient and must be imported for the feed industry (USDA, 2014).

The Northwest is the poorest among the seven regions in Viet Nam. Because the topography is high and mountainous, paddy rice production is insufficient, and maize production has been increasing rapidly in recent years. Maize has become the main income source for farmers (Luckmann, Ihle, & Grethe, 2011). Maize is the top annual crop in the Northwest. The area and production in 2011 were 28.74% and 14.92%, respectively (GSO, 2012b).

Son La, one of the seven provinces in the Northwest, has the largest area and greatest production of maize. The planted area and production of maize in 2012 was 168.7 thousand ha and 667.4 thousand tons, respectively. However, the planted area and production of paddy rice had only 48.2 thousand ha and 162.9 thousand tons, respectively (GSO, 2013). Maize is cultivated continuously all year, whereas paddy rice is cultivated mostly in the spring.

Son La Province and the Northwest are also facing many problems. The transportation system is poorly developed. Many ethnic groups live together with low education and traditional customs. The ratio of poverty in the Northwest in 2013 was 58.7%, which is high compared with the ratio for the entire country: 17.2% (GSO, 2012a). The annual crop plots are highly fragmented and very small. Farmers work as individuals and do not want to join co-operatives. Therefore, applying scientific and advance technologies here is a significant challenge for farmers and the government.

Although rice and maize have an important role in the Northwest social-economic structure, there are few studies regarding the efficiency of rice and maize crop production. To our knowledge, this article is the first study to estimate farm-level technical and scale efficiencies and identify the factors influencing them. This study estimated farm efficiency using a nonparametric method with a smooth bootstrap procedure to avoid measurement errors and data noise. After obtaining the technical and scale efficiencies, a Tobit regression method was used to determine the factors influencing these efficiencies.

There have been several studies regarding the efficiency of agriculture in Viet Nam. Khai and Yabe (2011) used stochastic production frontier analysis (SFA) with the Cobb-Douglas production function to measure the technical efficiency (TE) of rice production. Rios and Shively (2005) used a standard DEA method to first estimate technical and cost efficiencies and then a standard Tobit regression to identify the factor effect on these efficiencies of coffee farms in Dak Lak Province, Viet Nam. Linh (2012) is the first author to use a smooth bootstrapped DEA method to estimate technical and scale efficiencies of rice farms in the first step. Linh (2012) then used a standard and weight Tobit to determine the factors influencing technical efficiency. However, Linh (2012) used the Vietnam Household Living Standard Survey 2003-2004 (VHLSS 2004) for 8,813 households in all of Vietnam and only for rice farms. Dao and Lewis (2013) estimated the technical and scale efficiencies of annual crop farms in northern Vietnam using a DEA smooth bootstrap approach. However, Dao and Lewis (2013) have only used a DEA smooth bootstrap approach in the first step because the second step to identify factors was not completed. Dao and Lewis (2013) also did not focus on analyzing maize and rice farms in the Northwest.

This paper has two objectives. First, we estimate the technical and scale efficiencies of both rice and maize crops in Son La Province, Vietnam using a standard and smooth bootstrap DEA method. Second, we use a Tobit regression method to identify the factors influencing these efficiencies among farms.

The next section describes a short review of the methodology. Section 3 describes the main characteristics of the data. The results and discussion are presented in section 4. Finally, conclusions and several recommendations are provided.

2. Methodology

Efficiency can be estimated by two methods, namely, parametric and non-parametric. In the literature, most parametric approaches use the Stochastic Frontier Analysis (SFA) method, whereas the non-parametric approach uses Data Envelopment Analysis (DEA). The main difference between these two methods is based on the way that the production possibility frontier can be estimated. An advantage of the DEA method is that it does not require a functional form, whereas SFA requires a functional form.

DEA requires detailed information regarding all inputs and output data. Following the work of Coelli, Rahman, and Thirtle (2002); Farell (1957) and others, the four efficiencies often measured are technical, scale, allocative and cost. However, the allocative and cost efficiencies are strengthened by the availability of all inputs and output prices, which were difficult to collect when we conducted the field survey. Because the survey was conducted based on farmers' memories, sometimes they could not remember details. Therefore, this article focuses on measuring technical efficiency (TE) and scale efficiency (SE).

DEA has input orientations and output orientations. This study employed DEA to measure farm efficiency using an input orientation. We first used the smooth bootstrap procedure proposed by Simar and Wilson (2000) to estimate bias and the confident interval for technical efficiency (TE). We then used Tobit analysis to identify the exogenous factors affecting the estimated efficiency. We used the package FEAR developed by Wilson (2008) in the R platform and Stata software in the process.

2.1 Technical and Scale Efficiencies

The DEA production frontier is constructed and solved using linear programing techniques. Considering an ith farm out of a total of *n* farms, the input-based technical efficiency (TE) under variable return to scale (VRS) is calculated as:

$$TE_i = min_{\theta,\lambda}\theta_i, \text{ Subject to } Y_i \leq Y\lambda;\ \theta_i X_i \geq X_i;\ \lambda \geq 0;\ \sum_{i=1}^{n} \lambda_i = 1 \quad (1)$$

Where,

Y and X are the output and input vectors, respectively. The value of θ_i is the technical efficiency score for the *i = th* farm under VRS. In general, $0 \leq \theta_i \leq 1$, when $\theta_i = 1$, indicating that the farm is producing on the production frontier and is therefore technically efficient, whereas $\theta_i < 1$ shows that the farm is technically inefficient.

In case of a constant return to scale (CRS), we can easily impose it by deleting the convexity constraint $\sum_{i=1}^{n} \lambda_i = 1$ in Equation (1). Therefore, we can easily calculate scale efficiency (SE) by Coelli et al. (2002):

$$SE = TE_{CRS}/TE_{VRS} \quad (2)$$

We can also calculate the non-increasing to scale (NIRS) by replacing the convexity restriction $\sum_{i=1}^{n} \lambda_i = 1$ in Equation (1) to $\sum_{i=1}^{n} \lambda_i \leq 1$. In general, $0 \leq \theta_i \leq 1$ if SE = 1, and farms are considered to be scale efficient. If $TE_{VRS} = TE_{NIRS}$ and SE < 1, the farm is operating under a decreasing return to scale (DRS), or is "too large". If $TE_{VRS} \neq TE_{NIRS}$ and SE < 1, the farm is operating under an increasing return to scale (IRS) or is "too small". Finally, if $TE_{VRS} = TE_{CRS}$, the farm is operating under a CRS (Coelli et al., 2002).

2.2 Bootstrapping in DEA

DEA is a deterministic method, and one of its disadvantages is that no statistical noise is assumed in the analysis. Although DEA methods have been widely applied to date, many researchers have completely ignored the statistical noise in the estimators. This oversight can cause bias in the DEA estimates and mistaken results (Dao & Lewis, 2013). Simar and Wilson (1998, 2000) proved that bootstrapping is the best method to construct the statistical properties of DEA. In the bootstrap method, the data are simulated by resampling. Therefore, the data generating process (DGP) can be mimicked to the correct data generation. In the literature, several studies have applied the bootstrap method of Simar and Wilson (2000), such as Brummer (2001); Dao and Lewis (2013); Gocht and Balcombe (2006); Latruffe, Balcombe, Davidova, and Zawalinska (2005).

After using a smooth bootstrap procedure, we can check the biased DEA estimators and locate their confidence interval. The confidence interval is an important index to determine the exact results. The larger the variance, the more incorrect the efficiency results may be. Efron and Tibshinari (1993) have suggested that the number of iterations should be less than 1000 if researchers only want to estimate bias and standard deviation. Because we are more interested in confidence interval estimation, in our study, 2,000 bootstrap iterations were performed.

2.3 Variables Explaining the Efficiency Estimates

This article used Tobit analysis in the second stage to evaluate the factors influencing efficiency similar to most authors. Crop production is affected by many exogenous factors, such as household characteristics, household assets, extensions, and weather. Moreover, farms operating at optimal scale are assumed in constant return to scale (CRS) technical efficiency (Karimov, 2013). Therefore, this study selected the technical efficiency under variable return to scale (VRS) to become dependent variables in the Tobit model.

However, which index is suitable to use for analysis: the bias-corrected estimator or original technical efficiency? Simar & Wilson (2000) have suggested that using only bias-corrected estimates in the second step when,

$$\hat{\sigma}^2 < \frac{1}{3}(bias[\hat{\theta}])^2 \quad (3)$$

Where $\hat{\sigma}^2$ is the sample variance of the bootstrap values, and $\hat{\theta}$ is the uncorrected estimated score. In our study, using only bias-corrected estimates could not occur. Thus, we used the original technical efficiency score (TE_{VRS}) in the second stage.

The estimated coefficients in the Tobit regression models can not directly be used to interpret the results as the true marginal effect. This coefficient will affect the mean value of the dependent variable (Yi), given that it is observed and also affects the probability of the dependent variable being observed (Y^*_i) (Gujarati, 2011).

Therefore, similar to other studies, this study uses the marginal effects of all independent variables that are reported and calculated at the sample mean.

3. Data and Variables Used

The data used for this study originate from a survey conducted in six communes located in three districts of Son La Province, in northwest Vietnam. The survey was conducted from February to March 2014. The respondents were selected through a multi-stage sampling procedure. Three districts in the survey were based on three types of altitude: Mai Son < 1,000 m, Moc Chau 1,000 – 2,000 m and Phu Yen > 2,000 m. A total of 360 farm households from 12 villages were interviewed. Of this total, 292 households cultivate both rice and maize, 60 households cultivate only maize, six households cultivate only rice, and four conduct business. Therefore, we used the data of 292 household farms to compare the efficiency between the two main crops.

Output is measured as the gross income of rice and maize. Rice product is mostly used for self-consumption; maize product is used for both selling and self-consumption. Therefore, using profit or income from the rice and maize product index does not precisely show the technical and scale efficiencies. The inputs are planted land, family and hired labor, seed, fertilizer (including organic, NPK, nitrogenous and phosphate fertilizer), chemicals (including herbicide and pesticide) and other expenses (such as irrigation and transportation fees, etc.).

We also explain efficiency differences among farms using farm-specific variables. The selected variables are those most often used in the literature, such as the age and education of the household head, distance to the nearest market, access to credit, extension services, off-farm income and household assets (Binam, Tonyè, Wandji, Nyambi, & Akoa, 2004; Coelli et al., 2002; Linh, 2012).

Table 1 presents summary information regarding the variables used for rice and maize crops separately. The table shows that gross income from maize production is higher than rice production by 46,000 thousand vnd. The simple reason for this result is that all of the inputs of maize except other expenses are higher than rice. The planted rice crops are small by 0.32 ha, similar to the rice farm size in Bangladesh (Coelli et al., 2002), with an average size of only one-third of a hectare. The average size of a rice crop is equal to one-fifth the size of a maize crop. Most rice farms have only one or two plots. Because maize is cultivated on sloping land, it requires much more labor, seed, and fertilizer.

The farm-specific variables provide an overview of the farms' characteristics. The average age of the head of household is 44; the average education of the head of household is at the secondary level. The average family size is 5. Nearly 90% of farms use a national electricity source, and approximately 67% of farms have access to credit. The average distance to the nearest market is 9 km. Farmers have had contact with maize extensions more than rice. Few households have tractors and milling machines, and off-farm income is small: approximately 2 million Vietnamese dong per year.

Table 1. Descriptive variable used

Variable	Definition	Rice		Maize	
		Mean	SD	Mean	SD
Output and Inputs					
Output	Gross income (1,000vnd)	13,072.21	11,915.02	58,934.96	44,17566
Land	Cultivated land (ha)	0.32	0.44	1.48	1.10
Labor	Including family labor and hired labor (Man-days)	18.32	25.99	126.65	96.16
Seed	Total amount of seed (1,000vnd)	2,604.42	2,110.01	4,474.06	3,265.96
Fertilize	Total amount of fertilizer (1,000vnd)	2,638.31	2,356.90	9,146.38	6,485.98
Chemical	Total amount of chemicals (1,000vnd)	927.08	965.46	1,276.69	927.32
Others	Other expenses (1,000vnd)	190.34	819.79	117.89	766.24
Farm specific variables					
Age	Age of household head (years)	44.32	11.59	44.32	11.59
Edu	Schooling of household head (years)	6.62	3.22	6.62	3.22
Hhsize	Household members (person)	4.92	1.59	4.92	1.59
Sourele	Source of electricity of household, which takes 1 = use national source, 0 = otherwise	0.88	0.33	0.88	0.33
Dismark	Distance from household to nearest market (km)	9.20	7.11	9.20	7.11
Credit	Takes 1 if farmer has access to credit, 0 = not	0.67	0.47	0.67	0.47
Extent	Takes 1 if farmers received the information from extension services, 0 = none	0.41	0.49	0.94	0.24
Motor	Motorcycle of household, which takes 1 = have milling, 0 = otherwise	0.30	0.46	0.30	0.46
Tractor	Tractor machine of household, which takes 1 = have tractor, 0 = otherwise	0.22	0.41	0.22	0.41
Milling	Milling machine of household, which takes 1 = have milling, 0 = otherwise	0.28	0.48	0.28	0.48
Offic	Off-farm income of household (Million vnd)	1.99	10.39	1.99	10.39

Source: Own survey, 2014.

4. Results and Discussion

4.1 Technical and Scale Efficiencies

The standard DEA technical estimates under VRS and CRS are reported in Table 2. The results show that the majority of farms are inefficient in both the technical and scale efficiencies of rice and maize production.

The average technical efficiency score under VRS is 0.63 for rice crops and 0.54 for maize, with 34 rice and 19 maize crops fully efficient. These results suggest that, on average, farms can still maintain the same output performance with a decrease in the inputs by 37% for rice crops and 46% for maize crops. The results also indicate that technical efficiency under CRS and VRS in rice production is higher than in maize production, which may be partly because of the insufficiency and sloping of maize crop lands.

The mean scores of scale efficiency for rice and maize production are the same: 0.89. This conclusion indicates that farm size is much less important in changing technical efficiency. The last three rows of Table 2 show the percentages of farms that have constant return to scale (CRS), decreasing return to scale (DRS) and increasing return to scale (IRS). Overall, farms are mostly under increasing return to scale in both rice and maize crops, with scores of 84.25% and 77.40%, respectively. These results mean that farms are "too small" and may need to increase their scales. Only 11 rice crops and 10 maize crops are producing at optimal scale.

The technical efficiency score for both crops under VRS is lower than the scale efficiency score. This conclusion indicates that the technical inefficiency of maize and rice farms is mainly affected by management rather than the operating scale. This finding is similar to the result of Karimov (2013), which estimated the efficiency of potato and H-W melon crops on Uzbekistan farms. This result implies that farmers must focus more on improving the management of crop production, and rice farms must also increase their scale efficiency.

Table 2. Frequency distribution of technical and scale efficiency estimates in a pooled sample

	Rice crop			Maize crop		
	TE_{CRS}	TE_{VRS}	SE	TE_{CRS}	TE_{VRS}	SE
Mean	0.56	0.63	0.89	0.48	0.54	0.89
Std.dev	0.20	0.22	0.14	0.21	0.21	0.15
Minimum	0.15	0.16	0.21	0.07	0.14	0.28
Maximum	1.00	1.00	1.00	1.00	1.00	1.00
<60 %	60.96	51.72	4.45	73.97	67.10	6.85
60-69 %	12.67	9.93	6.51	11.30	12.00	4.79
70-79 %	13.01	12.33	6.51	6.51	8.20	6.16
80-89 %	7.19	8.90	17.81	2.40	3.80	14.05
90-100 %	6.17	17.12	64.73	5.82	8.90	68.15
IRS %			84.25			77.40
DRS %			11.99			19.18
CRS %			3.76			3.42

Source: Own survey, 2014.

4.2 Smooth Bootstrap Results

To improve the robustness of the results and realizing that the standard DEA method may have biases in estimating efficiency scores, we used a smooth bootstrapping method. The results of bias-corrected TE_{VRS} are reported in the sixth column of Table 3. The confidence intervals (CIs) of bias-corrected TE_{VRS} are also shown in the seventh and eighth columns.

Table 3. Technical efficiency estimates using smooth bootstrap method

Variables	Sample	Initial TE_{CRS}	Initial TE_{VRS}	% of farm with $TE_{VRS} = 1$	Bias-Corrected TE_{VRS}	Lower bound	Upper bound
Rice crop							
Pooled sample	292	0.56	0.63	11.64	0.55	0.50	0.62
Mai Son	93	0.61	0.68	16.10	0.58	0.52	0.66
Moc Chau	87	0.69	0.76	26.40	0.66	0.59	0.75
Phu Yen	112	0.62	0.69	19.60	0.59	0.53	0.68
Maize crop							
Pooled sample	292	0.48	0.54	6.50	0.46	0.42	0.52
Mai Son	93	0.65	0.75	26.90	0.66	0.59	0.74
Moc Chau	87	0.50	0.60	9.20	0.51	0.45	0.58
Phu Yen	112	0.51	0.58	8.04	0.49	0.44	0.56

Source: Own survey, 2014.

The results show that the mean of bias-corrected TE results are lower than the initial scores, and no farms have a full technical efficiency score. Similar results are found in Linh (2012) for a single bootstrap and Olson and Vu (2009) for a double bootstrap. Comparing efficiency scores in the location category of rice crops, farmers in the Moc Chau district have a higher TE score (0.66 compared with 0.58 and 0.59). The difference between the initial and bias-corrected efficiency scores is also highest in this group (0.76 compared with 0.66). However, with maize crops, farmers in the Mai Son district have the highest initial and bias-corrected TE scores compared with the Moc Chau and Phu Yen districts. These results indicate that rice crops in Moc Chau and maize crops in Mai Son have much more efficient farmers in the sample.

Karimov (2013) has suggested that authors must use bias-corrected efficiency scores to recommend policy. This suggestion is based on the distance between the initial and bias-corrected TE. For detail, the initial TE_{VRS} for the pooled sample suggests that rice and maize farms could decrease their inputs by 37% and 46%, respectively. If full efficiency was achieved, the bias-corrected TE_{VRS} suggests decreasing inputs for rice and maize production approximately 45% and 54%, respectively.

Using the smooth bootstrap method, the width of the 95% confidence intervals is 0.12 for rice production and 0.10 for maize production. This finding is similar to Karimov (2013) and Latruffe et al. (2005) for the single bootstrap procedure. This result indicates that farms could be more inefficient if we used a confidence interval index rather than the single estimated point. For example, in the location category, the mean TE_{VRS} of maize crops in the Moc Chau district shows that, on average, farms could obtain the same level of output by reducing 40% of their inputs. However, the confidence interval explains that inputs could be reduced from 42% to 55%. If we do not use the DEA bootstrap method, farms that were originally identified as lying on the production frontier may, in fact, lie below it.

4.3 Factors Explaining Efficiencies

Table 4. Results of Tobit regression

Variable	Rice		Maize	
	TE_{VRS}	SE	TE_{VRS}	SE
Constant	0.6507***	0.8009***	0.4402***	1.1088 ***
	(0.0951)	(0.0520)	(0.0861)	(0.0581)
Age	-0.0022*	0.0009	-0.0020*	-0.0015*
	(0.0013)	(0.0007)	(0.0012)	(0.0008)
Edu	0.0029	-0.0002	-0.0029	-0.0032
	(0.0049)	(0.0027)	(0.0045)	(0.0029)
Hhsize	0.0029	-0.0051	0.0215**	-0.0088
	(0.0096)	(0.0052)	(0.0088)	(0.0059)
Sourele	0.1146**	0.0444	0.0764	-0.0137
	(0.0507)	(0.0277)	(0.0463)	(0.0311)
Dismark	-0.0013	-0.0039***	0.0018	-0.0028**
	(0.0021)	(0.0011)	(0.0019)	(0.0013)
Credit	-0.0245	0.0106	-0.0647**	-0.0048
	(0.0314)	(0.0171)	(0.0281)	(0.0189)
Extent	0.0158	0.0330**	0.1018 *	-0.0480
	(0.0293)	(0.0160)	(0.0565)	(0.0381)
Motor	-0.0111	0.0469	-0.0474	0.0144
	(0.0628)	(0.0349)	(0.0307)	(0.0206)
Tractor	-0.0062	0.0055	-0.0152	0.0072
	(0.0344)	(0.0187)	(0.0343)	(0.0230)
Milling	-0.0391	0.0512**	-0.0356	-0.0020
	(0.0379)	(0.0207)	(0.0286)	(0.0192)
Offic	0.0006	0.0003	0.0005	-0.0005
	(0.0014)	(0.0008)	(0.0013)	(0.0009)
Log likelihood	-39.42	161.33	4.01	134.68

Note. ***, **, *: Significance at the 1%, 5%, and 10% level, respectively.

Source: Own survey, 2014.

To explain the variation of technical and scale efficiencies, these scores were regressed concerning the farm characteristics using a Tobit regression model. The results of the Tobit model are presented in Table 4, and the partial effect of each factor is listed in Table 5.

The age of the household head has a significant and negative effect on the technical efficiency of rice crops and the technical and scale efficiencies of maize crops. If household heads are older, they do not want to change their cultivation methods, apply new technologies or expand scales of production. Younger heads of household are considered to be more flexible by adopting new knowledge and technology and increasing investment.

Source electricity has positive effects on the technical efficiency of both rice and maize crops. The results indicate that farms using a national electricity source have a higher technical efficiency than those that do not. Because a national electricity source is stable and powerful, farms can use it for crop activities, such as using pumps for irrigation.

Unexpectedly, the distance to market has a significant and negative impact on the scale efficiency of rice and maize crops. This finding shows that if the distance to market is nearer, the scale inefficiency of rice and maize crops will increase. Arable land for rice and maize crops is small and limited. Although natural conditions allow cultivation in two seasons per year, most farmers cultivate rice in one season. In addition, the custom of people upland is to relax after the harvest. Therefore, these farms will consume more if they are near the market and if they do not keep money to invest in increasing the scale of rice and maize production.

Table 5. Partial effects of the Tobit regression

Variable	Rice		Maize	
	TE_{VRS}	SE	TE_{VRS}	SE
Age	-0.0022*	0.0009	-0.0020*	-0.0015*
Edu	0.0029	-0.0002	-0.0029	-0.0032
Hhsize	0.0030	-0.0051	0.0215**	-0.0088
Sourele	0.1146**	0.0444	0.0764*	-0.0137
Dismark	-0.0013	-0.0039***	0.0018	-0.0028**
Credit	-0.0245	0.01062	-0.0647**	-0.0048
Extent	0.0158	0.0330**	0.1018*	-0.0480
Motor	-0.0111	0.0469	-0.0474	0.0144
Tractor	-0.0062	0.0055	-0.0152	0.0072
Milling	-0.0391	0.0512**	-0.0356	-0.0020
Offic	0.0006	0.0003	0.0005	-0.0005

Note. ***, **, *: Significance at the 1%, 5%, and 10% levels, respectively.

Source: Own survey, 2014.

Family members have a positive effect on the technical efficiency of maize crops. This effect can come from the fact that maize plots are mainly located on upland, with a slope more than 15°; thus, it costs time and money to use modern technology, such as a tractor for land preparation and a motor for transportation. In fact, farmers prepare land with animals or their hands. Thus, maize crops require more labor than rice crops. Family laborers will help farms save money and initiate production. This finding is inconsistent with Coelli et al. (2002) who found that larger families have a negative effect on the technical, allocative and cost efficiencies of modern Boro rice.

The insignificant effects of education, motors, tractors and off-farm income indicate that these factors have a low impact on different efficiencies. Experience may be a more important factor than the education of the household head, especially with the people in the highlands. Moreover, the average education of the household head is at the secondary level. Motors are mostly used for daily life rather than for cultivating activities. Rice plots are also small and fragmented; maize plots are located on sloping lands. Thus, tractors have no effect on either rice or maize efficiencies.

Finally, milling machines have a positive effect on the scale efficiency of rice crops. Farmers who have their own milling machine could improve the scale of their rice crop. Using milling machines should save time and costs for farmers and allow them to be proactive in their farm activities.

5. Conclusion

This study uses a smooth bootstrapping method to analyze the variability of DEA technical efficiency estimates and to correct for the inherent bias in the DEA method. This study use detailed survey data for 292 farms that cultivate both rice and maize crops in 12 villages in three districts in Son La Province, Vietnam. The study shows that the opportunity for both technical and scale inefficiencies of maize and rice crops is significant. The results indicate that the TE_{VRS} among farmers differs across districts. The bias-corrected point estimate of TE_{VRS} in rice and maize crops is 0.55 and 0.46, respectively. These numbers indicate that input levels could decrease 45% for rice and 54% for maize with the present levels of output. This result suggests continuing to improve the management of annual crop production and cultivation methods for farmers. In terms of the scale efficiency score, most rice and maize crops are producing under increasing return to scale. This score indicates that farm scales are mostly "too small". Therefore, co-operation in cultivation, crop diversity and the optimal use of rice plots are several suggestions for optimal farm production.

In the second step, a Tobit regression is used to explain variations in efficiencies among farms. The results indicate that a national electricity source is an important factor to improve the technical efficiency of both rice and maize farms. Thus, expanding a national electricity source is an important strategy for the government in the near future to increase social welfare.

Large families are likely to be more technically efficient on maize farms. Because maize farms are mostly cultivated through human power, more people will be helpful. Therefore, motors and tractors are insignificant in both efficiencies of both types of farms. An undesirable credit factor is found to have a negative impact on the technical efficiency of maize farms; the distance to the nearest market has a negative effect on the scale efficiency of both rice and maize crops. These factors may come from outdated customs, low education and farmers' life behavior. Therefore, policies for adjusting customs, knowledge and credit in appropriate ways should be emphasized. Extension services are considered to continue enhancement because they have a positive effect on the scale efficiency of rice crops and the technical efficiency of maize crops.

References

Binam, J. N., Tonyè, J., Wandji, N., Nyambi, G., & Akoa, M. (2004). Factors affecting the technical efficiency among smallholder farmers in the slash and burn agriculture zone of Cameroon. *Food Policy, 29*(5), 531-545. http://dx.doi.org/10.1016/j.foodpol.2004.07.013

Brummer, B. (2001). Estimating confidence intervals for technical efficiency: The case of private farms in Slovenia. *European Review of Agricultural Economics, 28*(3), 285-306.

Coelli, T., Rahman, S., & Thirtle, C. (2002). Technical, Allocative, Cost and Scale Efficiencies in Bangladesh Rice Cultivation: A Non-parametric Approach. *Journal of Agricultural Economics, 53*(3), 607-626. http://dx.doi.org/10.1111/j.1477-9552.2002.tb00040.x

Dao, G., & Lewis, P. (2013). *Technical efficiency of annual crop farms in Northern Viet Nam.* Retrieved from http://www.murdoch.edu.au/School-of-Management-and-Governance/_document/Australian-Conference-of-Economists/Technical-efficiency-of-annual-crop-farms-in-northern-Vietnam.pdf

Efron, B., & Tibshinari, R. J. (1993). *An Introduction to the Bootstrap. Monographs on Statistics and Applied Probability* (p. 57). New York: Chapman& Hall.

Farell, M. (1957). The Measurement of Productive Efficiency. *Journal of the Royal Statistical Society, 120*(Series A), 253-90.

Gocht, A., & Balcombe, K. (2006). Ranking efficiency units in DEA using bootstrapping an applied analysis for Slovenian farm data. *Agricultural Economics, 35*(2), 223-229. http://dx.doi.org/10.1111/j.1574-0862.2006.00155.x

GSO. (2012a). *Poverty profile.* General Statistics Office of Viet Nam.

GSO. (2012b). *Results of the 2011 Rural, Agricultural and Fishery Census.* General Statistic Office.

GSO. (2013). *Statistical Yearbook of Viet Nam 2013.* General Statistic Office of Viet Nam.

Gujarati, D. (2011). *Econometrics by Example* (p. 196). Palgrave Macmilan.

Karimov, A. (2013). Productive Efficiency of Potato and Melon Growing Farms in Uzbekistan: A Two Stage Double Bootstrap Data Envelopment Analysis. *Agriculture, 3*(3), 503-515. http://dx.doi.org/10.3390/agriculture3030503

Khai, H. V., & Yabe, M. (2011). Technical efficiency analysis of rice production. *Journal of ISSAAS, 17*(1), 135-146.

Latruffe, L., Balcombe, K., Davidova, S., & Zawalinska, K. (2005). Technical and scale efficiency of crop and livestock farms in Poland: does specialization matter? *Agricultural Economics, 32*(3), 281-296. http://dx.doi.org/10.1111/j.1574-0862.2005.00322.x

Linh, V. H. (2012). Efficiency of rice farming households in Vietnam. *International Journal of Development Issues, 11*(1), 60-73.

Luckmann, J., Ihle, R., & Grethe, H. (2011). Can Vietnamese upland farmers profit from high world market prices? A price transmission analysis. In *EAAE 2011 Congress Change and Uncertainty Challenges for Agriculture, Food and Natural Resources*.

Olson, K., & Linh, V. (2009). Economic efficiency in farm households: trends, explanatory factors, and estimation methods. *Agricultural Economics, 40*(5), 587-599. http://dx.doi.org/10.1111/j.1574-0862.2009.00400.x

Prime Minister of Vietnam. (2012). *Decision No. 432/QD-TTg on approving the Viet Nam Sustainable Development Strategy for the 2011-2020.* Vietnamese Government. Retrieved from http://www.chinhphu.vn/portal/page/portal/English/strategies/strategiesdetails?categoryId=30&articleId=10050825

Rios, A. R., & Shively, G. E. (2005). Farm size and nonparametric efficiency measurements for coffee farms in Vietnam. In *American Agriculture Economics Association Annual Meeting*. Providence, Rhode Island.

Simar, L., & Wilson, P. (1998). Sensitivity analysis of efficiency scores: How to bootstrap in nonparametric frontier models. *Mamagement Science, 44*(1), 49-61.

Simar, L., & Wilson, P. W. (2000). A general methodology for bootstrapping in non-parametric frontier models. *Jounal of Applied Economics, 27*(6), 779-802.

USDA. (2014). *Vietnam Grain and Feed Annual* (Vol. i).

Wilson, P. W. (2008). FEAR 1.0: A Software Package for Frontier Efficiency Analysis with R. *Socio-Economic Planning Sciences, 42*, 247-254.

World Bank. (2014). *The World Bank Data.* Retrieved December 18, 2014, from http://data.worldbank.org/indicator/NV.AGR.TOTL.ZS/countries

16

Adapting to the Impacts of Drought by Smallholder Farmers in Sekhukhune District in Limpopo Province, South Africa

Sylvester Mpandeli[1,2], Edward Nesamvuni[3] & Phokele Maponya[4]

[1] University of Venda, School of Environmental Sciences, Department of Geography and Geo-Information Sciences, Thohoyandou, South Africa

[2] Water Research Commission of South Africa, Gezina, South Africa

[3] University of Venda, Office the Academic Registrar, Thohoyandou, South Africa

[4] Agricultural Research Council, Vegetable and Ornamental Plant Institute, Pretoria, South Africa

Correspondence: Sylvester Mpandeli, Water Research Commission of South Africa, Gezina, South Africa. E-mail: sylvesterm@wrc.org.za

Abstract

Smallholder farmers have been affected by drought impacts for several years. Sekhukhune district is characterized by poor and unreliable rainfall, frequent droughts and periodic flooding most of the time. Due to low and unreliable rainfall the smallholder farmers in the Sekhukhune district are finding it difficult to obtain high crop yields. As result of unreliable rainfall the majority of the households in the district are food insecure. The drought impacts in the Sekhukhune district has affected smallholder farmers in different ways including economically, socially and the production. Sekhukhune district has been receiving lower rainfall due to the effects of high extreme climatic events, climate variability and change. The impact of lower rainfall has negative effects on the agricultural sector, resulting in decrease in agricultural activities, loss of livestock, shortage of drinking water, low yields and shortage of seeds for subsequent cultivation in the district. The lowest average annual rainfall recorded was 438 mm in 1992. Limpopo Province including the Sekhukhune district has been characterised by low rainfall and recurrent drought problems especially in 1981/1984, 1988/1989, 1991/92 and in the 2004 and these hinder agricultural production in the province. The majority of farmers in the Sekhukhune district in 1992 lost high volumes of crops and livestock due to shortages of water and because of drought problems during that year. It was highlighted by several experts that the drought impacts in the Sekhukhune district are not only affecting the crop and the livestock smallholders, it is also affecting the vegetation status in the district. The quality and status of vegetation can be severely impacted by drought periods. The combination of these factors, for example low rainfall, poor vegetation condition and a range of other constraints, heightened during droughts, unfortunately produces a range of additional stressors for farmers in the Sekhukhune district. Poor vegetation usually means poor grazing and therefore poor cattle condition. This can further translate into loss of livelihoods as poor cattle often receive poor market prices.

Keywords: drought impacts, extreme climatic events, climate variability, climate change, low rainfall, Sekhukhune district

1. Introduction

In the Southern African Development Community region (SADC), for example in Zimbabwe, many farmers suffer frequently from drought. Farmers have however developed varies strategies to cope with drought (Babi et al., 2005). Coping strategies refer to action or activities used by farmers or households usually over short time spans to survive when confronted with unanticipated livelihood failure (Babi et al., 2005). Corbett (1988) observed that coping strategies seek to balance present adversity and sometimes comprise of a variety of livelihoods as a drought progresses. The majority of people in South Africa are also affected by poverty, a high unemployment rate, recurrent drought and inequitable land distribution (World Bank, 1997; FAO, 1998; UNAIDS, 2000; NEPAD, 2001; UNAIDS, 2004; UNDP, 2004). These factors are major problems further constraining the livelihoods of ordinary South Africans, for example, the farming community. Extremes of climate, such as periods of drought and floods, exert additional pressures on livelihoods (UNDP, 2004). Most parts of South Africa were affected by flooding in the year 2000. Drought, for example, has also been a recurrent phenomenon occurring in the 1960s, 1980s, 1990s, 2002 and 2003 (Tyson & Dyer, 1978; Vogel et al., 2000; Adger, 2001; Kihupi et al., 2003). Drought frequently adversely affects agricultural production in various provinces including Limpopo, Free State, parts of Western Cape and Northern Cape (e.g. Sowetan, 2003). Limpopo Province has been experiencing high frequency of severe drought occurring in different districts including Sekhukhune. Recently it has been noted that the agricultural drought in Limpopo Province is creating problems not only for the smallholder farmers but also for the commercial farmers. During drought periods, farmers may also embark on a number of agricultural and technical activities such as adjust fertilizer input, adopt varieties that are tolerant to drought environments and plant crops that require less water. Other coping strategies used by farmers in South Africa include direct seeding. Farmers also use the zero tillage system in order to conserve soil moisture (Ellis, 1993; Hansen, 2005; Patt & Gwata, 2002). This method also requires less water and is good for early planting. Some smallholder farmers in Zimbabwe, for example, have been known to sell their livestock to compensate for a lack of income because of insufficient harvest (Phillips et al., 1999; Adger et al., 2002; Patt & Gwata, 2002). This, however, only usually occurs under severe periods of stress.

In South Africa one of the coping strategies most frequently used is for farmers to shift to crops that require less water such as sorghum (Annandale et al., 2002). The problem with such a strategy is that the majority of farmers in South Africa have very limited access to technology, market access and farm inputs (Vogel, 2000; Ziervogel & Downing, 2004).

While coping strategies usually refer to short-term activities, they can be expanded for use in the longer term. Adaptation strategies refer to all those responses to climatic conditions that may be used to reduce vulnerability (IPCC, 2001; Adger et al., 2002). Adger et al. (2002), argue that adaptation will allow a system to reduce risks associated with hazards by reducing its social vulnerability. Adaptation can be reactive, concurrent or anticipatory, spontaneous or planned (Smit et al., 1996; Smithers & Smit, 1997; Smit et al., 2000). The objectives of this paper are: (a) Evaluate and analyse how smallholder farmers in Sekhukhune district have been coping and adapting to impacts of drought. (b) Identify better adaptation strategies used by smallholder farmers in Sekhukhune district. (c) To improve the understanding of current drought risk–reduction and coping mechanisms in Sekhukhune district of Limpopo Province. (d) To recommend effective drought adaptation strategies in Sekhukhune district.

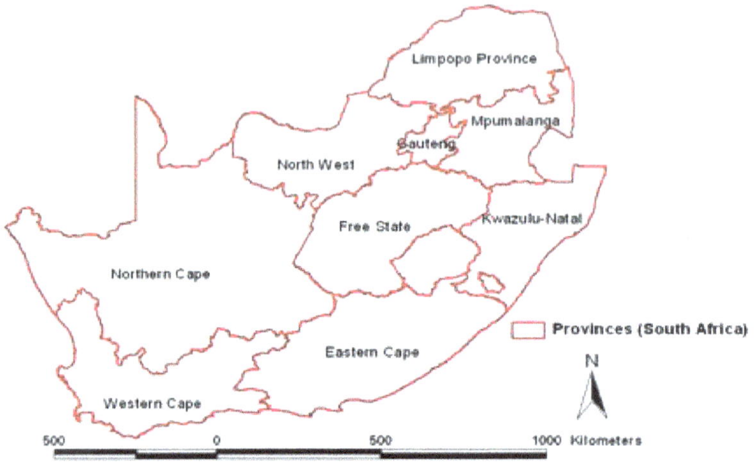

Figure 1a. Map of South Africa showing all 9 provinces

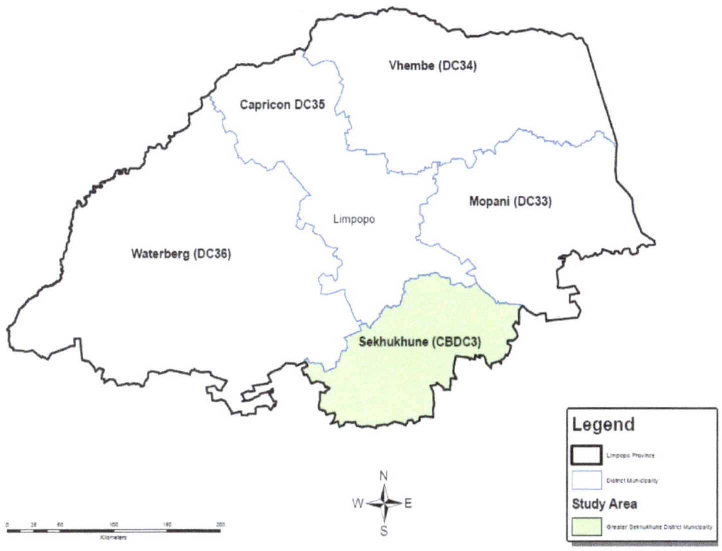

Figure 1b. Shows map of the Sekhukhune district in Limpopo Province

Source: ARC-ISCW (2014).

2. Study Site

According to Siambi et al. (2007), Greater Sekhukhune district is largely rural as 90% of the district population reside in the rural areas. Agriculture is considered the main contributor to employment and livelihoods. The Greater Sekhukhune district is rich with mineral deposits such as large reserves of platinum. The southern parts of the district have more agricultural potential than the rest of the district. The district average annual rainfall is approximately 560 mm. Subsistence or smallholder agriculture accounts for 70% of the farming activities in the district whilst the other 30% is commercial agriculture (Siambi et al., 2007). The Sekhukhune district is situated in semi – arid and it always experience water shortages most of the time. However, the majority of the commercial farmers depend on irrigation system for farming. Water shortages and poor rainfall distribution have been cited as one of the constraints hindering the growth and the development of the agricultural production in the district.

3. Materials and Methods

Two methods were used to analyze the results from this paper and this includes:

3.1 Mean Annual Rainfall Map

The rainfall data downloaded from the AgroMet databank at the Agricultural Research Council – Institute for Soil, Climate and Water (ARC-ISCW), South African Weather Service (SAWS) and ISCW weather stations from 1921–2013 with a recording period of 10 years and more. Regression analysis and spatial modelling were utilized during the development of the surface.

3.2 SPI Time Series

Rainfall GIS surfaces, covering South Africa, have been produced from data within the ARC-ISCW Climate Databank. The databank holds historical data from the South African Weather Service and the ARC-ISCW. Monthly rainfall GIS surfaces are produced from the historical rainfall data over the period from 1921–2013. The rainfall surfaces are produced as follows:

(i) Rainfall data from between 1200 and 3000 mechanical and automatic stations are extracted;

(ii) The long-term average rainfall for a specific month is used as a trend surface for interpolation;

(iii) Rainfall at a specific point is expressed as a percentage of the underlying rainfall trend surface;

(iv) The rainfall percentage values for a specific ten-day period are interpolated using the inverse distance weighting method.

The method results in a monthly rainfall surface that is true to the points where rainfall is recorded, but follows the climatology resulting from the influence of factors such as topography or distance from the ocean. The rainfall surfaces from 2003 onward are created through a combination of the abovementioned method and satellite rainfall estimates to supplement rainfall data over the South African plateau.

The resulting monthly rainfall surfaces are summarized to quaternary catchment (These are represented by +/- 1700 polygons in a GIS, covering the surface area of South Africa). For each catchment, the rainfall at monthly to 48-monthly time scale is transformed to Surface Precipitation Index (SPI) value for the catchment. The result is an SPI dataset for several time scales for each month since 1920. The SPI values can be used to study the time series of drought intensity classes by considering the traditional classification of the SPI ranges.

4. Results and Discussion

According to Mpandeli (2014), the average annual rainfall in the Sekhukhune district is less than 600 mm. The rainy season starts in late October in most areas of the Limpopo Province especially in the Sekhukhune district. The rainfall pattern peaks in January-February months, and thus when floods are also expected (Figure 2). As was reported by Mpandeli (2014), droughts across the Limpopo Province have impacted on crop & livestock production on top of their impact water supplies for irrigation activities, mining, and domestic use. It was also noted that the meteorological drought is the result of the negative deviation of rainfall from the mean and is normally the most common indicator for drought (Wilhite et al., 2000; Wilhelmi & Wilhite, 2002). Limpopo Province has high climatic variability & change and therefore farmers need to have seasonal climate forecast information or projections in advance in order for them to be able to plan their farming calendar accordingly. The Greater Sekhukhune district is well known to be dry – land area therefore smallholder farmers are expected to been use different coping and adaptation strategies including: (a) Plant crop that require less water such as groundnuts, lab lab, sorghum etc. (b) Some smallholder farmers in the area have adopted rain water harvesting and conservation agricultural principles in order to increase yields and income, for example, minimum tillage, zero tillage etc. (c) Livestock smallholder farmers in some of the areas in the district are also using the destocking especially during uncertainty periods. Other smallholder farmers in the district plant early mature crop varieties in order to counter act any natural vagaries.

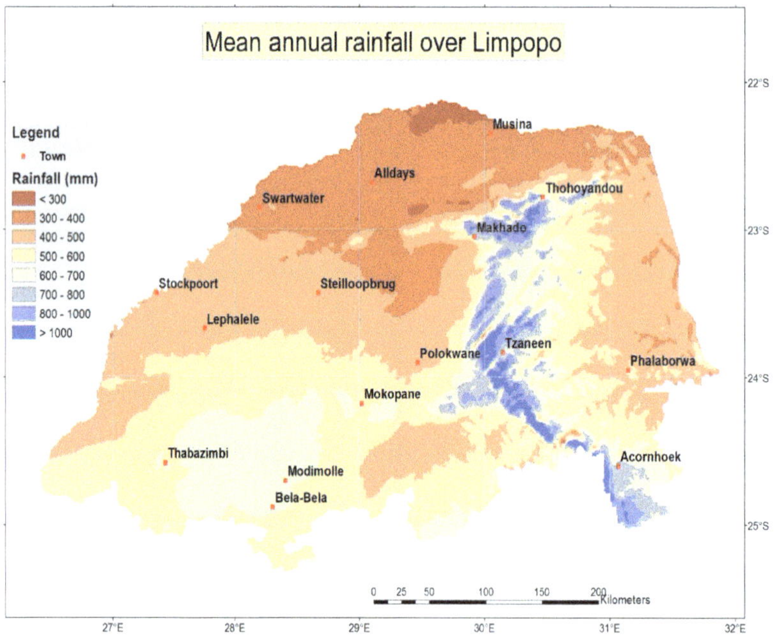

Figure 2. Mean annual rainfall map for Limpopo Province

Source: ARC-ISCW (2014).

Greater Sekhukhune district is different from the Vhembe district according to Mpandeli (2014) because Vhembe district is located in micro- climatic areas. In the Vhembe district, for example, the highest average monthly amount of rainfall received is normally from November to March every year (Mpandeli, 2006). Small holder farmers in the Greater Sekhukhune district are facing multi stress besides extreme climatic events, high climatic variability and change. These small holder farmers in the district also have to deal with the wild fire outbreaks which affect livestock grazing and also threaten human lives. The fact that the Greater Sekhukhune district is situated semi – arid area, any disturbances on the ecosystems especially changes in land use, diversion of water from local streams will have major consequences to the majority of the small holder farmers in the district. According to Greater Sekhukhune District Municipality report (2007) and Maponya (2013), the average annual rainfall in the Greater Sekhukhune district is 600 mm compared to other district such as Vhembe. According to Maponya (2013), Limpopo Province is a drought prone province which faces challenges of drought from time to time and Sekhukhune district is also not immune to drought impacts. Maponya (2013) further highlighted that due to this severe drought impacts across districts grazing and water for livestock and also irrigation activities were negatively affected and this also has an impact on the agricultural sector in general. According to Naidu (2003) and Maponya (2013), Limpopo Province was worst affected by drought in the past 8 years where dams were only 50% full compared to with 84% in the late 80s. This has severely affected the agricultural sector, smallholder farmers were most affected due to the fact that they are vulnerable and also low adaptive capacity. For the 15 years drought in the Sekhukhune district did not only affected the farming community, households, mining operations but have also have economic consequences (Greater Sekhukhune District Municipality report, 2011).

According to Mpandeli (2006), commercial farmers most of time have more wide choices during drought than subsistence and small-scale farmers. Due to the fact that the majority of the commercial farmers have strong financial backups, they also have good infrastructure especially irrigation systems such as Centre Pivot and drip irrigation system. Commercial farmers could easily switch their enterprise to a safer environment or to an area where they will be able to cope and adapt easily and most of the commercial farmers had enough reserves for their activities even if they are farming in new area. However, for subsistence and smallholder farmers, they have limited options due to the fact that these farmers do not have enough financial resources, they do not have collateral or security in case if they need loans from banks compared to the commercial farmers in the area. Besides the fact that the majority of the commercial farmers around Greater Sekhukhune district have good irrigation infrastructure, the majority of these farmers are literate, they have good technology and knowledge on how to apply the proper climate change adaptation strategy in order to deal with all type of agricultural risk including drought.

The high rainfall variability in Limpopo Province and some other parts of the country have led some farmers to use seasonal forecasts information for farm planning and decision making. Daily weather forecasts allow farmers to plan their daily activities. The daily weather forecast is provided by the SAWS through both the print and the electronic media and is published on line 24 hours.

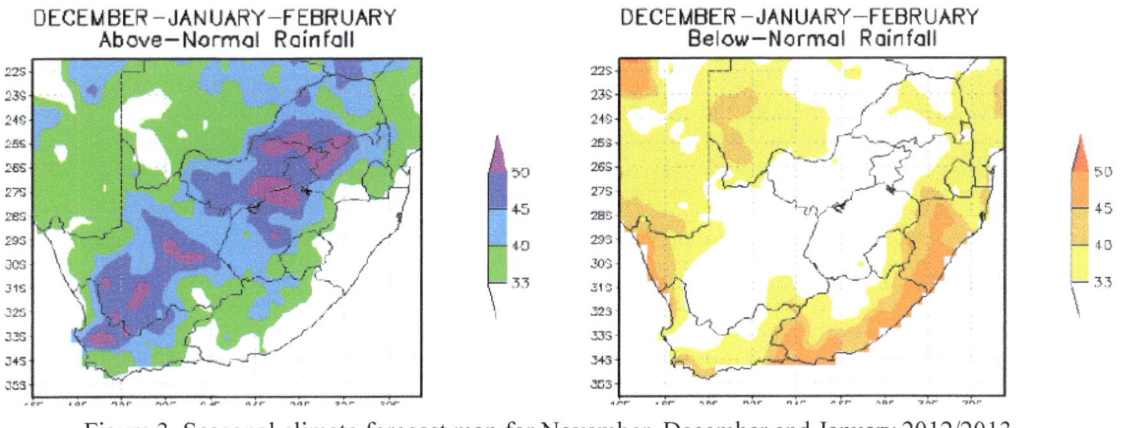

Figure 3. Seasonal climate forecast map for November, December and January 2012/2013

Source: SAWS (2014).

Forecast information that is useful to farmers is part of a process that includes an examination of the current needs, problems and context in which users operate. Forecasts, moreover, need to be expressed in the language of the users. Figure 3 shows that the above – normal rainfall in most parts of the Limpopo Province during December–January–February compared to other provinces such as North West and Northern Cape.

Seasonal climate forecasts are issued as probabilistic outlooks for the future usually for a coverage period of three months and with a rather broad spatial coverage (Figure 3). Conveying notions of 'probabilistic' information to a variety of users is not easy. It is important for users to understand that all seasonal forecast information or data are given as probabilities and not as deterministic. A probability forecast outlines how likely an event is likely to occur, as a percentage, and can assist farmers to be aware of the risks associated with weather and climate events. Figure 3 shows that the rainfall distribution in the Greater Sekhukhune district was below normal. These confirmed other studies done by Maponya (2013) which shows the rainfall distribution in Limpopo Province varies from area to area and district to district including Greater Sekhukhune. Figure 3 on the left hand side shows the probabilities for above – normal rainfall in most parts of Limpopo Province including Greater Sekhukhune district during December–January–February during 2012/13 (Figure 3). However, figure 6 on the right hand side shows that the probabilities for the below – normal rainfall in most parts of the Limpopo Province including Greater Sekhukhune district during December–January–February 2012/13.

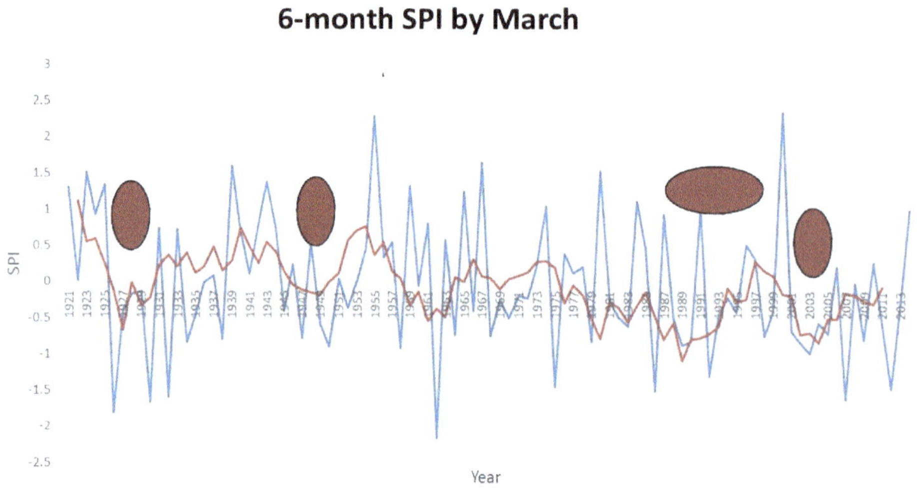

Figure 4. Shows the 6 month Surface Precipitation Index (SPI) for the Greater Sekhukhune district in Limpopo Province

As highlighted in earlier in this paper Greater Sekhukhune district has been affected by drought for number of years. Figure 3 shows that the Greater Sekhukhune district has been affected by the worst drought in 1926/1930/1932/1962/1975/1986/1992/2006/2012 (Figure 4). The worst drought record was in 1962, the majority of the smallholder farmers lost both the livestock and crops. According to Pachauri (2004) a decline in agriculture production would have serious implications on the rural populations and this would lead to food insecurity, reduced income and livelihoods of the poor of the poorest would also be affected. Some of the smallholder farmers decided to abandon their farming land and looked for employment opportunities in the non agricultural sector. The impact of lower rainfall in 1926/1930/1932/1962/1975/1986/1992 had negative impacts on the agricultural sector. According to Mpandeli and Maponya (2014), low rainfall resulting in decreases in agricultural activities including shortage of drinking water, loss of both livestock and crops and also lack of grazing capacity. However, figure 4 also shows that Greater Sekhukhune district also recorded the highest rainfall between 1921 and 2014. The highest rainfall recorded it was in 1923/1940/1955/1967/1980/2000 (Figure 4). During these periods the majority of people in the district lost their properties, roads were also damaged, and some of the people also lost their lives during these periods, smallholder farmers lost high production due to high rainfall distribution, lack of climate advisory information, poor access to Early Warning Information (EWS).

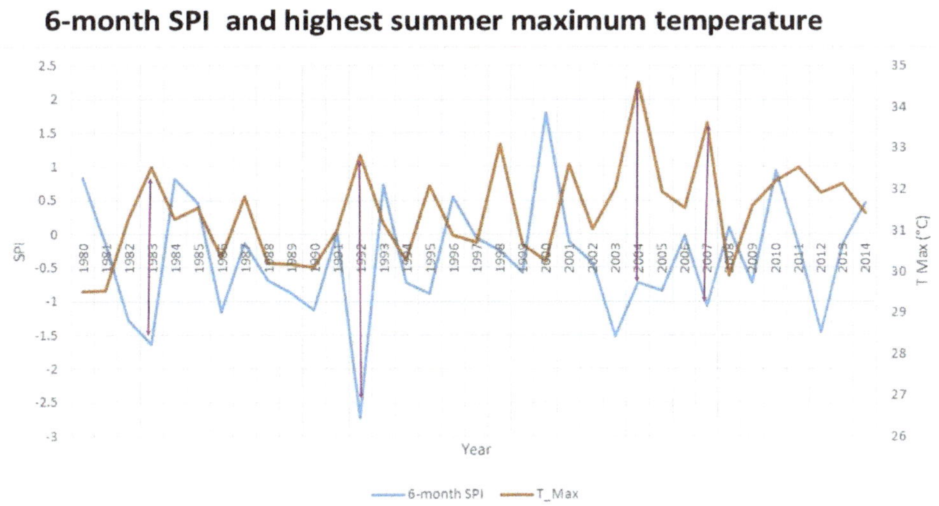

Figure 5. Shows the highest maximum temperature in Sekhukhune district, Limpopo Province

Figure 5 shows the relationship between the SPI and the maximum temperature from 1980 to 2014. The lowest temperature recorded was in 1980/1980/1983 and 1992 during that periods the majority of the smallholder farmers in the Greater Sekhukhune district experienced frost, hail etc. Figure 5, shows that in 1992 Limpopo Province experienced one of the worst droughts, smallholder farmers lost large number of livestock and crops. This had serious implications on the income of the smallholder farmers across the province including Greater Sekhukhune district. Similar trends of temperature effects were also depicted by Mpandeli (2014) in the Vhembe district. Due to high extreme climatic events and climatic variability and change smallholder farmers in the Greater Sekhukhune district should be able to access climate advisory information all the time due to poor rainfall distribution in the district. Small holder farmers in the Greater Sekhukhune district should be use different types of climate change adaptation strategies in order for them to be able to increase crop yields including: (a) Changing farming practices, (b) Use early matured crop varieties, (c) Use of drought resistance seeds etc. Figure 4 shows that the highest temperature recorded was in 2004 which was 34.5 degrees followed by 2007 which was 33.2 degrees. Greater Sekhukhune district recorded the highest rainfall in year 2000 and the year 2010 (Figure 5). More than 1100 mm of rainfall was recorded during those periods. According to Mpandeli and Maponya (2014) the majority of smallholder farmers lost large number of livestock and crops, the majority of the smallholder farmers in the district did not have good agricultural production. During that periods, the year 2000 and year 2010, the majority of the smallholder farmers were relying on their extended families for support and also on the non agriculture sector like mining, local municipality for survival etc.

In the Limpopo Province, the condition of the vegetation is a major concern for livestock farmers. Limpopo Province is characterised by different kinds of veld types, especially in the morphology, palatability and the nutritive value. Sweet veld grass is found around the Thohoyandou area in the Vhembe district and the sweet veld grows in frost-free and low-lying areas (Acocks, 1998). Sourveld grows mostly in the Tshakhuma area; these types of grass usually grow in a high rainfall area.

The quality and status of vegetation can be severely impacted by drought periods. One way to monitor the condition of the vegetation is to use the Normalised Difference Vegetation Index (NDVI). A relationship between rainfall and vegetation status exists, usually expressed as an index the NDVI. This index generally shows the presence and absence of green vegetation (Figure 6). Since green vegetation reflects more infrared radiation than visible radiation, the higher the value of NDVI, the higher the vegetation activity. An example of the relationship between periods of drought and vegetation can be observed in Figure 6.

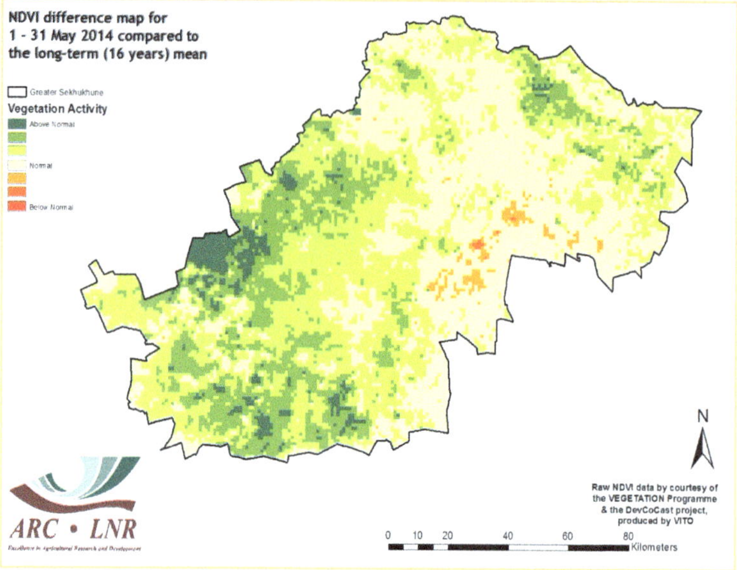

Figure 6. Normalised Difference Vegetation Index map for May 1, 2014 to May 30, 2014, compared to the long-term mean for 16 years

Source: ARC-ISCW (2014).

NDVI can thus be displayed and interpreted spatially over time. For example, from Figure 6, one can infer that the rainfall distribution was very low and sparse in Sekhukhune district compared to other parts of the Limpopo Province for the period 01 May 2014 to 31 May 2014. The green colour in most parts of Limpopo Province shows that the vegetation activity was higher than normal over this period. The green colour shows that there was higher vegetation than normal. In areas where the green colour dominates, there was good rainfall distribution (Figure 6).

The combination of these factors, for example high rainfall, good vegetation condition and a range of other available opportunities, heightened by figure 6, shows those smallholder farmers in the Sekhukhune district had good forage production and also managed to obtain good yields. According to Mpandeli (2014) poor vegetation usually means poor grazing and therefore poor cattle condition. This can further translate into loss of livelihoods as poor cattle often receive poor market prices. In the district such as Vhembe, older farmers indicated that drought has been a recurring process since the 1950s and this it was confirmed by results from figure 6. Some farmers stated that in 1992 they had severe drought problems resulting in a shortage of grazing and crop failures in many parts of the district (Bakali, pers. comm., 2004; Mpandeli, 2006, 2014). Livestock was slaughtered and farmers had to sell their livestock at prices as low as R5 per unit. The drought impacts in 1992 increased farming debt and affected food exports in the province (Koch, pers. comm., 2005; Mpandeli, 2006). According to Mpandeli (2006, 2014) farmers in the Vhembe district farmers describe drought based on the following factors such as food and feed shortages, low rainfall, decrease in water availability, dying vegetation and animals etc. These factors were also highlighted by O' Farrell et al. (2009). According to O' Farrell et al. (2009) farmers described drought in terms of the way it affected the agricultural systems.

5. Conclusions and Recommendations

Assessing the importance of climate variability is important in a drought-prone area such as Sekhukhune district in Limpopo Province. To improve drought management, risk reduction and responses of farmers in the Sekhukhune district were probed. A better understanding of the role drought plays in communities in Sekhukhune district, as well as strategies that may be used to best mitigate drought, were examined. Limpopo Province offers a range of opportunities as well as challenges for farmers. The results from this paper show that the changes in temperature and precipitation in the Greater Sekhukhune district in Limpopo Province could probably affect the accessibility, availability and the distribution of water resources in the area. The impact of drought in the Greater Sekhukhune district is not affecting the smallholder farmers or the agriculture sector only but other sectors such as water, mining and tourism are also affected. The Greater Sekhukhune district has been experiencing several types of droughts such as hydrological, agricultural and meteorological droughts for several decades. However, smallholder farmers in the Greater Sekhukhune district have been using several climate change adaptation strategies including, plant crops that requires less water, plat drought resistant crops, adjust agronomic practices and plant matured crop varieties..

References

Acocks, J. P. H. (1998). *Veld Types of South Africa: Memoir of the Botanical survey of South Africa* (No. 57, 3rd ed., p. 146). Department of Agriculture and Water Supply, Pretoria.

Adger, W. N. (2001). Scales of governance and environmental justice for adaptation and mitigation of climate change. *Journal of International Development, 13*, 921-931. http://dx.doi.org/10.1002/jid.833

Adger, W. N., Kelly, P. M., Winkels, A., Huy, L. Q., & Locke, C. (2002). Migration, remittances, livelihood and trajectories and social resilience in Vietnam. *Ambio, 31*, 358-366. http://dx.doi.org/10.1579/0044-7447-31.4.358

ARC-ISCW. (2014). Agricultural Research Council – Institute for Soil, Climate and Water.

Annandale, J. G., Jovanovic, N. Z., Mpandeli, N. S., Lobit, P., & du Sautoy, N. (2002). Two Dimensional Energy Interception and Water Balance Model for Hedgerow Tree Crops. *Paper presented at Southern African Society of Horticultural Sciences congress at Cedara, Kwazulu Natal, 15-17 January 2002.*

Babi, D. D. (2005). Assessing the vulnerability of FADAMA (flood plain) agriculture to climate variability and change. *A case study of semi-arid Northern Nigeria.*

Bakali. (2004). Agricultural scientist, personal communication.

Corbett, J. (1988). Famine and households coping strategies. *World Development, 16*, 1099-1112. http://dx.doi.org/10.1016/0305-750X(88)90112-X

Ellis, F. (1993). *Peasant Economics: Farm Households and Agrarian Development* (p. 309). Cambridge, UK: Cambridge University Press.

FAO (Food and Agriculture Organization). (1998). Agricultural Policies for Sustainable Use and Management of Natural Resources in Africa. *Report presented at 20th Regional Conference for Africa, February 16-20, 1998.* Addis Ababa, Ethiopia.

Greater Sekhukhune District Municipality (GSDM). (2007). *2006/2007 IDP Review.* Retrieved December 15, 2009, from http://www.sekhukhune.gov.za

Greater Sekhukhune District Municipality (GSDM). (2011). *2010/2011 IDP Review.* Retrieved January 15, 2012, from http://www.sekhukhune.gov.za

Hansen, J. W., Marx, S., & Weber, E. (2004). The role of climate perceptions, expectations and forecasts in farmer decision making: the Argentine Pampas and South Florida, *International Research Institute for Climate Predictions Technical Report, 04-01.* Palisades, New York.

Intergovernmental Panel on Climate Change Report (IPCC). (2001). Impacts, Adaptation and Vulnerability. *Contribution of the Working Group II to the Third Assessment Report on the Intergovernmental Panel on Climate Change.* UK: Cambridge University Press.

Kihupi, N., Kingamkono, R., Dihenga, H., Kingamkono, M., & Rwamugira, W. (2003). Integrating indigenous knowledge and climate forecasts in Tanzania. In C. Vogel & K. O'Brien (Eds.), *Coping with climate variability: The use of seasonal climate forecasts in Southern Africa* (pp. 155-169). Burlington, Vermont, USA: Ashgate.

Koch. (2005). Agricultural project manager, personal communication.

Maponya, P. I. (2013). Climate change and agricultural production in Limpopo: Impacts and Adaptation Options (Ph.D thesis, p. 108). Environmental Management, University of South Africa, Pretoria, South Africa.

Mpandeli, N. S. (2006). *Coping with climate variability in Limpopo Province* (Unpublished PhD thesis). University of Witwatersrand.

Mpandeli, N. S., & Maponya, P. I. (2014). Constraints and challenges facing the small scale farmers in Limpopo Province. *Journal of Agricultural Science, 6*(4), 135-143. http://dx.doi.org/10.5539/jas.v6n4p135

Mpandeli, N. S. (2014). Climate Risks Using Seasonal Climate Forecast Information in Vhembe District in Limpopo Province, South Africa. *Journal of Sustainable Development, 7*(5), 68-81. http://dx.doi.org/10.5539/jsd.v7n5p68

Naidu, E. (2003). *Large parts of South Africa face water restrictions.* Sunday Independent, South Africa.

NEPAD. (2001). The New Partnership for Africa's Development. *NEPAD* (Unpublished memorandum, October 2001).

Siambi, M., Simalenga, T., Mpandeli, N. S., Liphadzi, K., Mailula, N., Mkhari, J., & Ramugondo, R. (2007). Increased food security and income in the Limpopo Basin through integrated crop, water and soil fertility options and public-private partnerships. *Proceedings of the CGIAR Challenge Program on Water and Food.* Addis Ababa, Ethiopia.

Smithers, J., & Smit, B. (1997). Human adaptation to climate variability and change. *Global Environmental Change, 7*, 129-146. http://dx.doi.org/10.1016/S0959-3780(97)00003-4

Smit, B., McNabb, D., & Smithers, J. (1996). Agricultural adaptation to climate variation. *Climate Change, 33*(1), 7-29. http://dx.doi.org/10.1007/BF00140511

Smit, B., Burton, I., Klein, R. J. T., & Wandel, J. (2000). An anatomy of adaptation to climate change and variability. *Climatic Change, 45*, 223-251. http://dx.doi.org/10.1023/A:1005661622966

SAWS. (2014). South African Weather Service.

Sowetan. (2003). *Drought takes its toll in Limpopo Province* (p. 23, May 5, 2003). Gauteng: Sowetan.

Patt, A., & Gwata, C. (2002). Effective seasonal climate forecast applications: Examining constraints for subsistence farmers in Zimbabwe. *Global Environmental Change, 12*, 185-195. http://dx.doi.org/10.1016/S0959-3780(02)00013-4

Pachauri, R. K. (2004). Climate change and its impacts on development: The role of the IPCC assessments. *Climate and Development, IDS Bulletin, 35*(3). Sussex: Institute of Development Studies.

Phillips, J., Makaudze, E., Unganai, L., Makardho, J., & Cane, M. (1999). Current and potential use of climate forecasts for farm management in Zimbabwe. *Paper presented at the Society for Applied Anthropology Meetings*. Tuscon, AZ.

Tyson, P. D., & Dyer, T. G. J. (1978). The predicted above-normal rainfall of the seventies and the likelihood of droughts in the eighties in South Africa. *South African Journal of Science, 74*, 372-377.

UNAIDS. (2000). *Report global estimates of HIV/AIDS* (p. 32).

UNAIDS. (2004). *Report on the global AIDS epidemic* (p. 63).

UNDP. (2004). *United Nation Development Programme in South Africa* (Press release).

Vogel, C., Laing, M., & Munnik, C. (2000). Drought in South Africa. In D. Wilhite (Ed.), *Drought Volume 1, A Global Assessment, Routledge Hazards and Disasters Series* (pp. 348-366). London: Routledge. http://dx.doi.org/10.1080/03736245.2000.9713700

Vogel, C. (2000). Usable science: an assessment of long-term seasonal forecasts among farmers in rural areas of South Africa. *South African Geographical Journal,* 82, 107-116.

World Bank. (1997). Towards Environmentally Sustainable Development in Sub-Saharan Africa. *Findings: Africa Region* (p. 78). World Bank, Washington D. C., USA.

Ziervogel, G., & Downing, T. E. (2004). Stakeholder networks: improving seasonal climate forecasts. *Climatic Change, 65*, 73-101. http://dx.doi.org/10.1023/B:CLIM.0000037492.18679.9e

Wilhite, D. (2000). Drought as a natural hazard: Concepts and definitions. In D. Wilhite (Ed.), *DroughtVolume 1: A Global Assessment, Routledge Hazards and Disasters Series* (Vol. 1, pp. 3-18). London: Routledge.

Wilhelmi, O. V., & Wilhite, D. A. (2002). Assessing Vulnerability to Agricultural Drought: A Nebraska Case Study. *Natural Hazards, 25*, 37-58. http://dx.doi.org/10.1023/A:1013388814894

17

Control of Micronutrients Availability in Soil and Concentration in Rice Grain through Field Water Management

Juan Damian Marques Fong[1], Tsugiyuki Masunaga[1] & Kuniaki Sato[1]

[1] Faculty of Life and Environmental Science, Shimane University, Matsue, Japan

Correspondence: Tsugiyuki Masunaga, Faculty of Life and Environmental Science, Shimane University, 1060 Nishikawatsu, Matsue, Shimane 690-8504, Japan. E-mail: masunaga@life.shimane-u.ac.jp

Abstract

A pot experiment was conducted under different water managements, namely control (continuous flooding condition), 2,3,4WAH (water was drained at two to four weeks after heading), and FCap (the soil moisture was kept at field capacity level, to maintain aerobic soil condition). We examined the influence of the establishment of aerobic soil condition at different timing of rice post-heading stage. The results showed that water managements changed soil redox potential and pH that were known to be responsible of micronutrients availability in soil. Transition from flooding condition to aerobic condition during rice post-anthesis stage, reduced pH toward its initial value as well as Fe availability in soil. Although soil was re-oxidized in 2,3,4WAH treatments, which reduced Fe availability in soil from the level of control treatment, concentration of Fe in grain did not differ between control and drained treatments. Correlation analyses exhibited that the reduction of Fe availability in soil under aerobic condition in 2,3,4WAH and FCap enhanced Zn and Cu availabilities in soil and increased concentration of Zn, Cu and Mn in grain. The results point up toward that, enhancement of micronutrient availability in soil during post-heading stage are strongly associated with its grain micronutrient concentration. Establishment of aerobic condition from two weeks after heading increased the grain Zn, Cu and Mn concentration by 11, 20 and 35 percent respectively over the values in flooding condition, without negative effects on yield parameters. Unlike control and 2,3,4WAH treatments, FCap that kept aerobic soil condition performed high concentration of grain micronutrients but its low yield.

Keywords: grain filling stage, micronutrients concentration, soil availability, rice, water management

1. Introduction

Micronutrient deficiencies in soil have been identified as one of main factors affecting crop yield, food quality and human health (Yang et al., 2007; Alloway, 2008). In recent years, emphasis on agriculture, environment, demographic growth and micronutrient malnutrition has increasingly taken center stage in policy discussions on food security in developing countries. The inhabitants of these countries are prone to suffer from undernourishment, caused by inadequate food intake or intake of foods of poor nutritional quality (Bain et al., 2013). New approaches have been applied in achieving sustainable improvements in the micronutrient status of vulnerable populations. One of them is the "genetic biofortification" which has been accepted as a cost-effective and sustainable strategy in order to ameliorate the malnutrition (Cakmak, 2008; Meenakshi et al., 2007). The development of new improved crops varieties, through advanced breeding techniques as well as the screening of germplasm, has allowed for identification of varieties with high potential of accumulating micronutrient in edible parts. Some of the major successes have been observed in crops such as wheat, maize, beans, sweet potato, cassava and rice.

Rice (*Oryza sativa* L.) is the main staple crop for more than half of the world population (FAO, 2013). In spite of its high consumption, the low content of micronutrient in rice required to meet whole dietary elements necessary to keep a balanced diet in humans, calls for concern. Rice as a major cereal therefore has become a priority for current research works. New findings suggest that micronutrients content in rice grain depends not only on genotypic variables, but also on environmental factors, such as the endogenous soil mineral contents, soil properties (Cakmak, 2008; Sperotto et al., 2013) water management (Dobermann & Fairhurst, 2000; Liao et al., 2013) and climate conditions (Najafi-Ghiri et al., 2013).

"Agronomic biofortification" through the application of fertilizer (inorganic or organic), modification of cultivation systems, soil management and new irrigation strategies have proved effective in enriching micronutrients content in rice grain by controlling the availability of soil micronutrients for plant (White & Broadley, 2009). Unlike genetic biofortification, the agronomic biofortification has shown to be a short-term approach, more accessible for developing countries which are facing the burden of malnutrition and where the advanced technologies (e.g. plant breeding and genetic engineering) still remain unreachable (Cakmak, 2008).

Since soil micronutrient availability is highly associated with the contents in plant and grain quality; its availability has become the key limitation factor to productivity, stability and sustainability of rice production in many countries such as Australia, China, India and United States (Bell & Dell, 2008). Studies conducted by Yang et al. (2007) in China, reported that such deficiencies in soil overlapped areas where population exhibits health micronutrients imbalance, affecting more than 40% of the population. Usually those deficiencies are afforded by the applications of chemical fertilizer (edaphic and foliar) or organic amendment. The application of Zn fertilizer has had a positive response in wheat (*Triticum aestivum* L.), (Cakmak, 2008), sorghum (*Sorghum bicolor* L.) and rice (Gao et al., 2012). Iron is another important micronutrient for plant and human health. Major Fe deficiencies have been reported mainly in sandy and calcareous soils (Brady & Weil, 2014). However, under such condition application of inorganic Fe fertilizer often results in ineffective practice due the conversion to ferric form (Rengel et al., 1999). Besides that, the micronutrients content in grain could be controlled by genetic factor meaning that the application of fertilizer even in excess could not increase the content in grain and would rather trigger toxicity symptoms. Likewise micronutrient demand for physiological processes and plant uptake rates could also vary in function of different growing stages and environmental condition (Maschner, 2012).

Rice as well as other crops needs essential micronutrients in very small amount compared to macronutrient. Almost all the soil contains enough micronutrients to support the plants growth demands, however, micronutrients availability often is governed by some soil properties (White & Broadley, 2009). For example, changes in soil redox potential and pH are the most important properties that could affect micronutrient availability in paddy (Tao et al., 2007; Brady & Weil, 2014). Likewise, soil organic matter and its decomposition processes have a significant and direct impact on the availability of micronutrients (Marschner & Rengel, 2007). Other factors such as synergistic and antagonist interactions among micronutrients and essential elements could also often affect micronutrient uptake by crops (Fageira, 2002).

In order to increase the contents of micronutrients in grain, understanding the dynamics of micronutrient at different growing stages is also very important. Some studies have pointed out that the period between panicle initiation to maturity is considered as high demand for grain micronutrient accumulation. For instance, concerning Zn uptake, Jiang et al. (2007) reported that 36% of total Zn in rice grain was taken up directly from soil between anthesis and maturity stages, similar patterns has been described in wheat during grain filling stage (Pearson & Rengel, 1994). For Fe, as Inoue et al. (2009) observed, Fe-trasnporter gene OsYSL15 was highly expressed during the anthesis and seed formation stage, these stages probably are the active Fe uptake period of rice. Copper (Cu) on the other hand is essential in flower fecundation (Dobermann & Fairhurst, 2000). Therefore, its demand should significantly increase at anthesis.

With regards to above factors influencing the micronutrient availability in soil, plant uptake and the content in rice grain, little information is available about the effect of water management over micronutrient concentration in grain during grain filling stage. Since changes in soil water condition are strongly associated with micronutrient availability and plant uptake; shifting soil water condition could be a suitable approach to control soil micronutrient availability and to improve the nutritional value of rice in developing countries where farmers do not have easy accesses to fertilizers or other costly approaches. Therefore, the aim of this present study was to assess the effect of water management on the micronutrient contents in rice grain and availability of soil at grain filling stage.

2. Materials and Methods

The effects of water management on soil properties and micronutrients content in rice grain, were evaluated through a pot experiment conducted in a greenhouse at Shimane University (35°29′N 133°04′E), from 2011-2012 during rice growing season in Japan (May 15th/Oct 15th). Plastic bucket (Ø = 24 cm × 25 cm height) were filled with 7.5 kg of air dry "Gray Lowland" soil Typic Fluvaquent (USDA, 2010). The general soil characteristics are presented in Table 1. The semi-dwarf (cv. Koshihikari) rice variety was used in this study.

2.1 Experimental Design

The pot experiment was laid out in a randomized block design of five water irrigation treatments (Figure 1) with four replications. Control was flooded from transplanting to harvest. The heading stage (50% of the panicle exertion) was the benchmark for the next customized irrigation treatments where flooded-irrigation was stopped during the post-heading at four week after heading "4WAH," three week after heading "3WAH," and two weeks after heading "2WAH," and kept in a soil level approximately 35±3% (v/v) moisture, equivalent to ±80% of its field capacity level. The last treatment was kept at the same moisture condition, 80% of field capacity conditions "FCap" throughout growth period in order to mimic the conditions of upland systems. Treatments under flooding condition were ±5 cm water depth. Distilled water was used for irrigation to avoid micronutrient supply through it. Chemical fertilizer (N:P:K) at doses of 45:65:60 kg ha^{-1} respectively was applied based on crop demand as follows: N was split in four application, 1st-20% basal, 2nd-30% at tillering stage, 3rd-40% at panicle initiation and 4th-10% 15 days after panicle initiation (PI). P was applied 100% of dose basal. K was split in two application 50% basal and 50% at panicle initiation.

Table 1. Physical-chemical properties of Typic fluvaquent soil "Gray lowland" collected at paddy field in Shimane prefecture, Japan

Soil characteristics	Value
Sand (%)	22.8
Silt (%)	37.8
Clay (%)	39.4
pH	4.85
EC (mSm^{-1})	3.87
Bulk Density (g cm^{-3})	1.39
Total C (g kg^{-1})	17.6
Total N (g kg^{-1})	1.6
NH_4-N (mg kg^{-1})	4.91
NO_3–N (mg kg^{-1})	8.22
Available P (mg kg^{-1})	34.9

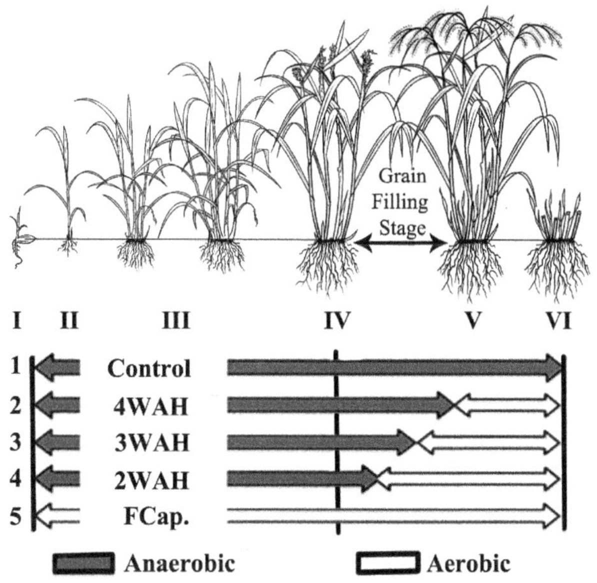

Figure 1. Schematic representation of randomized block design at with different irrigation treatments were applied during the growth season. (I) Sowing, (II) Transplanting, (III) Tillering, (IV) Heading, (V) Maturity Stage, (VI) Harvest. Period between (↔) IV and V was considered as grain filling stage

2.2 Plant and Rice Grain Chemical Analysis

Plant samples were collected at harvest. Flag leaves were first washed with tap water and twice with distilled water. Thereafter they were oven dried at 40 °C for 48 hours and dehusked in rotating rubber roll, (Satake THU35B, Japan). For micronutrient analysis, flag leaves and grain samples were ground into fine powder in agate grinding jars, using a mixer mill (MM200, Retsch GmbH, Haan, Germany). Grain and plant samples were oven-dried 12 hours at 80 °C. Subsequently 0.5 g of sample was digested in 2.5 ml HNO_3 within Teflon vessel. All samples were oven heated at 160 °C for 4 hours, kept resting overnight and diluted with distilled water up to 25 ml (Koyama & Sutoh, 1987). The concentrations of Fe, Zn, Cu and Mn were determined by Inductive Coupled Plasma Spectroscopy (ICPE-9000, Shimadzu, Japan).

2.3 Soil Sampling and Analysis

Soil samples were collected from 5, 10 and 15 cm depth in the bucket after harvest. They were immediately stored in ZIPLOC hermetic plastic bag and refrigerated in order to maintain original chemical soil properties. Portions of each sample was oven dried at 105 °C to measure the moisture content to calculate the necessary amount for all the subsequent analyses following standard analyses methods (Japan Soil Environment Analysis Methods Committee, 2003). The concentration of micronutrients (Fe, Zn, Cu and Mn was analyzed by mixing 10 g of soil with 20 ml of diethylene triamine pentaacetic acid (DTPA-TEA) extracting solution (Linsay & Norvel, 1978). The solution obtained was filtered and analyzed in ICPE-9000. Soil pH was determined by the soil-water ratio of 1:2.5 and electric conductivity (EC) soil-water ratio of 1:5 using electrode method (D-24 HORIBA, Japan). Redox potential was measured in situ from transplanting period to 25 DAT at 10cm depth using electrode method (PNR-41, DKK-TOA Co., Japan).

2.4 Statistical Analysis

To determine the influence of water management on soil properties, micronutrient concentration in soil, plant and rice grain accumulation, data were statistically analyzed by one–way analysis of variance (ANOVA). Significant differences among the treatments were determined by Turkey's honestly significant difference (HSD) test ($P < 0.05$) for multiple means comparison. Correlation analysis between grain micronutrient concentration and soil properties was done using Pearson Product Moment Correlation (PPMC). All statistical analysis was performed using IBM SPSS Statistic v20.0 (IBM SPSS, 2011. Chicago IL, USA).

3. Results

3.1 Soil Properties

Effects of flooding on soil redox potential (*Eh*) (Figure 2a) were observed from transplanting until early tillering stage (25 DAT). There was a sharp decrease over time in the *Eh* at the 10 cm depth, from +467 to -379 mVolt (mV). This suggested that depletion of O_2 and an increase in electron activity mediated by microorganisms respiration was highly marked in the first five days. However, after 17 days Eh values achieved an apparent stability, without significant changes. As a result of changes in the soil electrical conductivity in the flooded treatments, the pH increased from initial pH 4.85 (transversal dashed line) at transplanting (Figure 2b). In the control which was flooded throughout the growth stage, pH was 6.18 at the harvest stage, significantly higher than that of the other treatments that were drained after heading at 4, 3, 2 WAH. In the drained treatments, pH ranges declined between 5.54-5.58 without significant differences among them. Results show that pH acquires specific values at specific soil moisture. FCap treatment showed the lower pH of 5.24.

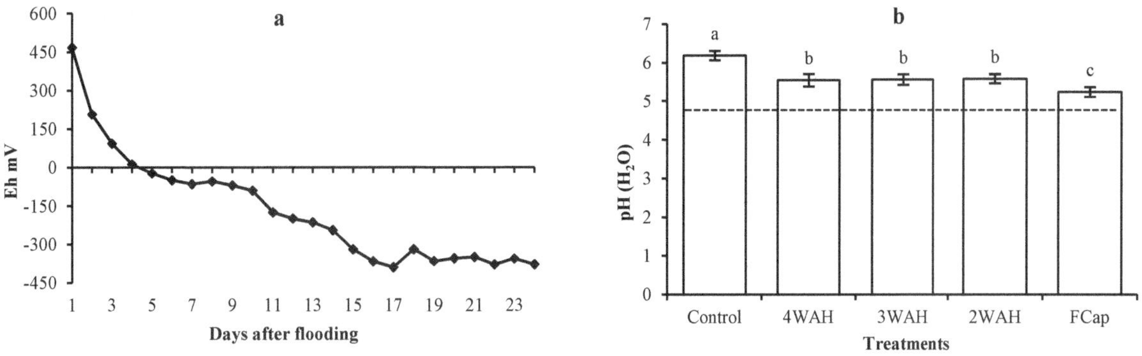

Figure 2. Influence of soil flooding on soil electrochemical properties. a) Soil Redox Potential; b) Soil pH(H_2O). [dashed line represent the pre-flood pH value]

3.2 Influence of Water Management on Grain Micronutrient Contents and Yield

One–way analysis of variance (ANOVA) showed that, there were statistically significant differences between means of micronutrient concentration in grain (P-value < 0.05) Table 2. Therefore resulted in rejection of the null hypothesis which stated that all the data came from groups with the same mean. In this case, subsequent analysis will help us to find a significant difference between pairs of means in at least one of the treatments.

Table 2. One–Way analysis of variance (ANOVA) for the concentration of micronutrients in grain

		Sum of Squares	df	Mean Square	F	Sig.
Grain-Fe Concentration	Between Groups	139.438	4	34.860	9.888	0.000
	Within Groups	123.385	35	3.525		
	Total	262.824	39			
Grain-Zn Concentration	Between Groups	204.186	4	51.046	11.431	0.000
	Within Groups	156.292	35	4.465		
	Total	360.477	39			
Grain-Cu Concentration	Between Groups	36.593	4	9.148	17.769	0.000
	Within Groups	18.020	35	0.515		
	Total	54.613	39			
Grain-Mn Concentration	Between Groups	3043.854	4	760.963	16.706	0.000
	Within Groups	1594.257	35	45.550		
	Total	4638.111	39			

The concentration of micronutrients in rice grain (Figure 3) differed significantly among irrigation treatments. The Fe concentration in grain ranged from 19.01 mg kg^{-1} to 13.95 mg kg^{-1}. 4WAH, Control and 2WAH showed the higher Fe concentration with no large significant differences, while 3WAH and FCap were 10% and 23% respecctively lower than Control. In contrast to Fe, Zn concentration in rice grain was more sensitive to changes in water management (Figure 3b). Treatments FCap and 2WAH showed values 35.40 mg kg^{-1} and 33.70 mg kg^{-1} respectively, representing 18% and 11% significantly higher than Control.

The concentration of Cu in grain clearly rose due the establishment of aerobic condition (Figure 3c). Treatments FCap and 2WAH showed higher concentrations of 4.82 and 4.11 mg kg^{-1} respectively without significant difference between them. Both treatments almost doubled their concentration in comparison with Control, 4WAH and 3WAH which were not significantly different. The trend observed for Mn grain concentration was not as we expected for the different water treatments. As soil anaerobic conditions were more prolonged, grain Mn concentration decreased. Values varied from 31.52 mg kg^{-1} to 55.68 mg kg^{-1}, treatments 2WAH and FCap rose their Mn concentration by 65-76% respectively higher than Control (Figure 3d).

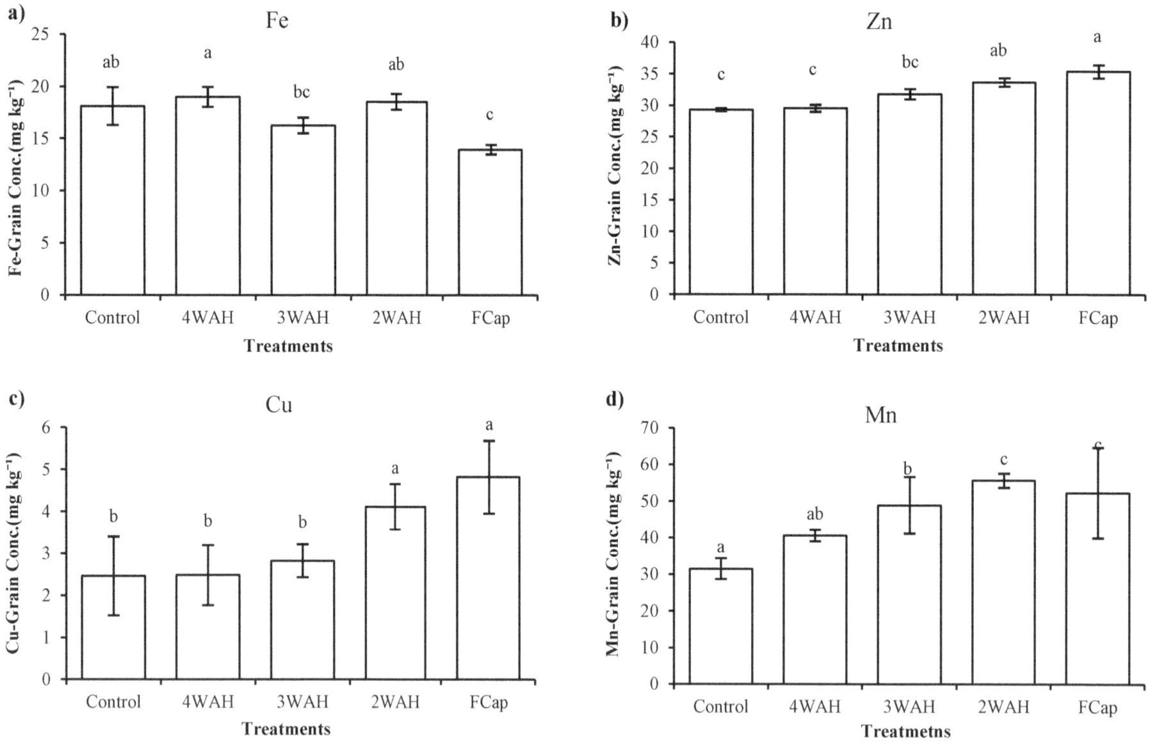

Figure 3. Concentration of micronutrients (Fe, Zn, Cu and Mn) in rice grain, influenced by water management

The effects of soil drainage on grain micronutrient concentration during early grain filling stage, around 75-80 DAT, was beneficial particularly from treatment 2WAH where the period of flooding was the shortest. That period was identified as high demand for micronutrient acquisition and accumulation. Despite of changes in soil properties, the transition from flooded to aerobic conditions increased grain yield and root density by 7% and 7.3 % respectively compared to the control (Table 3). Although aboveground plant biomass (leaf and stem weight) showed higher values for drained treatments , we observed that the increase was due the emergence of some new shoots after of the aerobic conditions were established. These tillers did not produce panicle at the time of harvest.

Table 3. Comparison of dry matter weight among different water managements. Values represent means of dry matter weight, on each part of the rice plant

	Control	4WAH	3WAH	2WAH	FCap.
Grain yield (g plant^{-1})	30.1^{b}	30.5^{ab}	31.3^{ab}	32.2^{a}	15.7^{c}
Leaf & Stems weight (g plant^{-1})	101.0^{b}	108.6^{ab}	115.9^{a}	119.4^{a}	72.0^{c}
Roots weight (g plant^{-1})	13.7^{a}	12.9^{ab}	13.9^{a}	14.7^{a}	12.5^{b}

Note. * Means followed by different lowercase letter represent the significant difference at $P < 0.05$ among treatments.

3.3 Soil Micronutrient Availability and Water Management

Soil micronutrient availabilities were clearly influenced by water managements. The Fe availability (Figure 4a) significantly decreased from Control to FCap. The decreasing change was more evident after the re-oxidation of treatments 4, 3, 2 WAH where the concentration was reduced by more than 20%. Treatments 3WAH, 2WAH and FCap showed a slight increase in Zn availability by 6, 11and 12% over Control, respectively, although differences were not statistically significant (Figure 4b).

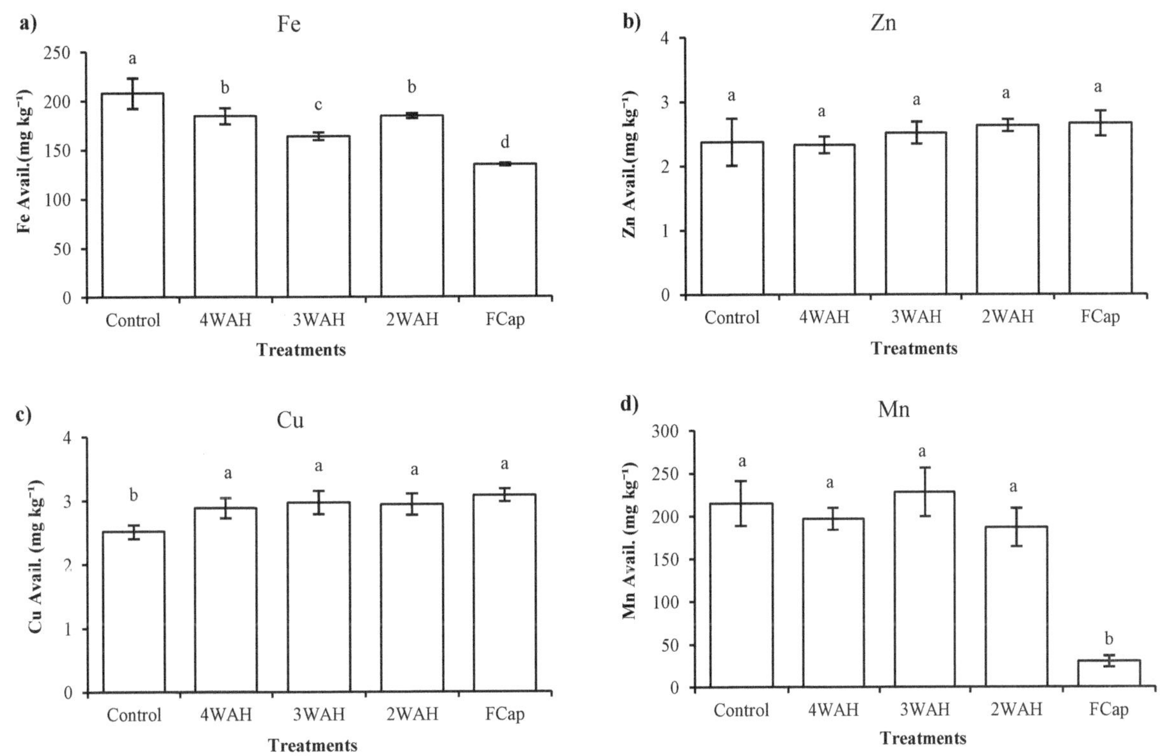

Figure 4. Micronutrients availability (Fe, Zn, Cu and Mn) in soil

Cu availability increased sharply as aerobic condition were established. Figure 4c shows how Cu availability increase from flooded condition to 4WAH in a short time. Cu availability reached its peak values as result of decreasing pH which is maintained across subsequent treatments. Availability of Cu did not show significant difference between the drained (4, 3, 2WAH) and FCap treatments. FCap and 2WAH were >20% significantly higher than Control. Likewise, Mn availability (Figure 4d) remained high across all the treatments regardless of flooding schedule. These treatments did not showed significant differences among them. Availability of Mn range from 187 mg kg^{-1} to 228 mg kg^{-1}. Control was six fold higher in available Mn than FCap.

4. Discussion

At first, for better understanding of the results of present study, we simply summarize the effect of Soil Eh and pH change with water management on micronutrient availability in soil. Soil flooding conditions caused modifications in soil pH and Eh. After its establishment, soil pores were water saturated and Eh started to decrease progressively (Figure 2a). The depletion of O_2, triggered new processes driven mainly by strict facultative anaerobic microbes that thrive under these environments. They take over further biochemical reactions, using substances other than O_2 as the terminal electron acceptor to metabolize organic compounds (Berkowitz, 2014). In the present study, the pH increase under flooding condition (Figure 2b), could be led by Eh reduction and the hydrogen consumption, during the reduction process of MnO_2 and $Fe(OH)_3$ and other transition elements demanded in microbes metabolism (Brady & Weil, 2014). The transition from flooding to aerobic condition during grain filling stage, reduced pH suggesting that establishment of aerobic condition could have affected the Eh and therefore pH could be progressively restored toward its initial value, as soil is re-oxidized (Zhou et al., 2014). Redox reactions (i.e. gain and losses of electrons) can define both the soil pH and the availability of micronutrients (White & Broadley, 2009).

4.1 Fe Dynamics

In our result, Fe availability (Figure 4a) was over the critical level in soil (5 mg kg^{-1}) for deficiency and below 300 mg kg^{-1} in leaf for toxicity (Dobermann & Fairhurst, 2000). The availability of Fe in soil was significantly higher in flooded than drained treatments, which implies that Fe availability in soil depends on soil Eh. Therefore water management becomes the major controlling factor to consider in promoting Fe plant uptake. Results also showed that even when soil became aerobic, Fe availability decreased at 2, 3 and 4WAH, but the grain Fe concentration was kept comparable in all the treatments except FCap. Moreover, transition from flooding to aerobic and re-oxidation of Fe^{2+} to less soluble form at 2WAH and the other drained treatments, did not affect substantially the concentration of Fe in grain. Remobilization of Fe from tissues to the grain may explain this phenomena. Studies conducted on rice by Sperotto et al. (2013) suggested that leaves were the main source to supply Fe to the seeds by remobilization from old tissues. Besides, release of organic acids (phytosiderophores) in rhizosphere might also contribute to Fe acquisition under aerobic condition and deficiency stress (Nozoye et al., 2013). Rice plant, during the tillering and reproductive stages is able to accumulate Fe as ferritin a "Fe-reserve protein" (Da Silveira et al., 2009). This Fe is remobilized via phloem by metal-transporters such as OsIRT1, OsIRT2, OsYSL15, and OsNRAMP5 which enhance their activity during grain filling stage (Bashir et al., 2013). In addition, the enhancement of Fe availability in soil, inhibited the availability of Zn and Cu in soil (Dobermann & Fairhurst, 2000) and the concentration of Zn, Cu and Mn in grain as showed correlation matrix (Table 4) therefore the re-oxidation of Fe to less soluble form from 2WAH, might facilitate the uptake of such micronutrients and improve the concentration in grain

4.2 Zn Dynamics

Regarding available Zn values, in soil were above the critical levels (0.8 mg kg^{-1}) for deficiency (Dobermann & Fairhurst, 2000). The concentration of Zn in grain as well as the concentration in soil was the highest in FCap, and also tended to increase by transition from Control, to 2, 3 ,4WAH (Figures 3b and 4b). These enhancements in grain positively correlated ($P < 0.01$) with the Zn availability in soil (see Table 4). Although Zn availability in soil might be affected by several factor (Sadeghzadeh, 2013) apparently, the reestablishment of aerobic condition, pH declining and reduction of Fe to less available form were the major factor controlling Zn availability. The timing in which irrigation was stopped, also played an important role for Zn acquisition. Unlike FCap, which achieved high grain Zn but low yield, 2WAH showed the best performance in terms of Zn grain concentration and grain yield (Table 3). The period in which soil was re-oxidized coincided with grain filling stage, which has been identified as high demand period not only for the uptake and remobilization of Zn but also other essential elements (Pearson & Rengel, 1994; Bashir et al., 2013) Many researchers found that Zn is translocated from old tissues to flag leaf and afterwards remobilized to the grain (Rengel et al., 1999; Wu et al., 2010). However, the higher percentage of Zn moved into the grain is taken up directly by root, during post-heading stage (Jiang et al., 2007; Sperotto et al., 2013). Although the availability in soil did not show significant differences with control, 2WAH responded more positively than other treatments. In comparison with 2WAH, treatments 3 and 4WAH did not increase their Zn grain concentration even when aerobic condition could have facilitated availability of Zn. Under flooding condition, Zn can be precipitated as ZnS and $Zn(OH)_2$ or reduced its transport and root uptake by HCO_3 which is the predominant anion at such condition (Dobermann & Fairhurst, 2000). The small enhancement in root density (Table 3) observed in 2WAH, also could be another positive factor that facilitated a larger soil contact area (Yang et al., 2004) and therefore, increase Zn uptake by plant.

4.3 Cu Dynamics

Available Cu in soil is held mainly as a free cupric ion (Cu^{2+}) which is the preferred form for plant (Hoang et al., 2008). In our results the Cu behavior in soil and grain was broadly similar with the trend observed in Zn. Results showed that the water condition at this period in which soil was re-oxidized at early grain filling stage, i.e. two weeks after heading, was determining factor controling the Cu contents in grain. Figures 3c and 4c, showed a close relationship between soil Cu availability and grain concentration. Furthermore, the shifting from anaerobic to aerobic condition promote enhancement of Cu solubility and bioavailability as well as better root uptake (Liao et al., 2013). On the other hand the lower Cu availability observed in treatments where anaerobic condition was kept longer, could be ascribed to the formation of Cu-ferrite ($Cu_2Fe_2O_4$), organic matter bound and formation of insoluble CuS (Dobermann & Fairhurst, 2000). Soil Zn availability did not significantly increase when the aerobic conditions were established, but Cu did (Figures 4b and 4c). This suggested that Cu is able to increase in availability easier than did Zn. Garnett et al. (2005) reported that more than 60% of Cu in grain was absorbed at filling stage, which explains the importance of timing in water management to increase grain Cu concentration. Enhancement of Cu availability is advisable in soils where Cu do not exceed toxic level for plant and humans (Xu et al., 2006).

4.4 Mn Dynamics

Despite of establishment of aerobic condition, scheduled at 2, 3 ,4 WAH, soil Mn availability remained high (Figure 4d) without significance changes compared with control. Since Mn is reduced at higher potential (+400~ -100mV) than Fe (+200~ -200 mV) (Kyuma, 1997), the Mn availability in soil can be kept stable longer than Fe in drained treatments of which the moisture condition changed. Although Fe and Mn take active parts in redox reactions in soil and both are able to increase their availability after soil flooding, presumably, high Fe availability observed in soil, adversely inhibited Mn uptake by roots, because of the antagonist interaction in soil (Dobermann & Fairhurst, 2000; Millaleo et al., 2010; Das, 2014). Alam (1985) reported that a concentration of Mn in leaves and stems of 7 weeks plants were increased when Fe application ranged 0-30 mg kg^{-1} in soil, but above 30 mg kg^{-1} application reduced Mn concentration in plant. Other studies conducted by Pearson and Rengel, (1994) concluded that Mn concentration in flag leaf and other active leaves increased during grain filling stage, but remobilization to the grain was not carried out, regardless of the leaf concentration. Suggesting that, continuous uptake from soil seems to be the main source of grain Mn. However its behavior still remains unclear, therefore further investigations are needed to elucidate this issue.

4.5 Relationship among Soil, Plant and Grain Micronutrients Composition.

The effect of water management on soil micronutrients availability and its relationship with the concentration in grain and flag leaf are shown in Table 4. The availability of Fe, Zn and Cu in soil significantly correlated with their respective concentrations grain. Soil pH, redox reactions, plant physilogy (Growth stage) and rhyzosphere activity are the main factors controlling the availability in soil and concentration of micronutrients in grain (Dobermann & Fairhurst, 2000; Gao et al., 2012; Marchner, 2012) mediated by the transition from flooding to aerobic condition. Mn detected in soil did not correlate with Mn concentration in flag leaf and showed a negative correlation with grain Mn ($p < 0.05$). Fe availability in soil correlated negatively with the concentration of Zn and Cu in plant and grain ($p < 0.01$). This observation suggested that availability of Fe in soil could be attributed to antagonistic interaction, namely inhibition of Zn, Cu and Mn uptake (Dobermann & Fairhurst, 2000) Negative effects of Fe excess on Cu and Mn shoots concentration have been observed in *Indica* varieties (Yoshihara et al., 2010). Concentration of Fe, Zn, Cu and Mn in flag leaf correlated positively with their respective concentration in grain. Moreover, concentration of Zn, and Cu in grain and flag leaf correlated significantly ($p < 0.01$) among each other. Studies conducted by Ishimaru et al. (2005) revealed that the expression of ZIP′s gene family are actively related with the Zn and Cu uptake and transport system. The same mechanism may control the behavior of Zn and Cu in plant, which brought the similar results of Zn and Cu in the present study.

Table 4. Correlation matrix between soil micronutrients availability and micronutrient concentration in flag leaf and grain

		Soil Micronutrient				Flag leaf Micronutrient				Grain		
		Fe	Zn	Cu	Mn	Fe	Zn	Cu	Mn	Fe	Zn	Cu
Soil	Zn	-0.33*										
	Cu	-0.77**	0.43**									
	Mn	0.71**	-0.47**	-0.47**								
Flag leaf	Fe	0.20	0.32	0.13	0.06							
	Zn	-0.68**	0.41	0.38	-0.88**	-0.14						
	Cu	-0.74**	0.47*	0.61**	-0.81**	0.20	0.77**					
	Mn	-0.43	0.55*	0.59**	-0.41	0.68**	0.31	0.72**				
Grain	Fe	0.63**	-0.38*	-0.41**	0.55**	0.31	-0.69**	-0.61**	-0.27			
	Zn	-0.61**	0.39*	0.46**	-0.58**	0.31	0.73**	0.86**	0.76**	0.25		
	Cu	-0.57**	0.45**	0.51**	-0.66**	0.34	0.72**	0.93**	0.80**	0.24	0.87**	
	Mn	-0.53**	0.48**	0.59**	-0.34*	0.46*	0.28	0.64**	0.85**	0.24	0.72**	0.79**

Note. **and* represent significant correlation at $P < 0.01$ and $P < 0.05$ respectively.

5. Conclusions

Soil water management changes the soil Eh and pH, which control the availability of micronutrient in soil and consequently the concentration in rice grain. When those changes occur during grain filling stage, micronutrients such as Zn, Cu and Mn increase their concentration in the grain, likely due the reduction of pH and their interaction with other soil properties. Although water management did not exerted significant changes on soil Fe availability, its re-oxidation to less available form, promoted the increase of Zn, Cu and Mn in grain.

Although researchers have been focused to increase the concentration of micronutrient in grain through breeding strategies, if soil management in relation to water condition is not applied, those strategies inevitably will fail. Therefore establishment of aerobic condition starting from 2 weeks post-anthesis till maturity by proper water management such as intermittent irrigation, seems to be the most feasible strategy to increase the concentration of micronutrients in rice grain. Furthermore, this technique did not result in grain yield losses.

Our results exhibited the importance of soil environmental managements in realization of the biofortification program. Plant breeding strategies must work alongside agronomic biofortification techniques. In this particular case of rice, the combination between water management and plant breeding could be the most viable option in order to ensure the supply of micronutrients from the soil and overcome malnutrition in vulnerable populations.

Acknowledgements

This study was supported by Japan Society for the Promotion of Science (Grant-in-Aid Scientific Research No. 24405047). The author is highly grateful to the members of the Pedosphere Eco-Engineering Laboratory, Faculty of Life and Environmental Science, Shimane University.

References

Alam, S. M. (1985). Effects of iron and manganese on the growth of rice and on the contents of these elements in rice plants. *Agronomie.* Retrieved January 7, 2015, from https://hal.archives-ouvertes.fr/hal-00884775/document

Alloway, B. J. (2008). *Micronutrient deficiencies in global crop Production.* Springer Sci., Dordrecht.

Bain, L. E., Awah, P. K., Geraldine, N., Kindong, N. P., Sigal, Y., Bernard, N., & Tanjeko, A. T. (2013). Malnutrition in Sub – Saharan Africa: Burden, causes and prospects. *The Pan African Medical Journal, 15*, 120. http://dx.doi.org/10.11604/pamj.2013.15.120.2535

Bashir, K., Takahashi, R., Nakanishi, H., & Nishizawa, N. K. (2013). The road to micronutrient biofortification of rice: progress and prospects. *Frontier Plant Science, 4*, 1-7. http://dx.doi.org/10.3389/fpls.2013.00015

Bell, R.W., & Dell, B. (2008). *Micronutrients for Sustainable Food, Feed, Fibre and Bioenergy production.* International Fertilizer Industry Association IFA, Paris, France.

Berkowitz, B. (2014). Biologicaly and mediated transformation. In B. Berkowitz, I. Dror, & B. Yaron (Eds.), Contaminant geochemistry. *Interactions and transport in the Subsurface environment* (2nd ed., pp. 381-395). Springer.

Brady, N. C., & Weil, R. R. (2014). *The Nature and Properties of Soil* (14th ed.). Pearson Education Limited, USA.

Cakmak, I. (2008). Enrichment of cereal grains with zinc: Agronomic or genetic biofortification? *Plant Soil, 302*, 1-17. http://dx.doi.org/10.1007/s11104-007-9466-3

Da Silveira, V. C., Fadanelli, C., Sperotto, R. A., Stein, R. J., Basso, L. A., Santos, D. S., ... Fett, J. P. (2009). Role of Ferritin in the Rice Tolerance to Iron Overload. *Scientia Agricola, 66*, 549-555. http://dx.doi.org/10.1590/S0103-90162009000400019

Das, S. K. (2014) Role of Micronutrient in Rice Cultivation and Management Strategy in Organic Agriculture – A Reappraisal. *Agricultural Sciences, 5*, 765-769. http://dx.doi.org/10.4236/as.2014.59080

Dobermann, A., & Fairhurst, T. (2000). *Rice: Nutrient disorders and nutrient management.* Singapore and Los Baños: Potash & Phosphate Institute (PPI), Potash & Phosphate Institute of Canada (PPIC), and International Rice Research Institute (IRRI), Philipine.

Fageria, N. K. (2002). Influence of micronutrients on dry matter yield and interaction with other nutrients in annual crops. *Pesquisa. Agropecuaria Brasileira, 37*, 1765-1772 http://dx.doi.org/10.1590/S0100-204X2002001200013

Food and Agriculture Organization (FAO). (2013). *In feeding the world.* FAO Statistical Yearbook. Retrieved January 7, 2015, from http://www.fao.org/docrep/018/i3107e/i3107e.PDF

Gao, X., Hoffland, E., Stomph, T., Grant, C. A., Zou, C., & Zhang, F. (2012). Improving zinc bioavailability in transition from flooded to aerobic rice: A review. *Agronomy for Sustainable Development, 32*, 465-478. http://dx. doi.org/10.1007/s13593-011-0053-x

Garnett T. P., & Graham R. D. (2005). Distribution and Remobilization of Iron and Copper in Wheat. *Annals of Botany, 95*, 817-826. http://dx.doi.org/10.1093/aob/mci085

Hoang, T. C., Rogevich, E. C., Rand, G. M., Gardinali, P. R., Frakes, R. A., & Bargar, T. A. (2008). Copper desorption in flooded agricultural soils and toxicity to the Florida apple snail *(Pomacea paludosa)*: Implications in Everglades restoration. *Environmental Pollution, 199*, 338-347. http://dx.doi.org/10.1016/j.envpol.2007.09.024

Inoue, H., Kobayashi, T., Nozoye, T., Takahashi, M., Kakei, Y., Suzuki K., ... Nishizawa, N. K. (2009). Rice OsYSL15 is an iron-regulated iron(III)-deoxymugineic acid transporter expressed in the roots and is essential for iron uptake in early growth of the seedlings. *Journal of Biological Chemistry, 284*, 3470-3479. http://dx.doi.org/10.1074/jbc.M806042200

Ishimaru, Y., Suzuki, M., Kobayashi, T., Takahashi, M., Nakanishi, H., Mori, S., & Nishizawa, N. K. (2005). OsZIP4, a novel zinc-regulated zinc transporter in rice. *Journal of Experimental Botany, 56*, 3207-3214. http://dx.doi.org/10.1093/jxb/eri317

Japan Soil Standard and Measurement Methods Committee. (2003). *Japan Soil Standard and Measurement Methods* (6th ed.). Hakuyusha, Tokyo, Japan.

Jiang, W., Struik, P. C., Lingna, J., van Keulen, H., Zhao, M., & Stomph, T. J. (2007). Uptake and distribution of root-applied or foliar applied ^{65}Zn after flowering in aerobic rice. *Annals of Applied Biology, 150*, 383-391. http://dx.doi.org/10.1111/j.1744-7348.2007.00138.x

Koyama, T., & Sutoh, M. (1987). Simultaneous multi element determination of soils, plant and animal samples by inductively coupled plasma emission spectrophotometry. *Soil Science & Plant Nutrition, 58*, 578-585.

Kyuma, K. (1997). *Newest Soil Science*. Asakura, Tokyo Japan.

Liao, L., Xu, J., Peng, S., Qiao, Z., & Gao, X. (2013). *Uptake and Bioaccumulation of Heavy Metals in Rice Plants as Affect by Water Saving Irrigation*. Retrieved January 7, 2015, from http://maxwellsci.com/print/ajfst/v5-1244-1248.pdf

Lindsay, W. L., & Norvell, W. A. (1978). Development of DTPA soil test for zinc, iron, manganese and copper. *Journal of America Soil Science Society, 42*, 421-428.

Marschner, P. (2012). *Mineral Nutrition of Higher Plants* (3rd ed.). Academic Press, CA., USA.

Marschner, P., & Rengel, Z. (Eds.). (2007). *Nutrient Cycling in Terrestrial Ecosystems*. Springer Science & Business Media. Berlin, Heidelberg.

Meenakshi, J., Johnson, N. L., Manyong, V. M., DeGroote, H., Javelosa, J., Yanggen, D. R., ... Meng, E. (2007). How Cost-Effective is Biofortification in Combating Micronutrient Malnutrition? An *Ex ante* Assessment. *World Development, 38*, 64-75. http://dx.doi.org/10.1016/j.worlddev.2009.03.014

Millaleo, R., Díaz, M. R., Ivanov, A. G., Mora, M. L., & Alberdi, M. (2010). Mn as essential and toxic element for plants, accumulation and resistance mechanisms. *Journal of Soil Science Plant Nutrition, 10*, 476-494 http://dx.doi.org/ 10.4067/S0718-95162010000200008

Najafi-Ghiri, M., Ghasemi-Fasaei, R., & Farrokhnejad, E. (2013). Factors Affecting Micronutrient Availability in Calcareous Soils of Southern Iran. *Arid Land Research and Management, 27*, 203-215. http://dx.doi.org/10.1080/15324982.2012.719570

Nozoye, T., Nagasaka, S., Kobayashi, T., Takahashi, M., Sato, Y., Sato, Y., ... Nishizawa, N. K. (2010). Phytosiderophore Efflux Transporters Are Crucial for Iron Acquisition in Graminaceous Plants. *Journal of Biological Chemistry, 286*, 5446 - 5454. http://dx.doi.org/10.1074/jbc.M110.180026

Pearson, J. N., & Rengel, Z. (1994). Distribution and remobilization of Zn and Mn during grain development in wheat. *Journal of Experimental Botany*. Retrieved December 15, 2014, from http://jxb.boxfordjournals.org/content/45/12/1829.full.pdf

Rengel, Z., Batten, G. D., & Crowley, D. E. (1999). Agronomic approaches for improving the micronutrient density in edible portions of field crops. *Field Crops Research, 60*, 27-40. http://dx.doi.org/10.1016/S0378-4290(98)00131-2

Sadeghzadeh, B. (2013). A review of zinc nutrition and plant breeding. *Journal Soil & Science Plant Nutrition, 13*, 905-927. http://dx.doi.org/10.4067/S0718-95162013005000072

Sperotto, R. A., Ricachenevsky, F. K., Waldow, V. A., Müller, L. H., Dressler, V. L., & Fett, J. P. (2013). Rice grain Fe, Mn and Zn accumulation: How important are flag leaves and seed number? *Plant Soil Environment, 59*, 262-266. Retrieved February 28, 2015, from http://www.agriculturejournals.cz/publicFiles/98742.pdf

Tao, H., Dittert, K., Zhang, L., Lin, S., Römheld, V., & Sattelmacher, B. (2007). Effects of soil water content on growth, tillering, and manganese uptake of lowland rice grown in the water-saving ground-cover rice-production system (GCRPS). *Journal of Plant Nutrition & Soil Science, 170*, 7-13. http://dx.doi.org/10.1002/jpln.200625033

United States Department of Agriculture (USDA). (2010). *Keys To Soil Taxonomy* (11th ed.). United States Departament of Agriculture. Retrieved February 28, 2015, from http://www.nrcs.usda.gov/Internet/FSE_DOCUMENTS/nrcs142p2_050915.pdf

White, P. J., & Broadley, M. R. (2009). Biofortification of crops with seven mineral elements often lacking in human diets – iron, zinc, copper, calcium, magnesium, selenium and iodine. *Review. New Phytologist., 182*, 49-84. http://dx.doi.org/10.1111/j.1469-8137.2008.02738.x

Wu, C., Lu, L., Yang, X., Feng, Y., Wei, Y., Hao, H. L., ... He, Z. (2010). Uptake, translocation, and remobilization of zinc absorbed at different growth stages by rice genotypes of different Zn densities. *Journal of Agriculture & Food Chemistry, 58*, 6767-6773. http://dx.doi.org/10.1021/jf100017e

Xu, J., Yang, L., Wang, Z., Dong, G., Huang, J., & Wang Y. (2006). Toxicity of copper on rice growth and accumulation of copper in rice grain in copper contaminated soil. *Chemosphere, 62*, 602-607. http://dx.doi.org/1.1016/j.chemosphere.2005.05.050

Yang, C., M., Yang, L., Yang, Y., & Ouyang, Z. (2004). Rice root growth and nutrient uptake as influenced by organic manure in continuously and alternately flooded paddy soils. *Agricultural Water Management, 70*, 67-81. http://dx.doi.org/10.1016/j.agwat.2004.05.003

Yang, X. E., Chen, W. R., & Feng, Y. (2007). Improving human micronutrient nutrition through biofortification in the soil–plant system: China as a case study. *Environmental Geochemistry & Health, 29*, 413-428. http://dx.doi.org/10.1007/s10653-007-9086-0

Yoshihara, T., Shoji, K., Kohno, Y., & Goto, F. (2010). Cross relationships of Cu, Fe, Zn, Mn, and Cd accumulations in common japonica and *Indica* rice cultivars in Japan. *Environmental and Experimental Botany, 68*, 180-187. http://dx.doi.org/10.1016/j.envexpbot.2009.10.006

Zhou, W., Lv, T. F., Chen, Y., Westby, A. P., & Ren, W. J. (2014). Soil Physicochemical and Biological Properties of Paddy-Upland Rotation: A Review. *The Scientific World Journal*. http://dx.doi.org/10.1155/2014/856352

18

Mapping of QTL for Color of Noodle Dough Sheet in a Doubled Haploid Population

H. Kang[1,2], C. S. Kang[2], H. Kim[2] & C. S. Park[1,3]

[1] Department of Crop Science and Biotechnology Chonbuk National University, Jeonju, Korea

[2] National Institute of Crop Science, RDA, Suwon, Korea

[3] Institute of Agricultural Sciences and Technology, Chonbuk National University, Jeonju, Korea

Correspondence: C. S. Park, Department of Crop Science and Biotechnology Chonbuk National University, Jeonju 561-756, Korea. E-mail: pcs89@jbnu.ac.kr

Abstract

This study was conducted for identification of the quantitative trait loci (QTL) associated with the color of flour and noodle dough sheets for the production of white salted noodles by analysis of a doubled haploid population from a cross between Korean wheat cultivars carrying different color values for the flour and noodle dough sheets. The DH population was evaluated in replicated field trials in 2011 and 2012 in upland conditions in Iksan, Korea. The color differences were measured as CIE-LAB *L*, *a* and *b* values. *Xgwm190* located on chromosome 5D explained 13.3%–22.8% of the phenotypic variation in the color of flour over the average of two years. *Xgwm190* also explained 11.8% and 24.1% of the phenotypic variation for the *L* and *b* values of the noodle dough sheets for the average of two years, respectively. *Xbarc81* and *Xgwm133* accounted for 7.9% and 9.8% of the phenotypic variation in the *b* value of flour produced in 2012 and the noodle dough sheets from flour produced in 2011, respectively. The putative marker for flour and noodle dough sheet color, *Xgwm190*, was applied to validate the relationship between genotypes of 24 Korean wheat cultivars. Fourteen of the Korean wheat cultivars contained genotype *a*, which represented homozygous to Keumkang at *Xgwm190*, showed lower *L* values of the flour and noodle dough sheets (91.6 and 83.5, respectively) than genotype *b* (93.6 and 85.1, respectively), which represented homozygous to Olgeuru at *Xgwm190*. *Xgwm190* could be used for the selection of wheat lines with bright color in Korean wheat breeding programs.

Keywords: wheat, quantitative trait loci (QTL), color, flour, noodle dough sheet

1. Introduction

The color of flour is a main consideration of determining the acceptability in noodles to consumers, albeit noodles are produced in wide ranges of color. Noodle color is mainly affected by the color of flour due to the simple formula and processing compared to bread-baking (Huo, 2001; Fu, 2008). There are two distinct categories of noodles based on the type of salt (Fu, 2008). White salted noodles, which are made from dough with flour, water and salt content based on the flour weight, are widely consumed in Korean and Japan (Fu, 2008). Bright color, ranging from white to creamy white, is desirable for white salted noodles, whereas a bright yellow color is preferable for yellow alkaline noodles (Huo, 2001; Fu, 2008).

Flour color includes complex attributes, such as pigments of starchy endosperm, pericarp and pigments in the aleurone of wheat grain, as well as the flour extraction ratio, particle size, protein and ash content, and microbial contamination (Miskelly, 1984; Mares & Campbell, 2001). The brightness of flour decreases with increase of the particle size as well as the protein, ash, damaged starch and tailing starch content, which causes dull color in white salted noodles (Huo, 2001; Fu, 2008). Slow discoloration is also important because the noodles are often stored raw for several days prior to cooking, during which discoloration can occur due to the effects of polyphenol oxidase, which is generally derived from the seed coat (Hatcher & Kruger, 1993; Fuerst et al., 2006). Therefore, execution of the correct milling procedures and inclusion of an appropriate range of protein content is important to maintain the brightness of white salted noodles (Huo, 2001; Fu, 2008). The factors related to the color of flour and the resulting noodles are genetically quantitative, often influenced by the environment and exhibit high genotype × environment interaction (Mares & Campbell, 2001; Davies & Berzonsky, 2003; Graybosch et al., 2004; Zhao et al., 2013).

Major QTL associated with the color of flour have been found on all chromosome arms using recombinant inbred lines, DH populations and the genetic resources of common wheat (*Triticum aestivum* L.) (Mares & Campbell, 2001; Kuchel et al., 2006; Zhang et al., 2009a, 2009b; Sadeque & Turner, 2010a; Bordes et al., 2011; Li et al., 2011; Tsilo et al., 2011; Roncallo et al., 2012; Zhao et al., 2013). The QTL for yellow pigments, polyphenol oxidase and carotenoid genes were also found to be associated with the flour and noodle color (Marles & Campbell, 2001; Raman et al., 2007; He et al., 2008; Zhang et al., 2009b; Sadeque & Turner, 2010b; Crawford et al., 2011; Martin et al., 2011; Crawford & Francki, 2013a, 2013b).

Korean wheat breeding programs have recently focused on the quality improvement of white salted noodles, and over 30 cultivars have been developed since the 1970s. However, few cultivars have been used in the production of white salted noodles because the consumers prefer cultivars which provide bright creamy color and soft texture of the cooked noodles. The objective of this study was to perform QTL analysis for the color of flour and noodle dough sheets using doubled haploid population crosses Keumkang and Olgeuru to gain an understanding of the genetic basis for improvement of the color in Korean wheat breeding programs. Keumkang, developed in 1997, is the most commonly grown cultivar in Korea, possessing good milling quality and medium dough strength, but its color is a little dull (Nam et al., 1998). Olgeuru is a soft red cultivar used for making white salted noodles, and has a brighter color than Keumkang (Nam et al., 1995).

2. Materials and Methods

2.1 Plant Materials

A population of 114 doubled haploid lines was derived from F_1 plants of Keumkang and Olgeuru crosses using the wheat × maize system according to the procedures of Inagaki and Mujeeb-Kazi (1995) at International Maize and Wheat Improvement Center (CIMMYT, El Batan, Mexico). The doubled haploid lines were grown at the Crop Experimental Farm of the National Institute of Crop Science, RDA (Iksan, Korea), for seed multiplication in 2009. Field trials were performed in randomized complete blocks with 3 replicates in upland conditions from 2011 to 2012 in Iksan, Korea. Twenty-four Korean wheat cultivars were also cultivated under the same conditions as the DH population in 2013. The seeds were sown on October 25, and each plot consisted of three 4 m rows spaced 25 cm apart. The plots were combine-harvested on June 20. Fertilizer was applied at 5: 7: 5 kg/10a (N: P: K) before sowing. Weeds, insects and disease were stringently controlled. No supplemental irrigation was applied. The grains from each plot were dried using forced air driers and bulked from replications to provide grains for color measurement.

2.2 Color of Flour and Noodle Dough Sheets

Wheat was milled with a Bühler experimental mill, according to the AACC International Approved Method 26-31.01. (2010). Two kilograms of wheat were conditioned overnight to reach 15% moisture content and then milled at a feed rate of 100 g/min with roll settings of 8 and 5 in break rolls, and 4 and 2 in reduction rolls. Noodle dough sheets for white salted noodles were prepared with 34% water absorption of the noodle dough according to the description of Park and Baik (2002). Flour (100 g, 14% mb) was mixed with a determined amount of sodium chloride solution in a pin mixer (National Mfg. Co., USA) for 4 min, with a head speed of 86 rpm. The concentration of sodium chloride solution for making noodles was adjusted to 2.0% sodium chloride. Crumbly dough was passed through the rollers of a noodle machine (Ohtake Noodle Machine Mfg. Co., Japan) running at 65 rpm with a 3-mm gap to form a dough sheet, and then the dough sheet was folded and put through the sheeting rollers. The folding and sheeting were repeated twice. The dough sheet was rested for 1 hr and then put through the sheeting rollers three times at progressively decreasing gaps of 2.40, 1.85 and 1.30 mm. After the last sheeting, the dough sheet was cut into 5 × 10 cm portions for measurement of the color. The flour and noodle dough sheet color was measured by colorimeter (CM-2002, Minolta Camera, Osaka, Japan) with an 11-mm measurement aperture. The color differences were recorded as CIE values, including *L* (lightness), *a* (redness-greenness), and *b* (yellowness-blueness).

2.3 PCR Conditions

Leaf tissue was collected from a single plant of each doubled haploid line after germination for two weeks, snap-frozen in liquid nitrogen and stored at -80 °C until needed. Genomic DNA was extracted from the young leaf tissue (100 mg) using a genomic DNA prep kit (Solgent Co., Korea), according to the manufacturer's instructions. DNA was quantified by Biodrop (Biodrop Ltd, UK). A total of 592 simple sequence repeats (SSRs) were used to survey the polymorphisms between the parents, and 140 polymorphic markers were used to genotype the DH populations. The PCR reactions were performed in a total volume of 15 μL containing 100 ng genomic DNA, 1X PCR buffer, 1.5 mM $MgCl_2$, 0.2 mM dNTPs, 1 unit *Taq* DNA polymerase, and 0.2 ug/ul each primer. After an initial denaturation at 94 °C for 5min, 10 cycles (decrease of 1 °C per cycle) were performed for 1 min at 94 °C, 1 min at 63 °C, and 1 min at 72 °C, followed by 35 cycles of 1 min at 94 °C, 1 min at 55 °C, and 1 min at 72 °C with a final extension at 72 °C for 10 min. The PCR products were analyzed with the QIAxcel Advanced system using a 12-capillary QIAxcel DNA high resolution cartridge (Qiagen Co., USA).

2.4 Statistical Analysis

Statistical analysis of the data was performed by SAS software (SAS Institute, NC, USA), utilizing the Fisher's least significant difference test (LSD), analysis of variance (ANOVA) and pair-wise t-test. Analysis of variance was conducted using the general linear model (GLM) procedure, with the genotype × year component as the error term. The GLM procedure was used for the randomized complete block (RCB) analysis. Sources of variation in the model were considered to be fixed effects. PROC GLM was used to estimate the relative contribution of the loci tested for the measured traits of the 114 DH lines. The mean of each marker allele type was compared by one-way ANOVA, and differences were considered to be significant at $P < 0.05$, unless otherwise specified. The pair-wise t test at $P = 0.05$ was conducted to compare means when F tests were significant. Pearson correlations between pairs of traits were determined from the least square means of DH lines separately at each year and over all environments, using the Corr procedure. Heritability (h^2) on a progeny mean basis was estimated from RCM analysis using the formula $h^2 = 1 - (M2/M1)$, proposed by Knapp et al. (1985), where $M1$ is the mean square of genotype and $M2$ is the mean square for genotype by environment interaction (G × E) in DH populations grown in different year.

2.5 QTL Analysis

A genetic map of the DH population using 140 SSR markers identified a partial genetic map spanning 2,340.7 cM over all 21 chromosomes. Individual maps for the initial QTL detection were constructed using MapDisto (http://mapdisto.free.fr). Composite interval mapping (CIM) analysis (Zeng, 1993, 1994) was performed using QTL Cartographer v2.5 for Windows (Wang et al., 2012), based on a model 6, Forward and Backward Regression method. The walking speed for all QTL was 2.0 cM, and the probability in stepwise regression was $P < 0.001$. Centimorgan units were calculated using the Kosambi mapping function. Empirical logarithm of the odds (LOD) value thresholds were estimated using 1,000 permutations at an experiment-wise $P < 0.01$ (Churchill & Doerge, 1994). The LOD scores for declaring QTLs were significant at 3.0. The additive value and coefficient of determination (R^2), based on the partial correlation of the putative QTL with the trait adjusted for cofactors in the multilocus model, were estimated based on the most significant dataset.

3. Results

Keumkang showed lower L value and higher a and b values of the flour than Olgeuru, but no significant differences were found in the color of the noodle dough sheet between the parental cultivars, because it was prepared with fixed water absorption in this study (Table 1). The frequency distributions of the color of the flour and noodle dough sheets were continuous, and transgressive segregation was observed in the DH populations based on the overall mean of two years (Figure 1). The year explained for the largest proportion of the variation among flour and noodle dough sheet color, although the color may have been influenced by both genotype and genotype by environment interactions (Table 2). However, the estimated heritability (h^2) of the flour and noodle dough sheet colors for in the DH population was 0.45% – 0.93%. The L value of flour was negatively correlated with the a and b values and a positive correlation was observed between the a and b values of flour (Table 3). The L value of the noodle dough sheet was also significantly correlated with the a and b values. In addition, the L and b values of the flour positively correlated with those of the noodle dough sheet, but no significant relationship was found for the a value between the parental cultivars.

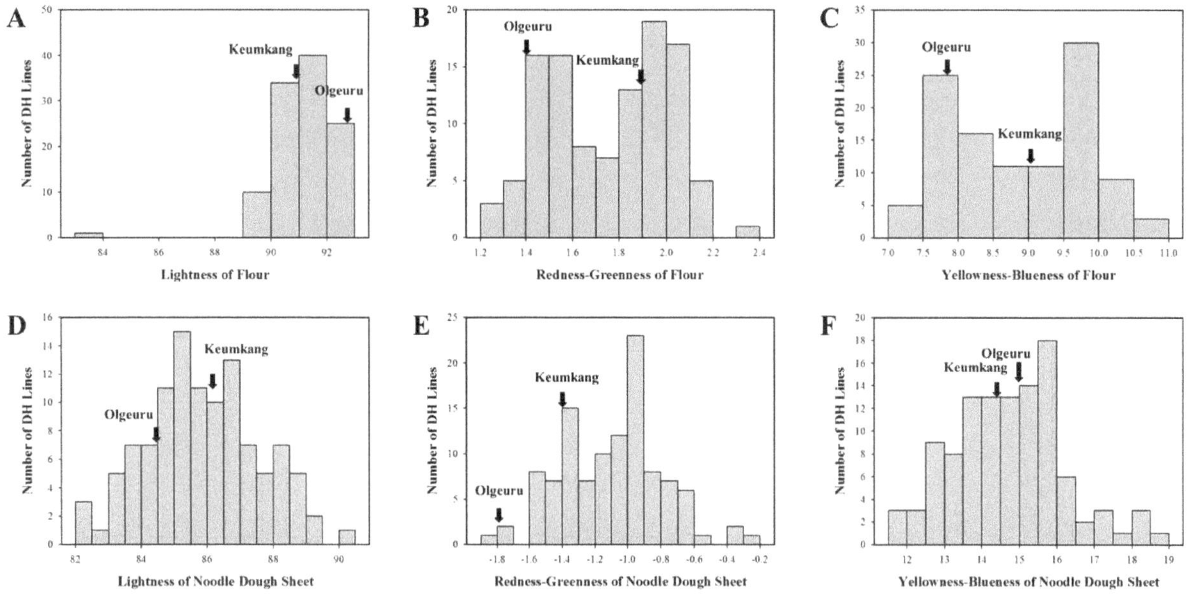

Figure 1. Frequency distributions of the lightness (A and D), redness-greenness (B and E) and yellowness-blueness (C and F) of flour and noodle dough sheets for white salted noodles, prepared from 114 DH populations. Arrows indicate mean values of the parental cultivars

Table 1. Means and ranges of flour and noodle dough sheet color for white salted noodles prepared from 114 DH populations grown two years in upland conditions

Traits*	Parents†		DH population		
	Keumkang	Olgeuru	Mean ± SD	Range	$h^{2‡}$
Flour-*L*	91.0b	92.8a	91.1 ± 1.2	83.0 – 92.7	0.45
Flour-*a*	1.9a	1.4b	1.8 ± 0.3	1.2 – 2.4	0.86
Flour-*b*	9.1a	7.9b	8.9 ± 1.0	7.0 – 11.0	0.93
Noodle-*L*	84.5a	85.8a	85.9 ± 1.7	82.2 – 90.1	0.45
Noodle-*a*	-1.1a	-1.8a	-1.1 ± 0.3	-1.8 – -0.3	0.46
Noodle-*b*	14.4a	14.9a	14.7 ± 1.5	11.7 – 18.8	0.60

Note. **L*, *a* and *b* of flour and noodle indicate the lightness, redness-greenness and yellowness-blueness of flour and noodle dough sheets for white salted noodles, respectively.

†Values followed by the same letters are not significantly different at $P < 0.05$.

‡Heritability (h^2) was estimated from the variance traits of the 114 DH populations.

Table 2. Analysis of variance of flour and noodle dough sheet colors for white salted noodles, prepared from 114 DH populations grown two years in upland conditions

Source of Variation	df	*F*-values					
		CIE value*					
		Flour			Noodle dough sheet		
		L	*a*	*b*	*L*	*a*	*b*
Year (Y)	1	21.92	677.98	4458.74	4445.00	5220.92	4385.68
P		<0.001	<0.001	<0.001	<0.001	<0.001	<0.001
Genotypes (G)	113	7.89	191.17	968.46	56.81	23.01	60.01
P		<0.001	<0.001	<0.001	<0.001	<0.001	<0.001
G × Y	113	4.34	27.29	69.32	31.29	12.52	23.76
P		<0.001	<0.001	<0.001	<0.001	<0.001	<0.001

Note. **L*, *a* and *b* of flour and noodle indicate the lightness, redness-greenness and yellowness-blueness of the flour and noodle dough sheets for white salted noodles, respectively.

Table 3. Pearson correlation coefficients among color of flour and noodle dough sheets for white salted noodles prepared from 114 DH populations grown two years in upland conditions

Parameters*	Flour-*L*	Flour-*a*	Flour-*b*	Noodle-*L*	Noodle-*a*
Flour-*a*	-0.46				
P	<0.001				
Flour-*b*	-0.52	0.84			
P	<0.001	<0.001			
Noodle-*L*	0.31	-0.48	-0.56		
P	<0.01	<0.001	<0.001		
Noddle-*a*	0.01	0.07	-0.13	0.37	
P	NS†	NS	NS	<0.001	
Noodle-*b*	-0.30	0.50	0.70	-0.82	-0.43
P	<0.01	<0.001	<0.001	<0.001	<0.001

Note. **L*, *a* and *b* of flour and noodle indicate the lightness, redness-greenness and yellowness-blueness of the flour and noodle dough sheets, respectively.

†NS = not significant.

Two QTLs, *Xbarc81* and *Xgwm190*, were found to be associated with the flour color, and were mapped to chromosomes 1B and 5D, respectively (Table 4 and Figure 2). *Xgwm190* explained 25.5%, 47.6% and 24.4% of the phenotypic variation (R^2) with LOD scores of 4.6, 9.1 and 9.1 for *L*, *a* and *b*, respectively, for the flour from wheat grown in 2012. *Xgwm190* also explained 22.2%, 13.3% and 22.8% of the R^2 with LOD scores of 5.2, 4.9 and 9.0 for each color value of flour in the average of two years. *Xbarc81* accounted for 7.9% of the R^2 with LOD scores of 3.4 for the *b* value of flour from wheat grown in 2012. *Xgwm190* had negative additive effects on the *L* value of flour, but positive additive effects on the *a* and *b* values. These results indicate that the Keumkang alleles reduced the lightness and increased the redness-greenness and blueness-greenness of the flour from the DH populations, demonstrating decrease of the *L* value of flour by 0.6 and increase of the *a* and *b* values of 0.2 and 0.5, respectively, in the average of two years.

Table 4. QTL for color of flour and noodle dough sheets for white salted noodles using composite interval mapping in 114 DH populations grown two years in upland conditions

Traits*/Marker	Chr.	LOD†	R^2 (%)‡	Additive effect§	Environment
Flour-*L*					
Xgwm190	5D	4.6	25.5	-0.7	2012
Xgwm190	5D	5.2	22.2	-0.6	Mean
Flour-a					
Xgwm190	5D	9.1	47.6	0.2	2012
Xgwm190	5D	4.9	13.3	0.1	Mean
Flour-*b*					
Xbarc81	1B	3.4	7.9	0.3	2012
Xgwm190	5D	9.1	24.4	0.5	2012
Xgwm190	5D	9.0	22.8	0.5	Mean
Noodle-*L*					
Xgwm190	5D	5.3	17.3	-0.8	2011
Xgwm190	5D	4.3	11.8	-0.6	Mean
Noodle-*a*					
Xgwm133	6B	3.0	9.8	0.1	2011
Noodle-*b*					
Xgwm190	5D	6.5	26.6	1.0	2011
Xwmc83	7A	3.4	23.9	-0.9	2012
Xgwm190	5D	4.6	24.1	0.7	Mean

Note. **L*, *a* and *b* of flour and noodle indicate the lightness, redness-greenness and yellowness-blueness of the flour and noodle dough sheets, respectively.

†Logarithm of the odds.

‡The ratio of phenotypic variation (R^2) explained by the overall mean of the dataset. The mean of each QTL was compared by one-way ANOVA at significance levels of $P \leq 0.001$.

§Additive effects of allele substitution. The units are those of the corresponding traits. Positive signs indicate that the Keumkang allele increased the quantitative trait, and vise-versa.

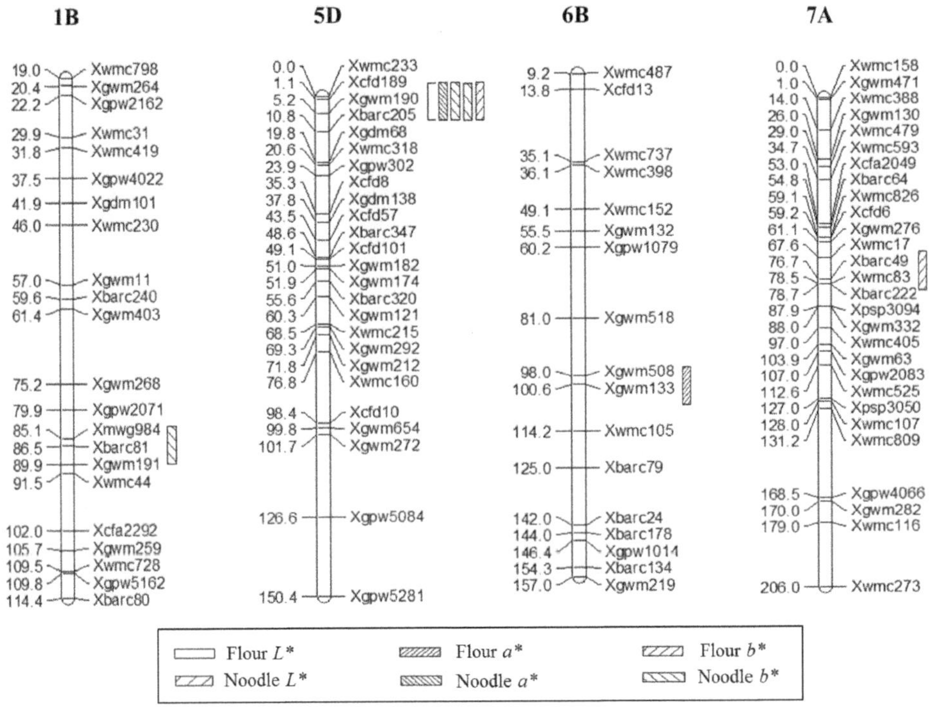

Figure 2. Linkage maps for the color of flour and noodle dough sheets for white salted noodles on chromosome 1B, 5D, 6B and 7A in 114 DH populations. Markers are indicated on the right, with genetic distances (cM) on the left. Different stripes of the markers indicate the color of flour and noodle dough sheets for white salted noodles, for which the symbols are located at the bottom of the figure

Xgwm190 was also found to be associated with the *L* and *b* values of the noodle dough sheets for white salted noodles for flour produced in 2011/2012 and the average of the two years. *Xgwm190* explained 17.3% and 11.8 % of the R^2, with LOD scores of 5.3 and 4.3 for the *L* value of the noodle dough sheet made with flour produced in 2011 and for the average of the two years, respectively. *Xgwm190* also explained 26.6% and 24.1% of the R^2, with LOD scores of 6.5 and 4.6 for the *b* value of noodle dough sheets produced from flour of wheat grown in 2012 and the average of two years, respectively. The additive effects of *Xgwm190* in *L* and *b* of noodle dough sheets were the same as those of flour. Another two QTLs found to be associated with the *a* and *b* values of the noodle dough sheet, *Xgwm133* and *Xwmc83*, were mapped to chromosomes 6B and 7A, respectively. *Xgwm133* explained 9.8% of the R^2 with LOD scores of 3.0 for the *a* value of the noodle dough sheets made from flour of wheat grown in 2011. *Xwmc83* explained 23.9% of the R^2 with LOD scores of 3.4 for the *b* value of noodle dough sheets produced from flour made in 2012. Similar to the other QTLs, *Xgwm133* had a positive additive effect on the *a* value of noodle dough sheet, whereas *Xwmc83* had a negative additive effect on the *b* value of noodle dough sheet.

The range of color in the Korean wheat cultivars was 90.6 – 94.2 for the *L* value, -1.4 – -0.7 for the *a* value and 6.9 – 12.0 for the *b* value in flour, with *L* values of 82.7 – 85.8, *a* values of -1.1 – 0.2 and *b* values of 11.8 – 18.3 for the noodle dough sheet for white salted noodles (Table 5). The 24 Korean wheat cultivars showed genotypes *a* and *b* at *Xgwm190*, a putative QTL for flour and noodle dough sheet color (Figure 3). Fourteen of the Korean wheat cultivars contained genotype *a*, which represented homozygous to Keumkang at *Xgwm190*, while the remaining 10 had genotype *b*, which indicated homozygous to Olgeuru. The Korean wheat cultivars carrying genotype *b* showed higher *L* values of both the flour and noodle dough sheets (93.6 and 85.1, respectively) than those with genotype *a* (91.6 and 83.5, respectively, Figure 4). No significant differences were found in the *a* and *b* values of flour and noodle dough sheets.

Table 5. Color of flour and noodle dough sheets for white salted noodles according to the genotype of Xgwm190 in 24 Korean wheat cultivars grown in upland conditions

Cultivar	*Xgwm190*[*]	CIE value[†]					
		Flour			Noodle dough sheet		
		L	*a*	*b*	*L*	*a*	*b*
Alchan	*a*	91.8	-0.9	9.3	83.5	-0.4	17.1
Anbaek	*a*	90.6	-1.2	10.8	83.2	-0.5	16.5
Baekjoong	*b*	93.6	-1.1	9.9	85.8	-0.8	13.5
Cheonggye	*b*	93.7	-1.0	8.2	85.0	-0.8	16.3
Dahong	*a*	92.6	-1.1	8.3	83.8	-0.8	16.8
Dajoong	*a*	91.5	-1.1	10.8	84.0	-0.4	16.6
Geuru	*a*	91.4	-1.1	10.2	84.2	-0.6	16.8
Gobun	*b*	93.5	-1.0	9.9	84.8	-0.3	17.2
Goso	*b*	94.2	-1.1	7.6	85.7	-0.4	14.7
Hanbaek	*a*	90.8	-0.7	9.7	82.7	0.2	12.1
Jeokjoong	*b*	93.3	-1.1	10.1	84.7	-0.8	14.3
Joa	*a*	93.0	-1.0	8.2	83.7	-1.0	15.6
Jokyung	*a*	91.4	-1.1	10.1	83.9	-0.8	14.8
Jonong	*a*	92.2	-0.7	6.9	83.3	-0.2	11.8
Jopoom	*a*	91.7	-1.1	9.8	83.1	-0.5	15.5
Keumkang	*a*	91.1	-0.8	9.6	83.3	-0.3	15.5
Namhae	*b*	93.9	-1.0	8.2	85.2	-0.7	16.3
Ol	*a*	92.3	-1.2	9.3	83.9	-1.1	18.3
Olgeuru	*b*	94.2	-1.0	7.8	85.4	-1.0	16.3
Suan	*b*	93.3	-0.8	8.8	85.4	-0.1	13.2
Sukang	*a*	90.6	-1.4	12.0	82.7	-1.1	17.3
Tapdong	*a*	91.3	-0.8	9.0	83.9	0.0	15.9
Uri	*b*	93.6	-1.0	8.5	85.0	-0.7	17.0
Younbaek	*b*	93.4	-1.1	10.5	84.3	-0.8	15.6
LSD[‡]	-	0.1	0.1	0.1	0.4	0.6	0.6

Note. [*]*a*, and *b* represent homozygous to Keumkang and Olgeuru, respectively.

[†]*L*, *a* and *b* of flour and noodle indicate the lightness, redness-greenness and yellowness-blueness of the flour and noodle dough sheets, respectively.

[‡]Least significant difference ($P < 0.05$).

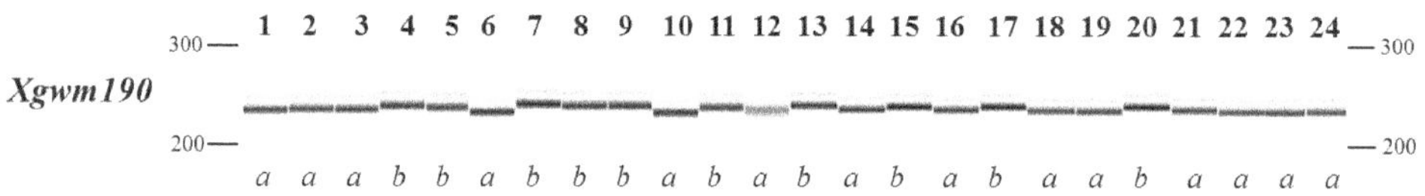

Figure 3. PCR products produced from QIAxcel system of QTL, *Xgwm190*, in 24 Korean wheat cultivars. 1, Ol; 2, Geuru; 3, Dahong; 4, Cheongkye; 5, Suan; 6, Tapdong; 7, Namhae; 8, Uri; 9, Olgeuru; 10, Alchan; 11, Gobun; 12, Keumkang; 13, Baekjoong; 14, Sukang; 15, Jeokjoong; 16, Hanbaek;17, Goso; 18, Anbaek; 19, Jopoom; 20, Younbaek; 21, Jonong; 22, Jokyung; 23, Dajoong; 24, Joa. *a* and *b* represent homozygous to Keumkang and Olgeuru, respectively

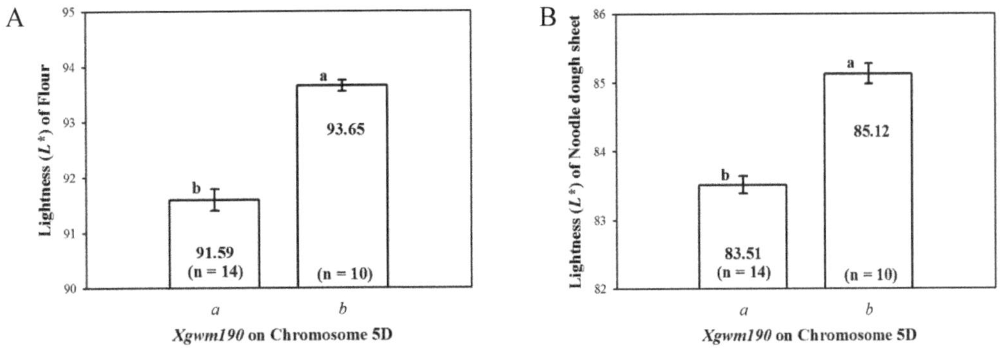

Figure 4. Difference in the lightness of flour (A) and noodle dough sheets for white salted noodles (B) according to the genotype of *Xgwm190* in 24 Korean wheat cultivars. Each bar represents the mean ± standard error, while *a* and *b* represent homozygous to Keumkang and Olgeuru, respectively. Bars with different letters above are significantly different at $p < 0.05$

4. Discussion

Color readings are usually expressed as CIE (Commission Internationale I'Eclairage) values, which are record as *L* value for lightness, *a* value for redness-greenness and *b* value for yellowness-blueness (Sun et al., 2002). Theoretically, pure white flour should have values of zero for both *a* and *b*, and one hundred for the *L* value (Sun et al., 2002). Therefore, flour or noodle dough sheets with low values of *L* and high positive values of a^* would result in a gray and dull color (Sun et al., 2002). Zhang et al. (2009b) proposed that high *L* and low *b* values of flour are very important to improve the noodle color for white salted noodles because the *b* value of noodle dough sheets for white salted noodles was significantly correlated with the *L* and *b* values of the wheat flour. Flour color is known to be influenced by both the bran and endosperm color of the grain (Baik et al., 1995; Mares & Campbell, 2001). The color of noodle dough sheets is also been affected by the physicochemical properties of the flour used, including flour yield, ash, protein, damaged starch content and protein quality (Feillet et al., 2000; Huo, 2001; Fu, 2008). These traits are also associated with polyphenol oxidase, which catalyzes the oxidation of phenols to produce dark-colored products, which are related to the discoloration of noodle dough (Baik et al., 1995; Feillet et al., 2000). There is generally a high correlation between the *b* value of flour and the yellow pigment content, which is primarily influenced by the levels of carotenoids (Mares & Campbell, 2001). Several enzymes in the carotenoid biosynthetic pathway, such as catalase (*Cat3-A1*), lycopene-e-cyclase (*e-LYC*), phytoene synthase (Psy) and Zeta (ζ)-carotene desaturase (ZDS), have been functionally associated with variation in the *b* value of flour (He et al., 2008; Howitt et al., 2009; Crawford et al., 2011; Zhang et al., 2011; Crawford & Francki, 2013a, 2013b).

Water absorption of the noodle dough affects the noodle color and sheeting, while the cutting and drying of noodles becomes difficult when there is deviation of the water absorption by over 2% from the optimum (Oh et al., 1986). The optimum water absorption has been determined by experienced personnel based on the handling properties of noodle dough in the industry, and a fixed absorption for making noodle dough has been used in a laboratory to evaluate noodle-making quality of wheat flours (Park & Baik, 2002). As shown in our previous report, Keumkang had higher protein, damaged starch content, SDS-sedimentation volume and water absorption of the mixograph than Olgeuru (Ahn et al., 2014). The color of the flour and noodle dough sheets are influenced by both genotype and environment, including factors such as growing conditions at the grain filling stage, milling process and the manufacturing process of noodles (Parker et al., 1998; Graybosch et al., 2004; Zhang et al., 2009). Heritability of the *L* and *b* values of flour and noodle dough sheets for white salted noodles was also found in DH lines of Australian wheat (Mares & Campbell, 2001). The flour and noodle dough sheet color showed high heritability and low genotype by environment interactions, which could allow useful QTL detection power in a small sample of environments (Narasimhamoorthy et al., 2006).

Major QTL associated with the color of flour have been detected on 17 chromosomes for the *L* value of flour: 1A, 1B, 2A, 2B, 2D, 3A, 3B, 3D, 4B, 4D, 5A, 5B, 5D, 6A, 6B, 7B and 7D; on 15 chromosomes for the *a* value of flour: 1A,1B, 2A, 2B, 3A, 3D, 4A, 4B, 4D, 5A, 5B, 5D, 6A, 6B and 7B; and on 14 chromosomes for the *b* value of flour: 1A, 1B, 1D, 2A, 2D, 3A, 3B, 4A, 4B, 4D, 5B, 5D, 6A, 6D, 7A, 7B and 7D (Mares & Campbell, 2001; Kuchel et al., 2006; Howitt et al., 2009; Zhang et al., 2009a, 2009b; Li et al., 2011; Tsilo et al., 2011; Crawford & Francki, 2013b; Zhao et al., 2013). The *Xbarc81* and *Xgwm190* QTLs detected herein were localized at 86.5 on chromosome 1B and at 5.2 cM on chromosome 5D. Zhang et al. (2009a) reported one flanking QTL, *Xcfd2226–Xwmc765*, detected at 8.0 cM on chromosome 5D for the *L* value of flour and two flanking QTLs, *Xbarc372–Xwmc412.2* and *Xwmc215–Xgdm63*, detected at 0.1cM and 9.0 cM on chromosomes 1B and 5D for the *a* value of flour, respectively, in a Chinese DH population. These two flanking markers located on chromosome 5D were closely located with the *Xgwm190* QTL detected in this study. Tsilo et al. (2011) reported one QTL, *Xbarc130*, detected at 0.0 cM on chromosome 5D that affected the *L* and *b* values of flour in recombinant inbred lines of hard red spring lines. This marker was near the hardness (*Ha*) locus on chromosome 5D, which is known to influence endosperm texture and the tightly linked puroindoline genes, *Pina* and *Pinb*, functioning as the *Ha* locus (Morris, 2002). Li et al. (2011) reported one flanking QTL, *Xbarc161–Xbarc144*, detected at 4.0 cM on chromosome 5D for the whiteness of flour in recombinant inbred lines of Chinese wheat, although QTLs with significant relation to flour color were not found on chromosomes 1B and 5D. Zhang et al. (2009b) reported that the rye secalin marker *Sec1*, which identified the 1RS rye chromosome and was associated with 1B/1R wheat-rye translocation, demonstrated an association with the *b* value of flour and was significantly related to the yellow pigment of wheat flour made from Chinese recombinant inbred lines. Zhao et al. (2013) also reported two flanking QTLs, *Xwmc419b–Xcfd20* and *Xubc880d–Xswes119a*, detected on chromosome 1B which affected the *b* value of flour from Chinese recombinant inbred lines. They also reported that one flanking QTL, *Xswes555b–Xswes558b*, and two flanking QTLs, *Xswes1061–Xswes340a* and *XwPt5505–Xgwm182*, detected on chromosome 5D affected the *L* and *a* values of the flour, respectively.

Mares and Campbell (2001) reported that several QTLs detected on chromosomes 1B, 2B, 3B, 5B and 7A in DH populations of Australian wheat lines affected the *L* and *b* values of noodle dough sheets for white salted noodles. Zhang et al. (2009b) reported two flanking QTLs, *Sec 1–HVM23* and *Xwmc809–YP7A*, detected on chromosome 1B and 7A, respectively, related to the color of noodle dough sheets for Chinese white salted noodles. This flanking marker was localized near the distal end of chromosome 7A, but *Xwmc83* detected herein was localized at 78.5 cM. A QTL for the color of noodle dough sheets for white salted noodles was located on chromosome 5D in this study, while QTLs for the *L* and *b* values of noodle dough sheets for yellow alkaline noodles were previously detected on chromosomes 5B, 5D and 7AL in DH populations of Canadian wheat lines (McCartney et al., 2006). Sadeque and Turner (2010b) found a major QTL for the *L* value of alkaline noodles on chromosome 2A, with additional QTLs on chromosomes 1A, 1B and 7A. The discoloration of Asian noodles is mainly influenced by polyphenol oxidase activity, which is controlled by two major genes, *Ppo-A1* and *Ppo-D1*, located on chromosome 2A and 2D (Martin et al., 2011). More specific molecular markers should be applied in Korean wheat breeding programs to improve the color of noodle dough sheets for white salted noodles. The discoloration of the noodle dough sheets and yellow pigments in the flour and noodles should also be considered to improve the acceptance of noodles in the market.

Increasing the brightness of flour and the end-use products has been the most important criteria in Korea, as well as other countries. However, it is a laborious, time-consuming and high-cost process to measure the color of flour and end-use products because large quantities of grains should be milled. Color related traits of flour and the end-use products have generally been evaluated on the advanced stages of breeding populations. Therefore, it is important to accurately predict the color to allow selection of wheat lines with bright color at the early stages of breeding programs. The putative QTL, *Xgwm190*, could be helpful to select wheat lines with bright color in the Korean wheat breeding population, although validation lines are limited and further fine mapping is needed for application of color-improving traits of flour and end-use products in Korean wheat breeding programs.

Acknowledgements

This work was carried out with the support of the “Cooperative Research Program for Agriculture Science & Technology Development (Project title: Establishment of quality criteria for high uniformity in end-use of Korean wheat cultivars, Project No. PJ011009), Rural Development Administration, Republic of Korea.

References

AACC International. (2010). *Approved Methods of Analysis* (11th ed., Methods 26-31.01). AACCI: St. Paul, MN, USA.

Ahn, J. H., Kang, C.-S., Jeung, J.-U., Baik, B.-K., Pena, R. J., & Park, C. S. (2014). Effect of allelic variations at the *Glu-D1*, *Glu-A3*, *Glu-B3* and *Pinb-D1* loci on flour characteristics and bread loaf volume. *International Food Research Journal, 21*, 1177-1185.

Baik, B.-K., Czuchajowska, Z., & Pomeranz, Y. (1995). Discoloration of dough for oriental noodles. *Cereal Chemistry, 72*, 198-205.

Bordes, J., Ravel, C., Le Gouis, J., Lapierre, A., Charmet, G., & Balfourtier, F. (2011). Use of a global wheat core collection for association analysis of flour and dough quality traits. *Journal of Cereal Science, 54*, 137-147. http://dx.doi.org/10.1016/j.jcs.2011.03.004

Churchill, G. A., & Doerge, R. W. (1994). Empirical threshold values for quantitative trait mapping. *Genetics, 138*, 963-971.

Crawford, A. C., & Francki, M. G. (2013a). Chromosomal location of wheat genes of carotenoid biosynthesis pathway and evidence for a catalase gene on chromosome 7A functionality associated with flour B* colour variation. *Molecular Genetics and Genomics, 288*, 483-493. http://dx.doi.org/10.1007/s00438-013-0767-3

Crawford, A. C., & Francki, M. G. (2013b). *Lycopene-ε-cyclase* (*e-LCY3A*) is functionally associated with quantitatively trait loci for flour b* colour on chromosome 3A in wheat (*Triticum aestivum* L.). *Molecular Breeding, 31*, 737-741. http://dx.doi.org/10.1007/s11032-012-9812-x

Crawford, A. C., Stefanova, K., Lambe, W., McLean, R., Wilson, R., Barclay, I., & Francki, M. G. (2011). Functional relationships of *phytoene synthase 1* alleles on chromosome 7A controlling flour colour variation in selected Australian wheat genotypes. *Theoretical and Applied Genetics, 123*, 95-108. http://dx.doi.org/10.1007/s00122-011-1569-9

Davies, J., & Berzonsky, W. A. (2003). Evaluation of spring wheat quality traits and genotypes for production of Cantonese Asian noodles. *Crop Science, 43*, 1313-1319. http://dx.doi.org/10.2135/cropsci2003.1313

Feillet, P., Autran, J. C., & Vernière, C. L. (2000). Pasta brownness: An assessment. *Journal of Cereal Science, 32*, 215-233. http://dx.doi.org/10.1006/jcrs.2000.0326

Fu, B. X. (2008). Asian noodles: History, classification, raw materials, and processing. *Food Research International, 41*, 888-902. http://dx.doi.org/10.1016/j.foodres.2007.11.007

Fuerst, E. P., Anderson, J. V., & Morris, C. F. (2006). Delineating the role of polyphenol oxidase in the darkening of alkaline wheat noodles. *Journal of Agricultural and Food Chemistry, 54*, 2378-2384. http://dx.doi.org/10.1021/jf0526386

Graybosch, R., Ames, N., Baenziger, P. S., & Peterson, C. J. (2004). Genotypic and environmental modification of Asian noodle quality of hard winter wheats. *Cereal Chemistry, 81*, 19-25. http://dx.doi.org/10.1094/CCHEM.2004.81.1.19

Hatcher, D. W., & Kruger, J. E. (1993). Distribution of polyphenol oxidase in flour millstreams of Canadian common wheat classes milled to three extraction rates. *Cereal Chemistry, 70*, 51-55.

He, X. Y., Zhang, Y. L., He, Z. H., Wu, Y. P., Xiao, Y. G., Ma C. X., & Xia, X. C. (2008). Characterization of *phytoene synthase* 1 gene (*Psy1*) located on common wheat chromosome 7A and development of a functional marker. *Theoretical and Applied Genetics, 116*, 213-221. http://dx.doi.org/10.1007/s00122-007-0660-8

Hou, G. (2001). Oriental noodles. *Advances in Food and Nutrition Research, 43*, 141-193. http://dx.doi.org/10.1016/S1043-4526(01)43004-X

Howitt, C. A., Cavanagh, C. R., Bowerman, A. F., Cazzonelli, C., Rampling, L., Mimica, J. L., & Pogson, B. J. (2009). Alternative splicing, activation of cryptic exons and amino acid substitutions in carotenoid biosynthetic genes are associated with lutein accumulation in wheat endosperm. *Functional and Integrative Genomics, 9*, 363-376. http://dx.doi.org/10.1007/s10142-009-0121-3

Knapp, S. J., Stroup, W. W., & Ross, W. M. (1985). Exact confidence intervals for heritability and heritability on a progeny mean basis. *Crop Science, 25*, 192-194. http://dx.doi.org/10.2135/cropsci1985.0011183X002500010046x

Inagaki, M. N., & Mujeeb-Kazi, A. (1995). Comparison of polyhaploid production frequencies in crosses of hexaploid wheat with maize, pearl millet and sorghum. *Breeding Science, 45*, 157-161.

Kuchel, H., Langridge, P., Mosionek, L., Williams, K., & Jefferies, S. P. (2006). The genetic control of milling yield, dough rheology and baking quality of wheat. *Theoretical and Applied Genetics, 112*, 1487-1495. http://dx.doi.org/10.1007/s00122-006-0252-z

Li, W.-H., Liu, W., Liu, L., You, M.-S., Liu, G.-T., & Li, B.-Y. (2011). QTL mapping for wheat flour color with additive, epistatic, and QTL× environmental Interaction effects. *Agricultural Sciences in China, 10*, 651-660. http://dx.doi.org/10.1016/S1671-2927(11)60047-3

Mares, D. J., & Campbell, A. W. (2001). Mapping components of flour and noodle color in Australian wheat. *Australian Journal of Agricultural Research, 52*, 1297-1309. http://dx.doi.org/10.1071/AR01048

Martin, J. M., Berg, J. E., Hofer, P., Kephart, K. D., Nash, D., & Bruckner, P. L. (2011). Allelic variation of polyphenol oxidase genes impacts on Chinese raw noodle color. *Journal of Cereal Science, 54*, 387-39. http://dx.doi.org/10.1016/j.jcs.2011.08.003

McCartney, C. A., Somers, D. J., Lukow, O., Ames, N., Noll, J., Cloutier, S., ... McCallum, B. D. (2006). QTL analysis of quality traits in the spring wheat cross RL4452 × 'AC Domain'. *Plant Breeding, 125*, 565-575. http://dx.doi.org/10.1111/j.1439-0523.2006.01256.x

Miskelly, D. M. (1984). Flour components affecting paste and noodle color. *Journal of the Science of Food and Agriculture, 35*, 463-471. http://dx.doi.org/10.1002/jsfa.2740350417

Morris, C. F. (2002). Puroindolines: the molecular genetic basis of wheat grain hardness. *Plant Molecular Biology, 48*, 633-647. http://dx.doi.org/10.1023/A:1014837431178

Nam, J.-H., Ha, Y.-W., Park, M.-W., Lee, S.-Y., Sung, B.-R., Park, H.-H., ... Park, K.-G. (1995). Early maturing, semi-dwarf and high yielding and good quality new soft wheat variety "Olgeurumil". *RDA Journal of Agricultural Science, 37*, 126-130.

Nam, J.-H., Song, H.-S., Park, H.-H., Ha, Y.-W., Park, M.-W., Park, K.-H., ... Kim, D.-H. (1998). A new high milling, early maturing, semi-dwarf and white grain wheat variety "Keumkangmil" with good bread quality. *RDA Journal of Agricultural Science, 40*, 81-87.

Narasimhamoorthy, B., Gill, B. S., Fritz, A. K., Nelson, J. C., & Brown-Guedira, G. L. (2006). Advanced backcross QTL analysis of a hard winter wheat × synthetic wheat population. *Theoretical and Applied Genetics, 112*, 787-796. http://dx.doi.org/10.1007/s00122-005-0159-0

Oh, N. H., Seib, P. A., Finney, K. F., & Pomeranz, Y. (1986). Noodles. V. Determination of optimum water absorption of flour to prepare oriental noodles. *Cereal Chemistry, 63*, 93-96.

Park, C. S., & Baik, B.-K. (2002). Flour characteristics related to optimum water absorption of noodle dough for making white salted noodles. *Cereal Chemistry, 79*, 867-873. http://dx.doi.org/10.1094/CCHEM.2002.79.6.867

Parker, G. D., Chalmers, K. J., Rathijen, A. J., & Langgridge, P. (1998). Mapping loci associated with flour colour in wheat (*Triticum aestivum* L.). *Theoretical and Applied Genetics, 97*, 238-245. http://dx.doi.org/10.1007/s001220050891

Raman, R., Raman, H., & Matin, P. (2007). Functional gene markers for polyphenol oxidase locus in bread wheat (*Triticum aestivum* L.). *Molecular Breeding, 19*, 315-328. http://dx.doi.org/10.1007/s11032-006-9064-8

Roncallo, P. F., Cervigni, G. L., Jensen, C., Miranda, R., Carrera, A. D., Helguera, M., & Echenique, V. (2012). QTL analysis of main and epistatic effects for flour color traits in durum wheat. *Euphytica, 185*, 77-92. http://dx.doi.org/10.1007/s10681-012-0628-x

Sadeque, A., & Turner, M. A. (2010a). QTL Mapping of flour color in a hexaploid wheat doubled haploid population using diversity array technology. *Thai Journal of Agricultural Science, 43*, 103-108.

Sadeque, A., & Turner, M. A. (2010b). QTL mapping of polyphenol oxidase (PPO) activity and yellow alkaline noodle (YAN) color components in an Australian hexaploid wheat population. *Thai Journal of Agricultural Science, 43*, 109-118.

Sun, X. D., Wang, L. K., Ren, H. B., & Lan, J. (2002). The application of tristimulus colorimeter in the determination of flour color. *Science and Technology of Cereals, Oils and Foods, 10*, 31-33.

Tsilo, T. J., Hareland, G. A., Chao, S., & Anderson, J. A. (2011). Genetic mapping and QTL analysis of flour color and milling yield related traits using recombinant inbred lines in hard red spring wheat. *Crop Science, 51*, 237-246. http://dx.doi.org/10.2135/cropsci2009.12.0711

Wang, S., Basten, C. J., & Zeng, Z.-B. (2012). *Windows QTL Cartographer 2.5.* Retrieved from http://statgen.ncsu.edu/qtlcart/WQTLCart.htm

Zeng, Z. B. (1993). The theoretical basis for separation of multiple linked gene effects in mapping quantitative trait loci. *Proc. Natl. Acad. Sci., 90*, 10972-10976. http://dx.doi.org/10.1073/pnas.90.23.10972

Zeng, Z. B. (1994). Precision mapping of quantitative trait loci. *Genetics, 136*, 1457-1468.

Zhang, C., Dong, C., He, X., Zhang, L., Xia, X., & He, Z. (2011). Allelic Variants at the *TaZds-D1* Locus on Wheat Chromosome 2DL and their Association with Yellow Pigment Content. *Crop Science, 51*, 1580-1590. http://dx.doi.org/10.2135/cropsci2010.12.0689

Zhang, K.-P., Chen, G.-F., Zhao, L., Liu, B., Xu, X.-B., & Tian, J.-C. (2009a). Molecular genetic analysis of flour color using a doubled haploid population in bread wheat (*Triticum aestivum* L.). *Euphytica, 165*, 471-484. http://dx.doi.org/10.1007/s10681-008-9756-8

Zhang, Y., Wu, Y., Xiao, Y., He, Z., Zhang, Y., Yan, J., ... Ma, C. (2009b). QTL mapping for flour and noodle colour components and yellow pigment content in common wheat. *Euphytica, 165*, 435-444. http://dx.doi.org/10.1007/s10681-008-9744-z

Zhao, Y., Sun, H.-Y., Wang, Y.-Y., Pu, Y.-Y., Kong, F.-M., & Li, S.-S. (2013). QTL mapping for the color, carotenoids and polyphenol oxidase activity of flour in recombinant inbred lines of wheat. *Australian Journal of Crop Science, 7*, 328-337.

19

The Effect of Heterogeneity Due to Inappropriate Tillage on Water Advance and Recession in Furrow Irrigation

Hamid Raeisi Vanani[1], Ali Reza Soltani Todeshki[2], Kaveh Ostad Ali Askari[3] & Mohammad Shayannejad[2]

[1] Irrigation and Drainage engineering, Isfahan University of Technology, Isfahan, Iran

[2] Water Engineering Department, Isfahan University of Technology, Isfahan, Iran

[3] Civil Engineering, Water Engineering, Lecturer of Islamic Azad University of Najafabad, Iran

Correspondence: Kaveh Ostad Ali Askari, Civil Engineering, Water Engineering, Lecturer of Islamic Azad University of Najafabad, Isfahan Province, Iran.

Abstract

In surface irrigation, uniformity of longitudinal slope and depth of tillage are very important factors in advance duration time. Inappropriate tillage and land preparation can cause uneven surfaces and non-uniform slopes in fields which is especially important in furrow irrigation due to their influence on advance times. The purpose of this study is to evaluate the effect of non-uniform longitudinal slope due to inappropriate tillage on advance and recession phases in furrow irrigation system, which plays an important role in irrigation uniformity and application efficiency. For this purpose 12 furrows, each 42 m long were made with different longitudinal slopes and a width of 0.6 meter. The number of irrigation was 5 with irrigation interval of 4 days and input discharge of 0.8 lps. Results showed that advance times are very different in furrows especially in the first irrigation and varied from 19min up to 50min. The values for recession time were 9 to 29 min. The results of these five irrigation events indicate that non-uniform slope has significant effect on advance and recession time. Therefore with attention to the water crisis and the need to increase efficiency and uniformity of water distribution in surface irrigation, using modern machinery and paying more attention to accurate preparation of land are necessary in order to obtain uniform furrows in fields.

Keywords: non-uniform slope, advance, recession, furrow irrigation

1. Introduction

Tillage is performed for various purposes such as preparing the seed bed, increasing water and air penetration into the soil, burying plant residues under the soil, mixing types of fertilizers with soil, removing weeds, controlling and decreasing the population of pests and diseases, crust breaking, breaking the root growth limiting layers, soil humidity conservation etc (Brye et al., 2003; Marques da Silva et al., 2004). Due to the importance of the tillage issued, in various researches the effect of different methods used for performing such operations has been evaluated. For example, investigations indicate the in the dry-farming conditions, type of tillage can significantly influence the humidity conservation of the soil surface layer and increasing the yield of a crop like sunflower (Arora et al., 2011; Mizuba et al., 2001).

Decrease of runoff and erosion in protecting tillage compared with the traditional method especially at large scale, is amongst the advantages of such operations (Ramos et al., 2006; Tullberg et al., 2007). As a whole, tillage operations (like plowing) should be such designed and operated which provides the conditions for reducing the soil superficial density and increased effective porosity in water conserving (Rounsevell & Jones, 1993; Martin et al., 2004; Moreno et al., 2008), and for example in dry-farming increases the soil's capacity to conserve precipitations. Also, attention should be paid to soil displacement in the length unit in various tillage methods since regarding the relation of fertility and soil translocation, this peculiarity is different in various procedures; such that for example, the highest soil translocations has been seen under the condition of dense soil and the highest machineries movement speed and while crushing of clod has been observed (Allmaras et al., 1967). Soil surface roughness is one of the crucial peculiarities of the soil which is influenced by runoff and erosion.

This peculiarity in different soil humidity and due to using various tillage procedures and stages and also following two tillage operations performed via identical tools, can be very different (Hamza et al., 2005; Dexter et al., 2004) Results of the studies indicate that almost always the disking and harrowing of soil is accompanied by surface roughness (Darve et al., 2013; Unger et al., 1990). Regarding the importance of humidity while tillage is performed, the most important limiting factor for the time of performing such operations, is the soil humidity; sine tillage in a more humid soil, causes formation of large clods and damaging the soil structure. Also, in case the soil humidity is lower that the desired level, in addition to increased energy consumption, large clots is formed (Barzegar et al., 2004; Vidal Vazquez et al., 2006). The compliance of soil for performing tillage operations, indicate the time when such operations can be done for desired soil structure, the appropriate size of the aggregates and suitable form of the surface (Nakajima et al., 2014; Ozpinar et al., 2006). Due to the tillage impact on soil properties including roughness and penetration, the tillage procedure and depth is effective on the furrow irrigation hydraulic and the soil capacity for water storing; since for example, increased tillage depth from 150 mm to 250 mm can significantly increase the water penetration into the soil, duration of water progression, application efficiency, and the saturation hydraulic conduction. The reason for increased advance time due to increased plough depth is augmented surface roughness and penetration. In connection with increased tillage depth, the water retention capacity is also decreased and also a significant effect has been seen on decreasing ration of the excite runoff and dispersion uniformity of the one or two primary irrigations however; this decrease has been less in other irrigation turns (Allmaras et al., 1967; Mosaddeghi et al., 2009).

According the research performed in this connection, tillage procedure caused a significant difference only for the first irrigation in the outflow runoff, dispersion uniformity, application efficiency and the deep percolation. Increased tillage depth as well, caused faster germination due to high penetration in the first irrigation (Gomez et al., 2012; Gomez et al., 2005). Due to hydraulic importance of advance time in the surface irrigation and it effect on efficiency, it seems very crucial to investigate the impact of soil tillage on this very important peculiarity.

In this same connection, comparison of advance time in furrow irrigation for two types of the farm tillage including traditional tillage and non-tillage operations showed that water advance is slower in furrows without tillage at a specified time (Leys et al., 2010; Yonts et al., 2007). At the identical time too, the advance in the farm without tillage in the firmed furrows is over twice that of advance in the usual furrows of this farm. Also the advance period in the farm through the conventional tillage in the smooth furrows improved and decreased (Olson et al., 2008; Oguntunde et al., 2006). Anyway in all cases for a specified period of time, the level of water advance in the firmed furrows was more than in the usual (not-firmed) furrows that through increased pressure for firming the furrow, this level of advancement increased (Dalla Rosa et al., 2012). Also, in connection with advance it was revealed that the conservation tillage (decrease of depth and number of tillage operations and direct seed sowing inside the cover remained from the previous harvest) compared with the conventional tillage, penetration was increased and as a result water advance on the soil is decreased (Rashidi & Keshavarzpour, 2007). It is worth mentioning that one of the methods for increasing water advancement in the furrow and reducing the irrigation period, is compaction of it bed and stabilizing it and as a result augmentation of efficiency of irrigation without decrease of yield (Daniel, 2004). It is worth mentioning that density of weed due to decreased plowing can be one of the problems of this type of tillage and potentially effective on the water advance speed and yield during the farming season (Dexter et al., 2005).

One of the most important tillage operations which have as great influence on the advance stage is the land leveling operations. Precise and complete leveling is one of the important and effective components of management of water requirement of the farm. Research indicates that among various leveling procedures including the laser and conventional leveler, from the viewpoint of the volume of water consumed for irrigating the farm leveled through these two methods there exists a significant difference and that he layer leveler can, compare with the conventional method, 35% decrease the irrigation time (Berehe et al., 2013).

It should be noted that land leveling can increase uniformity of receiving irrigation water (Hajabbasi et al., 2000). Studies show that due to employing the laser leveling technology, irrigation efficiency is increased to about 20%. Also the laser leveling technology can be applied as an appropriate tool for increasing efficiency of the consumables, increase of yield at the surface unit, and facilitation of agricultural operations (decreased work force) (Berehe et al., 2012). Regarding the importance of influence of tillage operations (including leveling) on soil quality, researches revealed that the leveling operations cause main differences on the soil depth and its specifications all over the land. This issue in the vineyards caused a difference in the humidity conditions of the soil at various situations of the form (Dexter et al., 2001; Kroulik et al., 2009).

Comparison of land before and after leveling also showed that the superficial density of the soil, amount of clay and sand is significantly increased (Hajabbasi et al., 2000; Hoogmoed et al., 2003). The above said issue also influence the penetration level and as a result water advance. Also, the leveling operations have a significant effect on local distribution of properties related to fertility (nutrients) (Bandaranayake et al., 1998; Keller et al., 2007). Therefore, to decrease the negative effects of leveling on soil fertility, maximum leveling section, limited to 10 cm or less has been recommended (El-Awad et al., 2000; Hoogmoed et al., 2003). Regarding the above said issues and importance of tillage and water advance in the farm, in this research the effect of inappropriate tillage on progression and regression of water in the furrow irrigation has been evaluated.

2. Materials and Methods

To achieve the project's objectives, a farm located in the Industrial University of Isfahan was chosen. Some of the soil properties of this farm have been presented in Table 1.

Table 1. Some of the properties of the farm's soil in the 0-15 cm and 15-40 cm soil layers

Section	Depth (cm)	Soil particles (%)			Soil texture	Gravel (%)	Organic matter (%)	Bulk density (g/cm^2)	EC (ds/m)	pH	FC (%)	PWP
		Sand	silt	clay								
1	0-15	50.4	21.8	27.8	Loam sandy clay	38	<1	1.57	1.82	7.87	31.57	15
	15-40	52.3	20.3	27.4	Loam sandy clay	50	<1	1.68	1.97	7.84	33.71	15
2	0-15	54.2	18.8	27	Loam sandy clay	34	<1	1.78	0.98	8.29	20.87	15
	15-40	56.1	17.3	26.6	Loam sandy clay	43	<1	1.71	0.89	8.43	19.81	15

Executive operations of the project started with tillage. In this phase, plowing, disking and roller – harrowing on the land was done. Following roller harrowing, using mapping, a regular grilling was performed on the land and general slope of the farm and its up and down points was specified (Figure 1).

Figure 1. The plowing, crushing and reticulation

In the second phase, leveling operations were performed via the traditional tool (handbarrow "Zanbeh") and based on judgments and experience of the machine operator. After leveling and grading, non-wheel furrows in the width of 60 and depth of about 25 cm were established by the furrower machine and 42 m of which was selected for the test (Figure 2). To dig furrows parallel to each other, a straight line by the mapping camera was drawn using chalk on the farm border and the furrow establishing operations were performed parallel to this line. Following establishing the furrows, slope of each one was determined separately.

Figure 2. Leveling with traditional instruments and creating a furrow in two field section

Rate of inflow water to the furrows was selected to be 0.8 l/s and value of the water conveying pipe was calibrated for this amount of water conveyance. The test started through water conveyance to the furrows and the water progression process to the end of them was measured (Figure 3). The water regression period as well was determined by observing the first and the last regression point along each furrow. To eliminate the marginal effects, for each main furrow for which measurement was performed, two lateral furrows were considered.

Prior to any irrigation the longitudinal profile of the furrows was determined via mapping and five irrigations with an irrigation interval of four days in each were performed. The irrigation interval (flow interruption) in all the furrows was almost identical. Overall 60 furrow irrigation operations were performed in 12 main furrows and in 24 lateral furrows. It is worth mentioning that the test farm was culture free.

Figure 3. Water advance and recession in furrows

3. Results and Discussion

Diagram of the progression data obtained of the research from has been depicted in Figures 4 and 5.

As seen in the diagrams, general trend of progression interval modifications in all the furrows was identical however the progression in the first irrigation (regarding the difference between conditions of soil roughness and penetrability) compared with other irrigation intervals was significantly longer. This is acceptable regarding principles of flow hydraulic in the surface irrigation.

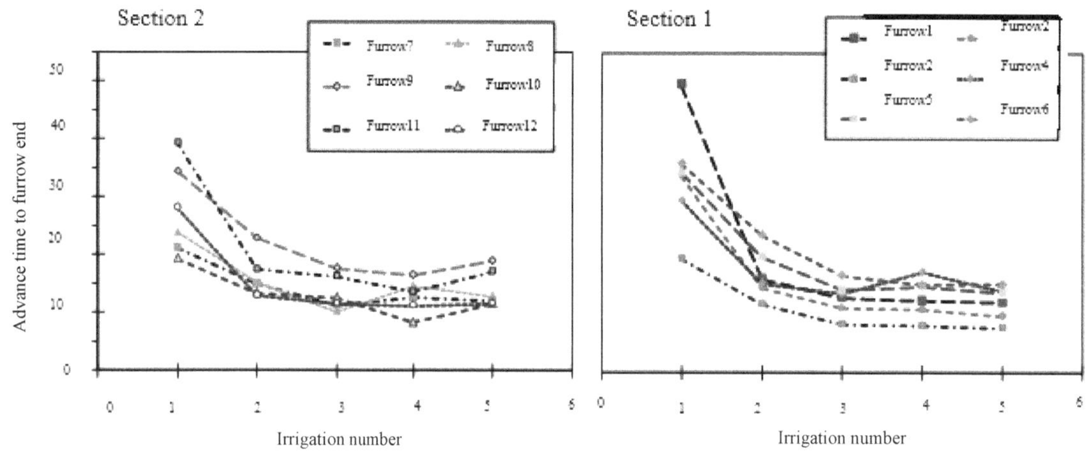

Figure 4. Furrow irrigation advance times

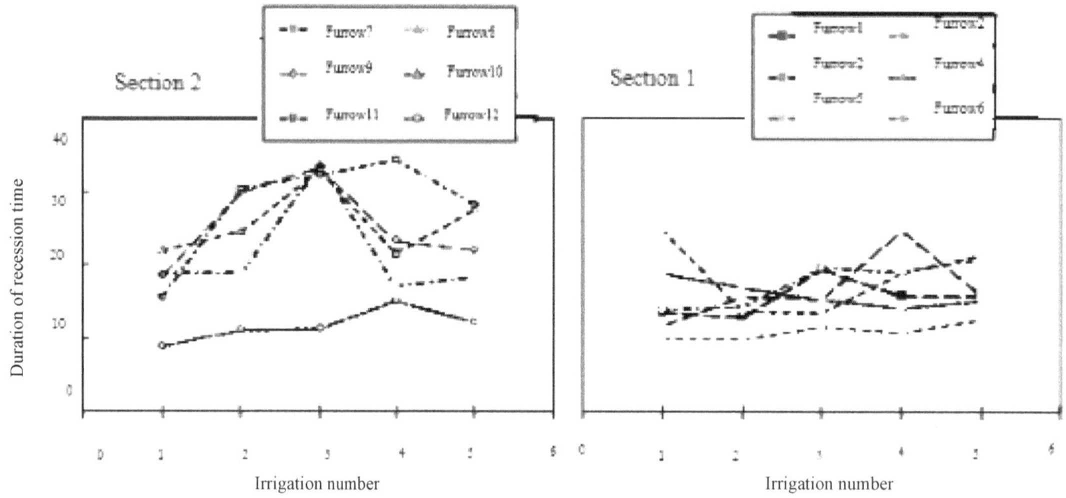

Figure 5. Time Chart recession and shifts in furrow irrigation

Difference of progression time in various furrows indicate that regarding similarity of the soil type, primary humidity and irrigation operations in all the furrows, the difference in "Surface roughness and the deep soil unevenness" of furrows (even after five irrigation intervals and partial modification of the flow bed), has had a considerable effect on the water progression time. Also, unevenness of the farm led to formation of water ditches in the regression phase in some of the longitudinal points causing disordered discharge and regression phase leading to development of waterlogging conditions. In the meantime, part of the heterogeneity in the regression phase is due to penetration differences arising from plow depth and furrow establishment depth differences. Figure indicates fluctuations of the regression time in the 12 understudied furrows. Interval of regression is the time interval between the first and the last regression in each furrow. As it can be seen in each furrow, not only no particular trend is seen in various irrigation shifts in each furrow, but even in different furrows this specification much varies. Since the length of the period for the regression to happen in very much dependent on the farm conditions (longitudinal profile of the furrows), fluctuations and unevenness of the surface of the furrows can be considered as a reason for this even.

To specify the extent of progression and regression modifications, minimum and maximum values of the two above said peculiarities in different irrigation shifts and in the tested furrows are presented in Table 2. According to Table 2, the progression and regression values in all irrigation shifts change in a vast extent of time. Also, a descending trend is evident in the minimum and maximum progression data during the irrigation shifts which can be attributed to the partial stabilization of bed of the furrows during the irrigation shifts. No special trend however, is observed in the regression duration peculiarity.

Table 2. The minimum and maximum period time of recession, the shift forward and irrigation (min)

Number of irrigation	Advance time		Recession time	
	Minimum	Maximum	Minimum	Maximum
1	19.4	49.7	9	28.5
2	11.9	23.7	10	30.5
3	8.4	17.8	11.4	33.9
4	8.2	17.4	11	34.4
5	7.8	19.1	12.2	28.3

Drawing of the land roller harrowing on the progression and regression intervals (Figure 6) indicate that influenced by the primary tillage management (as counter plugging toward the center of the land), a larger mass of soil was accumulated in the center of the land. Evidently this makes precise leveling after roller harrowing a difficult task. On one side, leveling via the traditional method (compared with new technologies like laser leveling) and exclusive use of visual leveling (without concurrent control by the camera) in spite of the worker's care and sufficient experience had made difficult establishment of identical furrows. This issue indicates that in spite of experience and attention of the leveling machine in charge, still such operations suffer errors mostly writing from human sources. Therefore, employment of modern technologies like laser leveling, regarding its precision is of special importance.

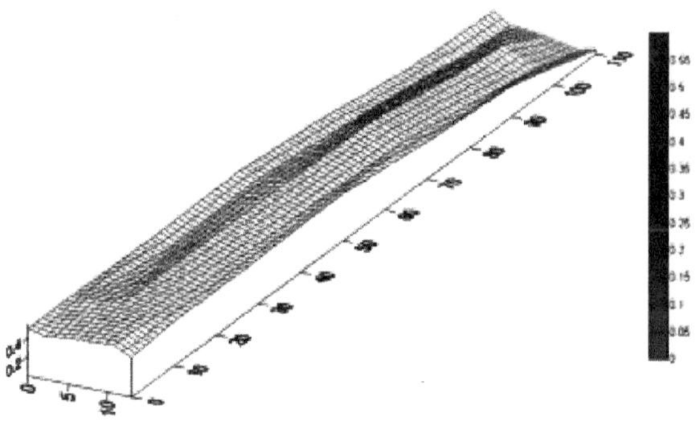

Figure 6. Soil surface pressing view

Also, the primary tillage being not deep due to hardness of the soil layer, inappropriate humidity and containing pebbles, caused that in some parts of the farm, the furrower machine dependent on its tolerance, change the furrow depth and as a result longitudinal profile of the furrows become uneven (Figure 7). On the other hand, inappropriate slope of the first few meters of the furrow as well, indicates that the beginning point of the furrower machine should be selected several maters farther than the beginning of the furrow; otherwise the machine will not have the require respite for achieving the required depth for establishing the furrows.

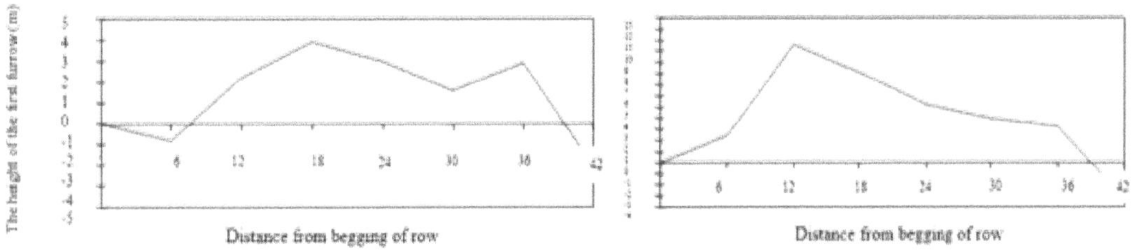

Figure 7. Two examples of longitudinal profile of furrows created by the leveling operation

Existence of particular profiles in the furrows (similar to Figure 7) caused increased cross section of the flow in the few meters of the first few entry meters as well; since water was devoid of the required hydraulic pressure for movement towards the end of the furrows which in turn caused accumulation of water layers and as a result erosion of the section soil at the beginning of the furrow. Therefore, surface and subsurface unevenness of soil was one of the factors indirectly influencing the water progression.

Farm studies from observation also indicate that in addition to inappropriate primary tillage, it's in sufficient depth and conventional leveling were of major factors contributing to unevenness of the farm surface. Inefficiency of machinery, imbalance Frame of the furrower device and its disturbed adjustment along the course of movement (Figure 8) are also other contributing factors in unevenness of sorrows. The above said factors that caused the established furrows to have different slopes and longitudinal profiles compared to each other.

Also, with regard to the mapping information, it was revealed that the situation of the longitudinal profile of the furrows is not constant and is different in subsequent irrigations (Figure 9). This issue can influence progression and regression since regarding Figures 4 and 5, the progression time as well as duration of regression in a furrow at various irrigation turns has had different behaviors compared with other furrows. The reason might be that the soil erosion behavior in furrows with uneven slope causes different progressions and regressions in various irrigation turns.

Figure 8. Furrows system and its problems during labor

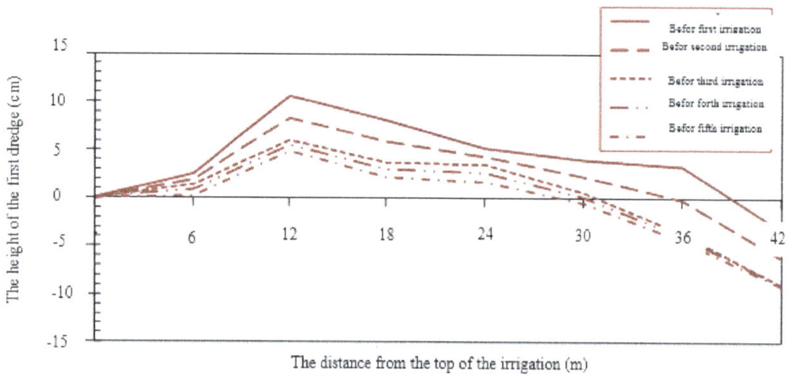

Figure 9. An example of a longitudinal profile changes during shifts furrow irrigation

Through investigation of the land tillage, existence of frequent large a hard colds even after their disking was revealed (Figures 10a and 10b). This observation indicates insufficiency of the disking machine and inappropriate humidity at the time of operations. Existence of the same clods plays and important role in the progression time in the first irrigations. It should be noted that large soil particles like these clods have no farming value and often cause problems for soil management and irrigation (Lampurlanes & Cantero-Martinez,

2006). Also, the pulverized soil particles in the conventional tillage system (used in this project) is trans located due to surface irrigation and is accompanied by aggregation of particles in the deep parts of the land. This aggregation of small particles is followed by surface crusting and water erosion (Arvidsson & Bolenius, 2006) which was also evident in this project (Figure 10c). This event influences the penetration process followed by impact on progression.

a) b) c)

Figure 10. a) Remain gravels after pressing, b) furrow creation, c) soil crust after irrigation

4. Conclusion

The actual timing of the tillage operations and the soil conditions under which they are performed can influence water progression and regression in the furrows. Results of this research reveal that conventional tillage employed in majority of farms cause differences in hydraulic specifications of the advance and recession time in the furrow irrigation.

5. Recommendations and Suggestions

Since condition of majority of farms resembles the farm investigated in this project and also considering the fact that actually our country is suffering a severe water crisis, it is necessary that experts in the agriculture and promotion sectors consider this issue in order to increase the irrigation efficiency and sustained production and instruct the farmers, according to the scientific findings of complications of the conventional tillage and let them know the necessity of employing modern tools and new methods appropriate for tillage and paging much attention regarding tillage of the culture bed.

References

Allmaras, R. R., Burwell, R. E., & Holt, R. F. (1967). Plow-layer porosity and surface roughness from tillage as affected by initial porosity and soil moisture at tillage time. *Soil Sci. Soc. Am. Proc., 31*, 550-556. http://dx.doi.org/10.2136/sssaj1967.03615995003100040033x

Arora, V. K., Singh, C. B., Sidhu A. S., & Thind, S. S. (2011). Irrigation, tillage and mulching effects on soybean yield and water productivity in relation to soil texture. *Agr. Water. Manag., 98*, 563-568. http://dx.doi.org/10.1016/j.agwat.2010.10.004

Arvidsson, J., & Bolenius, E. (2006). Effects of soil water content during primary tillage- laser measurements of soil surface changes. *Soil Till. Res., 90*, 222-229. http://dx.doi.org/10.1016/j.still.2005.09.005

Bandaranayake, W. M., Butters, G. L., Hamdi, M., Prieksat, M., & Ellsworth, T. R. (1998). Irrigation and tillage management effects on solute movement. *Soil Till. Res., 46*, 165-173. http://dx.doi.org/10.1016/S0167-1987(98)00029-4

Barzegar, A. R., Hashemi, A. M., Herbert, S. J., & Asoodar, M. A. (2004). Interactive effects of tillage system and soil water content on aggregate size distribution for seedbed preparation in Fluvisols in southwest Iran. *Soil Till. Res., 78*, 45-52. http://dx.doi.org/10.1016/j.still.2004.01.002

Berehe, F. T., Melesse, A. M., Fanta, A., & Alamirew, T. (2013). Characterization of the effect of tillage on furrow irrigation hydraulics for the Dire Dawa Area, Ethiopia. *Catena, 110*, 161-175. http://dx.doi.org/10.1016/j.catena.2013.06.003

Berehe, F. T., Fanta, A., Alamirew, T., & Melesse, A. M. (2012). The effect of tillage practices on grain yield and water use efficiency. *Catena, 100*, 128-138. http://dx.doi.org/10.1016/j.catena.2012.08.001

Brye, K. R., Slaton, N. A., Savin, M. C., Norman, R. J., & Miller, D. M. (2003). Short-term effects of land leveling on soil physical properties and microbial biomass. *J. Soil. Sci. Soc. Am., 67*, 1405-1417. http://dx.doi.org/10.2136/sssaj2003.1405

Dalla Rosa, J., Cooper, M., Darboux, F., & Medeiros, J. C. (2012). Soil roughness evolution in different tillage systems under simulated rainfall using a semivariogram-based index. *Soil Till. Res., 124*, 226-232. http://dx.doi.org/10.1016/j.still.2012.06.001

Daniel, N. M. (2004). *Modeling the effects of rainfall intensity and deep chiseling on infiltration and runoff within DRAINMOD for alluvial soils*. PhD Dissertation submitted to Louisiana state university and agricultural and mechanical college.

Darve, F., Freitas, H., & Moldes, A. (2013). Short-term effects of land leveling on irrigation-related some soil properties in a clay loam soil. Hindawi publishing corporation. *J. Sci. World,* 1-16. http://dx.doi.org/10.1155/2013/187490

Dexter, A. R., Czyz, E. A., Birkas, M., Diaz-Pereira, E., Dumitru, E., Enache, R., ... Simota, C. (2005). The optimum and the range of water content for tillage- further developments. *Soil Till. Res., 82*, 29-37. http://dx.doi.org/10.1016/j.still.2005.01.005

Dexter, A. R., & Birkas, M. (2004). Prediction of soil structures produced by tillage. *Soil Till. Res., 79*, 233-238. http://dx.doi.org/10.1016/0022-4898(79)90022-3

Dexter, A. R., & Bird, N. R. A. (2001). Methods for predicting the optimum and the range of soil water contents for tillage based on the water retention curve. *Soil Till. Res., 57*, 203-212. http://dx.doi.org/10.1016/S0167-1987(00)00154-9

El-Awad, S. E. D. A. G. (2000). Effects of irrigation interval and tillage systems on irrigated cotton and succeeding wheat crop under a heavy clay soil in the Sudan. *Soil Till. Res., 55*, 167-173. http://dx.doi.org/10.1016/S0167-1987(00)00112-4

Gomez, E., & Garland, J. L. (2012). Effects of tillage and fertilization on physiological profiles of soil microbial communities. *App. Soil. Ecol., 61*, 327-332. http://dx.doi.org/10.1016/j.apsoil.2011.10.008

Gomez, J. A., Vanderlinden, K., & Nearing, M. A. (2005). Spatial variability of surface roughness and hydraulic conductivity after disk tillage: implications for runoff variability. *J. Hydrol., 311*, 143-156. http://dx.doi.org/10.1016/j.jhydrol.2005.01.014

Hajabbasi, M. A., & Hemmat, A. (2000). Tillage impacts on aggregate stability and crop productivity in a clay-loam soil in central Iran. *Soil Till. Res., 56*, 205-212. http://dx.doi.org/10.1016/S0167-1987(00)00140-9

Hamza, M. A., & Anderson, W. K. (2005). Soil compaction in cropping systems: a review of the nature, causes, and possible solutions. *Soil Till. Res., 82*(2), 121-145. http://dx.doi.org/10.1016/j.still.2004.08.009

Hoogmoed, W. B., Cadena-Zapata, M., & Perdok, U. D. (2003). Laboratory assessment of the workable range of soils in the tropical zone of Veracruz, Mexico. *Soil Till. Res., 74*, 169-178. http://dx.doi.org/10.1016/j.still.2003.06.001

Keller, T., Arvidsson, J., & Dexter, A. R. (2007). Soil structures produced by tillage as affected by soil water content and the physical quality of soil. *Soil Till. Res., 92*, 45-52. http://dx.doi.org/10.1016/j.still.2006.01.001

Kroulik, M., Kumhala, F., Hula, J., & Honzik, I. (2009). The evaluation of agricultural machines field trafficking intensity for different soil tillage technologies. *Soil Till. Res., 105*, 171-175. http://dx.doi.org/10.1016/j.still.2009.07.004

Lampurlanes, J., & Cantero-Martinez, C. (2006). Hydraulic conductivity, residue cover and soil surface roughness under different tillage systems in semiarid conditions. *Soil Till. Res., 85*, 13-26. http://dx.doi.org/10.1016/j.still.2004.11.006

Leys, A., Govers, G., Gillijns, K., Berckmoes, E., & Takken, I. (2010). Scale effects on runoff and erosion losses from arable land under conservation and conventional tillage: The role of residue cover. *Hydrol., 390*, 143-154. http://dx.doi.org/10.1016/j.jhydrol.2010.06.034

Martin, E. C., Adu-Tutu, K. O., McCloskey, W. B., Husman, S. H., Clay, P., & Ottman, M. (2004). Conservation tillage effects on infiltration and irrigation advance times in Arizona cotton. Arizona Cotton Report (Vol. 142, pp. 7-30). University of Arizona, Tucson, AZ. Retrieved from http://hdl.handle.net/10150/198126

Marques da Silva, J. R., Soares, J. M. C. N., & Karlen, D. L. (2004). Implement and soil condition effects on tillage-induced erosion. *Soil Till. Res., 78*, 207-216. http://dx.doi.org/10.1016/j.still.2004.02.009

Mizuba, M. M., & Hammel, J. E. (2001). Infiltration rates in fall-seeded winter wheat fields following preplant subsoil tillage. *J. Soil. Water. Conserv., 56*(2), 133-137.

Moreno, R. G., Diaz Alvarez, M. C., Alonso, A. T., Barrington, S., & Requejo, A. S. (2008). Tillage and soil type effects on soil surface roughness at semiarid climatic conditions. *Soil Till. Res., 98*, 35-44. http://dx.doi.org/10.1016/j.still.2007.10.006

Mosaddeghi, M. R., Morshedizad, M., Mahboubi, A. A., Dexter, A. R., & Schulin, R. (2009). Laboratory evaluation of a model for soil crumbling for prediction of the optimum soil water content for tillage. *Soil Till. Res., 105*, 242-250. http://dx.doi.org/10.1016/j.still.2009.08.005

Nakajima, T., & Lal, R. (2014). Tillage and drainage management effect on soil gas diffusivity. *Soil Till. Res., 135*, 71-78. http://dx.doi.org/10.1016/j.still.2013.09.003

Oguntunde, P. G., Ajayi, A. E., & Giesen, N. V. D. (2006). Tillage and surface moisture effects on bare-soil albedo of a tropical loamy sand. *Soil Till. Res., 85*, 107-114. http://dx.doi.org/10.1016/j.still.2004.12.009

Olson, N. C., Schmalle, G., Adekola, L., Gulliver, J. S., & Nieber, J. L. (2008). *Improvements in infiltration rates of compacted soil with tillage and compost.* International low impact development conference, ASCE. http://dx.doi.org/10.1061/41009(333)84

Ozpinar, S. (2006). Effects of tillage systems on weed population and economics for winter wheat production under the Mediterranean dryland conditions. *Soil Till. Res., 87*, 1-8.

Ramos, M. C., & Martınez-Casasnovas, J. A. (2006). Impact of land levelling on soil moisture and runoff variability invineyards under different rainfall distributions in a Mediterraneanclimate and its influence on crop productivity. *J. Hydrol., 321*, 131-146. http://dx.doi.org/10.1016/j.jhydrol.2005.07.055

Rashidi, M., & Keshavarzpour, F. (2007). Effect of different tillage methods on soil physicals properties and crop yield of Watermelon (Citrullus vulgaris). *ARPN J. Agric. and Biol. Sci., 2*(6), 1-6.

Rounsevell, M. D. A., & Jones, R. J. A. (1993). A soil and agroclimatic model for estimating machinery workdays: the basic model and climatic sensitivity. *Soil Till. Res., 26*(3), 179-191. http://dx.doi.org/10.1016/0167-1987(93)90043-O

Tullberg, J. N., Yule, D. F., & McGarry, D. (2007). Controlled traffic farming-from research to adoption in Australia. *Soil Till. Res., 97*(2), 272-281. http://dx.doi.org/10.1016/j.still.2007.09.007

Unger, P. W., Fultonand, L. J., & Jones, O. R. (1990). Land leveling effects on soil texture, organic matter content, and aggregate stability. *J. Soil. Water. Cons., 45*(3), 412-415.

Vidal Vazquez, E., Miranda, J. G. V., Alves, M. C., & Paz Gonzalez, A. (2006). Effect of tillage on fractal indices describing soil surface microrelief of a Brazilian Alfisol. *Geoderma., 134*, 428-439. http://dx.doi.org/10.1016/j.geoderma.2006.03.012

Yonts, C. D., & Eisenhauer, D. E. (2007). Firming irrigation furrows to improve irrigation performance. *Irrigation engineering, irrigation operations and management.* University of Nebraska-Lincoln.

20

Distribution and Extraction of Heavy Metals in Soil and Their Accumulation in *Brassica oleracea* L. after Long Term Wastewater Irrigation

Faridullah[1], Farid Ul Haque[1], Alias Bin Abdullah[2], Muhammad Irshad[1], Arif Alam[3] & Akhtar Iqbal[1]

[1] Department of Environmental Sceinces, COMSATS Institute of Information Technology Abbottabad, Pakistan

[2] Faculty of Arts and Social Sciences, Department of East Asian Studies, University of Malaya, Kuala Lumpur, Malaysia

[3] Department of Development Studies, COMSATS Institute of Information Technology Abbottabad, Pakistan

Correspondence: Faridullah, Department of Environmental Sciences, COMSATS Institute of Information Technology Abbottabad, Pakistan. E-mail: faridullah@ciit.net.pk

Abstract

Wastewater irrigation has become a common practice especially in third world countries. Over the period of time population growth has resulted in increased domestic and industrial wastes. People produce huge quantities of vegetables and crops yields with wastewater irrigation without knowing its effects on soils, plants and ultimately on consumers. Therefore, a study was carried out to compare accumulation of heavy metals (Fe, Cu, Pb, Zn, and Cr) in wastewater irrigated soils with rain fed soils. Water soluble, and total extractable heavy metals were determined. The contents of water soluble, exchangeable and total plant essential elements (K, Ca, and Mg) were also determined. Soil samples from three different layers (0-30 cm, 30-60 cm and 60-100 cm depth) were collected from both wastewater irrigated field and rain fed field. Results indicated that water soluble heavy metals varied in soil in order Fe> Cr > Pb > Zn > Cu irrespective of the depth and irrigation management. Total heavy metals in all layers of soil were noted as Cr > Zn > Pb > Cu > Fe for wastewater irrigated field and Cr > Zn > Pb > Fe > Cu for rain-fed field. On the other hand the concentrations of water soluble essential elements varied as K > Mg > Ca for both rain-fed and wastewater irrigated soils. The study clearly indicated that wastewater irrigation caused heavy metal accumulation in both soils and plants. The use of wastewater for agriculture may be economically productive due to abundance of nutrients present in it but have adverse effects on soil, plant and ultimately its consumers. Although heavy metals in plant were found within the standard limits, however, continuation of such practices for a longer period of time may escalate their levels beyond the safe limits.

Keywords: heavy metals, wastewater irrigation, cabbage

1. Introduction

Wastewater is used for irrigation because of its availability without cost and the scarcity of fresh water. Excessive use of the wastewater for irrigation not only contaminates soil but also the plants (REF). Heavy metals leach down the surface and subsurface soils with repeated applications of wastewater and result in gradual increase in the concentrations of heavy metals in vertical soil column. These heavy metals are taken up by the plants thus posing potential hazard to the human health because of transmission into consumer chain (Burn et al., 2001). The excessive use of heavy metals thus depletes essential nutrients in the body causing several physical and psychological abnormalities. Trace elements requirement is met either through food or water; therefore balanced amount of these metals should essentially be present in food sources. In trace amounts, heavy metals are essentially required for living organisms; however, in higher concentrations they are toxic. The use of wastewater and sewage sludge on agricultural soil has become a common practice in the developing countries, as a result of which these toxic metals can be transferred and concentrated into plant tissues from the soil (Alloway, 1995). At higher concentrations, some metals have strong toxic effects and are regarded as environmental pollutants (Nedelkoska & Doran, 2000; Chehregani et al., 2005). In soils contaminated with heavy metals, plant growth can be inhibited by metal absorption. However, some plant species are able to accumulate fairly large amounts of metals without showing stress, which represents a potential risk for animals and humans (Oliver, 1997).

Potential danger from metal accumulation by plants grown on such soil is becoming an increasing problem throughout the world. This has created a demand for an intensive research effort aimed at predicting the availability of heavy metals in the soil (Nriagu, 1991). Studies have been carried out to analyze heavy metals both in plants and soils. However, distribution of heavy metals in three soil layers of 0-30 cm, 30-60 cm and 60-100 cm depth and their accumulation in cabbage (*Brassica oleracea* L.) after irrigation with wastewater in Abbottabad, Khyber Pukhtunkhwa province of Pakistan has not been investigated earlier. The objective of the present study was to investigate distribution and extractability of heavy metals in soil irrigated with domestic wastewater of Mirpur village, Abbottabad and its surroundings and hospital wastewater of Ayub Medical Complex, Abottabad and to determine the accumulation of heavy metals in cabbage (*Brassica oleracea* L.) irrigated with wastewater.

2. Materials and Methods

2.1 Study Area

General area of study was Abbottabad, Pakistan, urbanizing at a very high population influx rate in recent years. This increase in population posed multifaceted challenges not only for the environment but also for its habitats. The study area was carried out in two adjacent fields, one of which being irrigated by mixture of domestic wastewater of Mirpur and hospital wastewater of Ayub Medical Complex, Abbottabad for cabbage plant (*Brassica Oleracea* L.). The other adjacent field was rain-fed with the same plants cultivated in it.

2.2 Soil and Plant Sampling

Soil sampling was done from 0-30 cm, 30-60 cm and 60-100 cm depths in order to assess the distribution and leaching of heavy metals (Fe, Cu, Cr, Pb and Zn) in vertical soil column. Plant samples were also collected and lab experiments were performed to find heavy metal accumulation in soils and cabbage plant.

2.3 Soil Analysis

Soil samples were air dried, crushed, and sieved (< 0.5 mm) to ensure homogeneity and digested in a mixture of duplicate acids (HNO_3 and $HClO_4$). Total elements i.e. potassium (K), calcium (Ca) magnesium (Mg), iron (Fe), zinc (Zn), copper (Cu), chromium (Cr), lead (Pb) in the extract of digested soil samples, were determined by atomic absorption spectrophotometer (IBSRAM, 1994). Sample weighing 0.25 g was digested with 5 mL concentrated $HClO_4$ by gradual heating over a hot plate for 1 h. After drying 20% HNO_3 was added to the sample and it was heated again for 1 h. The solution was diluted to 50 mL with deionized water and passed through a 0.22 μm filter. Ammonium acetate soluble cations (Ca, Mg and K) were extracted with 1 M NH_4OAc, adjusted to pH 7.0 (Thomas, 1982). The soil samples weighing 3 g were placed in a 100 mL centrifuge tube. 25 mL NH_4OAc solution was added and was shaken mechanically for 1 h. The supernatant was separated from the soil by centrifugation at 3000 rpm for 10 min. Supernatant is filtered into a 100 mL volumetric flask. The extract was made up to the volume with deionized water and used for determination of the ammonium acetate soluble cations, Ca, Mg, and K and above mentioned heavy metals. The contents were determined using an atomic absorption spectrophotometer.

The above mentioned procedures were used to determine the extracted heavy metals with 1 M NH_4OAc (Thomas, 1982). Soil texture was determined by the pipette method (Gee & Bauder, 1986). Soil pH and electrical conductivity (EC) were measured in soil-water (1:5; w:v) suspensions.

2.4 Plant Analysis

Plant tissues were dried in oven on 65 °C for one hour, crushed, and sieved (< 0.5 mm) to ensure homogeneity and digested in a mixture of duplicate acids (HNO_3 and $HClO_4$). Total elements i.e. potassium (K), calcium (Ca) magnesium (Mg), iron (Fe), zinc (Zn), copper (Cu), chromium (Cr), lead (Pb) in the extract of digested plant samples, were determined by atomic absorption spectrophotometer (IBSRAM, 1994). Sample weighing 0.25 g was digested with 5 mL concentrated $HClO_4$ by gradual heating it over a hot plate for 1 h. After drying 20% HNO_3 was added to the sample and it was heated again for 1 h. The solution was diluted to 50 mL with deionized water and passed through a 0.22 μm filter. The total elemental concentrations of cations and other elements (K, Ca, Mg and above mentioned metals) in plant tissue were determined on an atomic absorption spectrophotometer. Treatments were replicated thrice.

2.5 Water Analysis

All the analytical procedures used were the standard methods for water and wastewater analysis (APHA, 2005).

Wastewater samples were analyzed prior to and after the treatment with used 40% H_2O_2. The BOD was measured by using standard method (APHA, 2005), COD was determined by closed reflux colorimetric method using digester (HACH-LTG 082.99.40001) (APHA, 2005). The wastewater sample, digestion solution and sulphuric acid were digested in vials for two hours at 150 °C. After digestion, absorbance was measured at wavelength 605 nm in a spectrophotometer (LOVIBOND tintometer GMBH, 44287 DORTMUND). The pH meter (HANNA, HI-991003) was used for pH determination. H_2O_2 was measured according to Tetra Pak Technical data manual by using hydrometer and temperature. The H_2O_2 value was obtained after connecting temperature and hydrometer reading on third scale of H_2O_2 W/W

2.6 Statistical Analysis

The data collected during the studies were statistically analyzed using the Stat View software (SAS, 1999). A probability level of $P < 0.05$ was considered significant, and means were separated by Fisher's least significant difference (LSD) test.

3. Results

3.1 Behaviour of Different Elements in Soil

Concentrations of heavy metals in soils and plants in the area of study were found highly dependent on the source of irrigation. The results of the study indicated that water soluble K in the rain fed field shows an increasing trend with the soil depth (0–30 cm), (30–60 cm) and (60–100 cm), whereas in the wastewater irrigated field no significant differences were observed in different layers as shown in Figure1a. For rain fed field concentration of water soluble K was minimum 19.97 mg kg^{-1} in top soil (0–30 cm) and noted maximum 23.14 mg kg^{-1} in sub soil having depth (60–100 cm). On the other hand concentration of water soluble K in the wastewater irrigated field was greater in top soil 28.10 mg kg^{-1} and lesser was investigated (27.12 mg kg^{-1}) in 60–100 cm depth. Concentration of extractable K in wastewater irrigated field was (0–30 cm) > (30–60 cm) > (60–100 cm) (Figure 1b). There is no significant difference between different layers for total K in rain-fed field as shown in Figure 1c. Total concentration of K in wastewater irrigated field varied for different layers in order of (60–100 cm) > (30–60 cm) > (0–30 cm). Generally the concentration of water soluble Ca was found higher in the wastewater irrigated field as compared to the adjacent rain fed field. No significant differences were observed for the water soluble Ca in the three layers of wastewater irrigated field whereas for the rain-fed field maximum water soluble Ca was noted from 30–60 cm depth as shown in the Figure 1d. Amount of extractable Ca was noted almost same in all soil columns in waste water irrigated field (Figure 1e). While comparing the total Ca in wastewater irrigated field and rain fed field, it was found that total Ca concentration in wastewater irrigated field was almost 1.5 times greater than the rain fed field. Concentration of total Ca in the rain fed field varied in different soil columns as 30–60 cm > 0–30 cm > 60–100 cm as shown in the Figure 1f. Concentration of Mg in all layers of the soils was noted almost same with no significant differences for both types of fields Figure 2a. However, the concentration of water soluble Mg was slightly lesser (175.40 mg kg^{-1}) in the top soil as compared to the subsoil layers Figure 2a. Extractable Mg varied in the order 60–100 cm > 30–60 cm > 0–30 cm in the rain fed field. Contents of total Mg in rain fed soil are greater as compared with the waste water irrigated soil. Regardless of irrigation source maximum concentration of extractable Mg was obtained from topsoil (0–30 cm) with 412.83 mg kg^{-1} and minimum was achieved from the bottom layer (30–60 cm) with 405.62 mg kg^{-1} (Figure 2b).

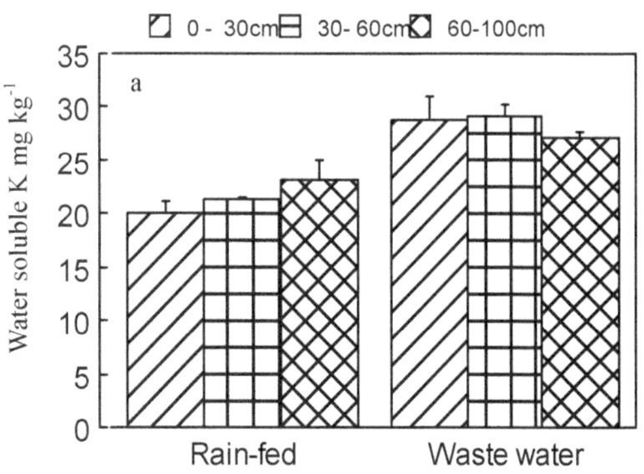

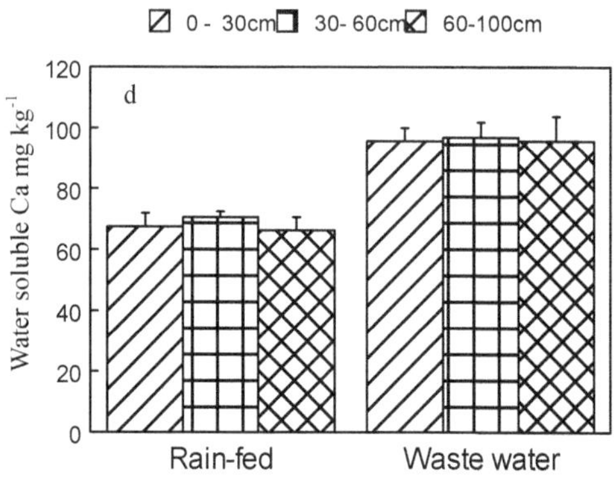

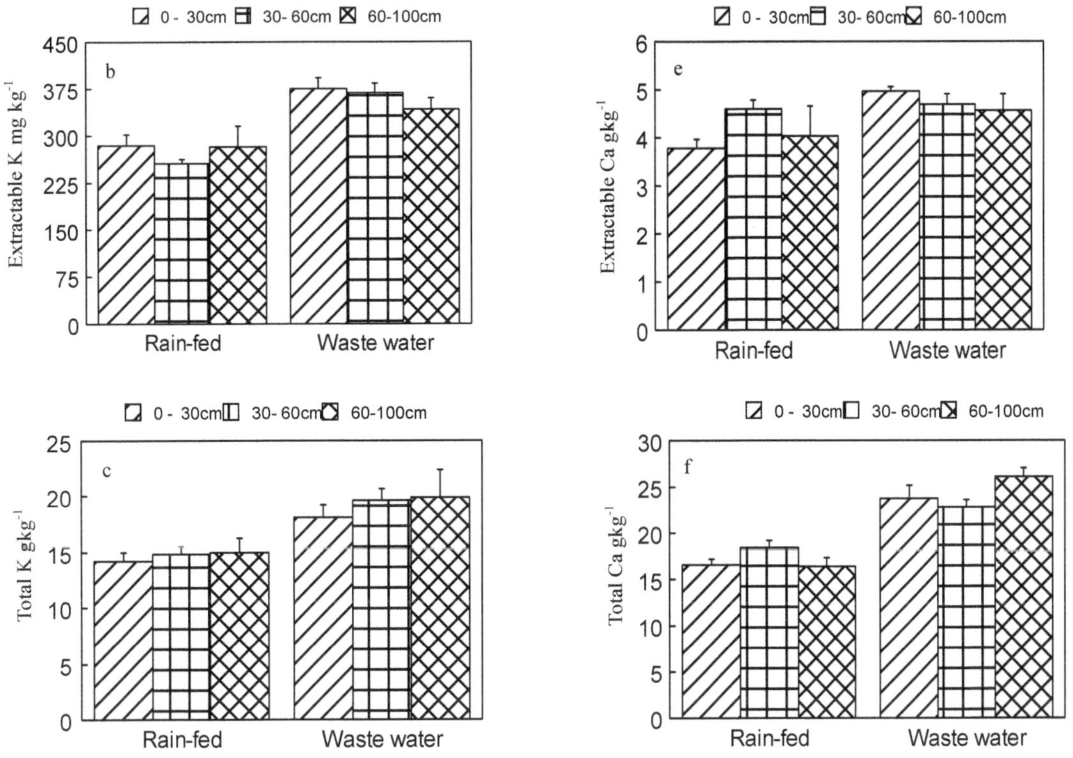

Figure 1. a, b and c show water soluble, extractable and total K (mg kg^{-1}) in rain fed and wastewater irrigated field. While d, e and f show water soluble, extractable and total Ca (mg kg^{-1}) in different soil layers in rain fed and wastewater irrigated field

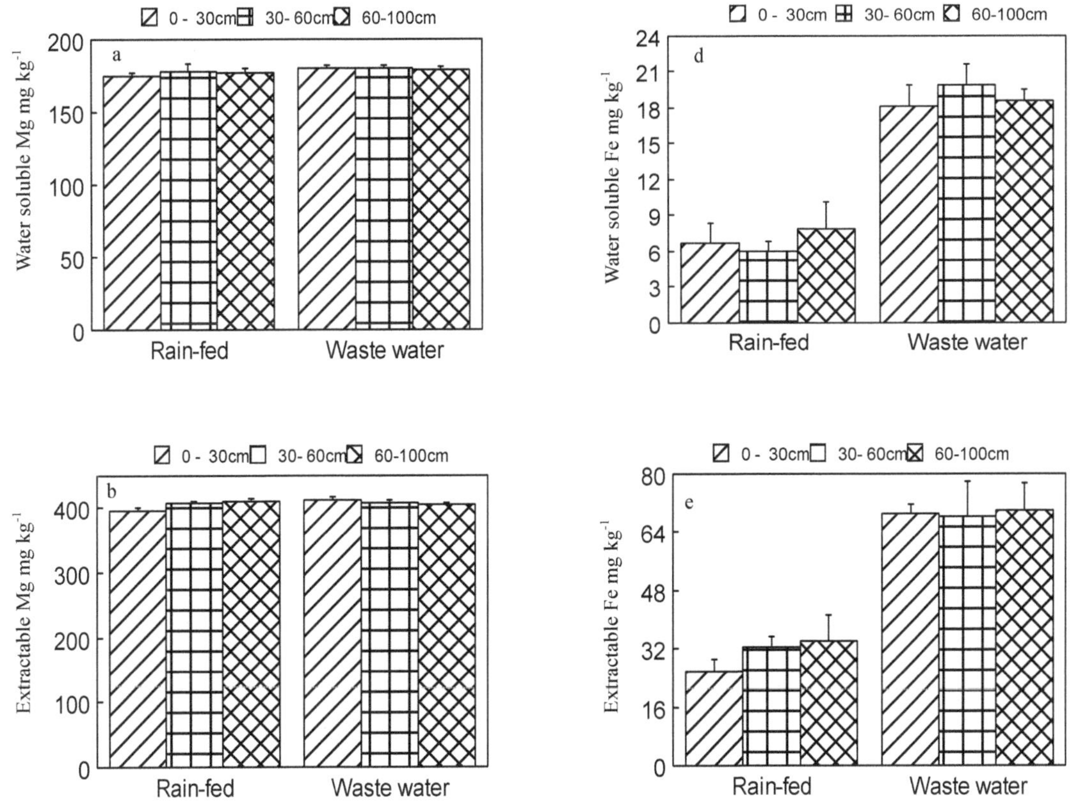

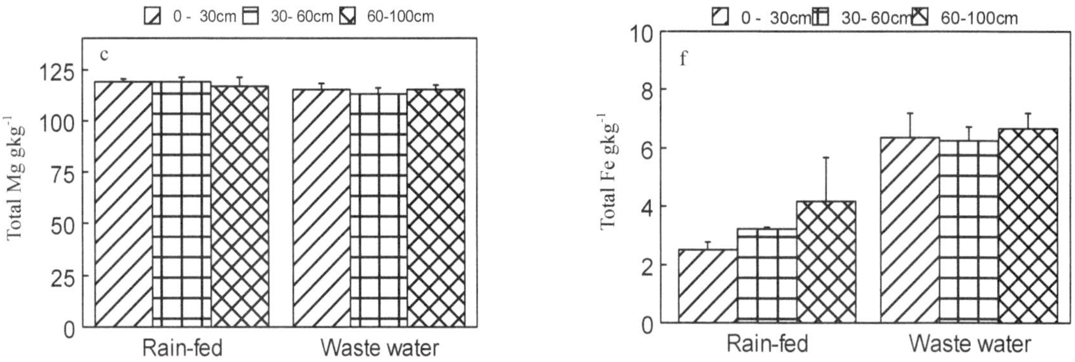

Figure 2. a, b and c show water soluble, extractable and total Mg (mg kg^{-1}) in rain fed and wastewater irrigated field. While d, e and f show water soluble, extractable and total Fe (mg kg^{-1}) in different soil layers in rain fed and wastewater irrigated field

In wastewater irrigated field water soluble Fe varied in order of 30–60 cm > 60–100cm > 0–30 cm whereas in rain fed field the concentration was 60–100 cm >0–30 cm > 30–60 cm (Figure 2d). Maximum concentration of extractable Fe in wastewater irrigated field found in the soil depth 60–100 cm (70.02 mg kg^{-1}) whereas the minimum value of extractable Fe was observed in the subsoil (68.27 mg kg^{-1}) as shown in Figure 2e. In the rain fed field the amount of extractable Fe was considerably less as compared with the adjacent wastewater irrigated field as shown in Figure 2e. In the wastewater irrigated field the maximum concentration of total Fe was found in the bottom soil layer (6.68 g kg^{-1}) and minimum (6.25 g kg^{-1}) amount was observed in the subsoil layer with depth 30–60 cm. In the wastewater irrigated field the total Fe varied in order (30–60 cm) < (0–30 cm) < (60–100 cm) whereas total Fe varied in the rain fed field as (0–30 cm) < (30–60 cm) < (60–100 cm), however, amount of total Fe in wastewater irrigated field was noted almost same in different depths of soil with non significant variations (Figure 2f).

Figure 3a shows that water soluble Zn greatly reduced in the rain fed field as compared to the wastewater irrigated field where slightly higher concentration was found in top layer of soil (0–30 cm) with 0.43 mg kg^{-1}. Amount of water soluble Zn as observed in the rain fed field varied in order 60–100 cm > 30–60 cm > 0–30 cm. Accurate measurements of the total metal contents in polluted soils are required to assess the potential risk of these areas. Higher total Zn contents were found in the wastewater irrigated field, whereas in the rain fed field the concentration of total Zn was decreased. In wastewater irrigated field total Zn concentration was observed higher with maximum amount of total Zn (79.93 mg kg^{-1}) in topsoil. Minimum amount of total Zn (59.83 mg kg^{-1}) was found in the soil layer having depth 30–60 cm. Concentration of total Zn was observed lesser (22.3 mg kg^{-1}) in rain fed field in top soil layer 0–30 cm while its maximum value was observed in the bottom soil layer as shown in Figure 3c.

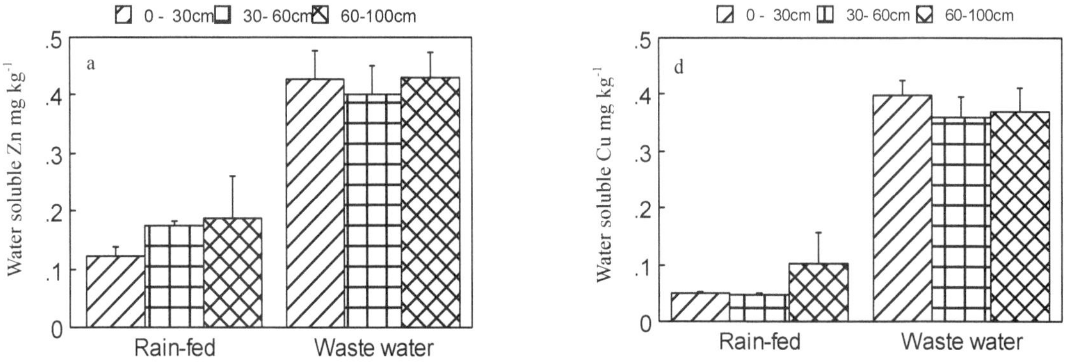

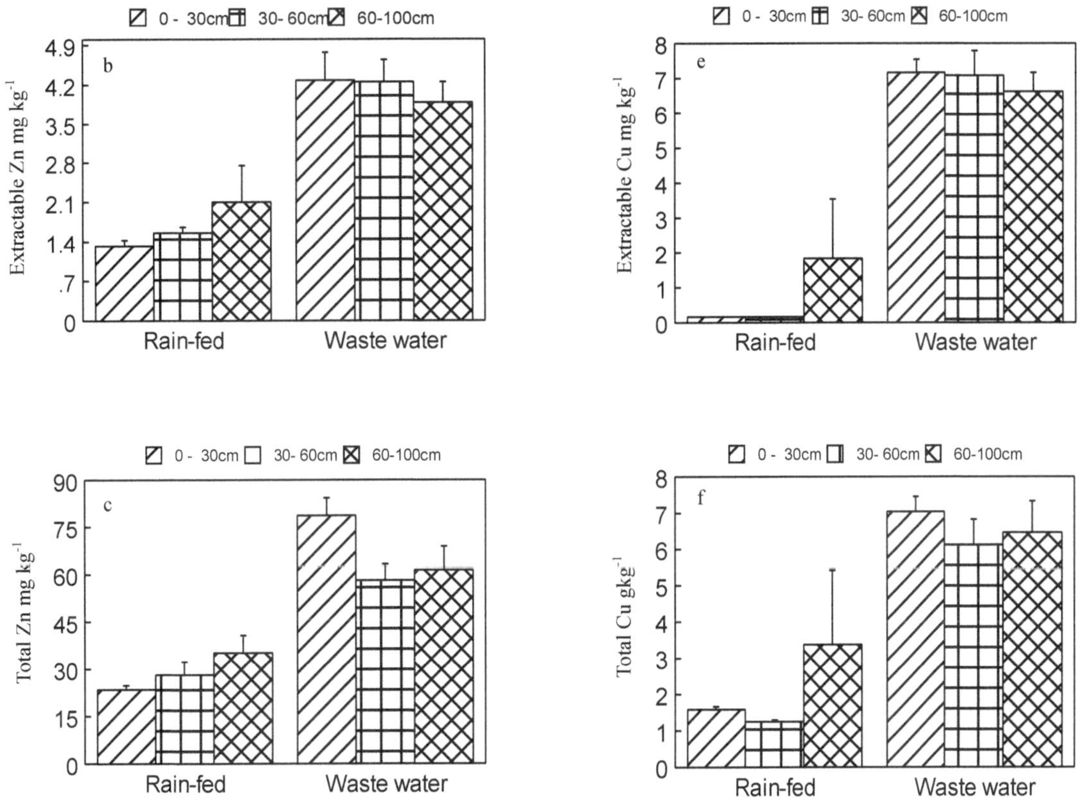

Figure 3. a, b and c show water soluble, extractable and total Zn (mg kg^{-1}) in rain fed and wastewater irrigated field. While d, e and f show water soluble, extractable and total Cu (mg kg^{-1}) in different soil layers in rain fed and wastewater irrigated field

Amount of water soluble Cu in rain-fed field is negligibly lesser in all soil columns and found more than three times higher in the wastewater irrigated field. Within the soil column in the wastewater irrigated field the concentration of water soluble Cu was found significantly higher in the top soil (0.39 mg kg^{-1}) and observed minimum (0.35 mg kg^{-1}) in the subsoil layer with depth 30–60 cm with marginal differences as shown in Figure 3d. In the rain fed field concentration of water soluble Cu was noted negligibly small (0.05 mg kg^{-1}) in the top soil and slightly higher (0.10 mg kg^{-1}) as shown by Figure 3d. The order of extractable Cu concentration in the wastewater irrigated field was noted as 0–30 cm > 30–60 cm > 60–100 cm. The concentration of extractable Cu in the rain fed field was negligible (0.17 mg kg^{-1}) in the topsoil and subsoil column but its concentration increased in the bottom layer to (1.8 mg kg$^{-1)}$ as shown in Figure 3e. The amount of total Cu in the wastewater irrigated field was observed higher and noted relatively lesser in the adjacent rain fed field. Total Cu contents in the wastewater irrigated field varied in order of 0–30 cm > 60–100 cm > 30–60 cm as shown by Figure 3f.

Although Cr is required in trace amounts for sugar and lipid metabolism and in larger amounts it can be toxic. Water soluble Cr contents in the wastewater irrigated field were found significantly higher as compared with the adjacent rain fed field. In wastewater irrigated field the amount of water soluble Cr in soil layer 60–100 cm was found increased to 6.17 mg kg^{-1} and subsoil layer (30–60 cm) water soluble Cr concentration was decreased to 5.61 mg kg^{-1} (Figure 4a). In the rain fed field maximum water soluble Cr contents (4.60 mg kg^{-1}) were found in the bottom soil layer (60–100 cm). Concentration of water soluble Cr varied in the rain fed field in order of 60–100 cm > 0–30 cm > 30–60 cm. The value of extractable Cr was found almost double in the wastewater field and lower in the rain fed field (Figure 4b). Total Cr was found increased in the wastewater irrigated field by an appreciable amount as compared to the rain fed field. Maximum total Cr (128.78 mg kg^{-1}) was found in the subsoil 30–60 cm while minimum concentration (122.42 mg kg^{-1}) was observed in the top soil column. Total Cr varied in the rain fed field in the order 60–100 cm > 30–60 cm > 0–30 cm (Figure 4c).

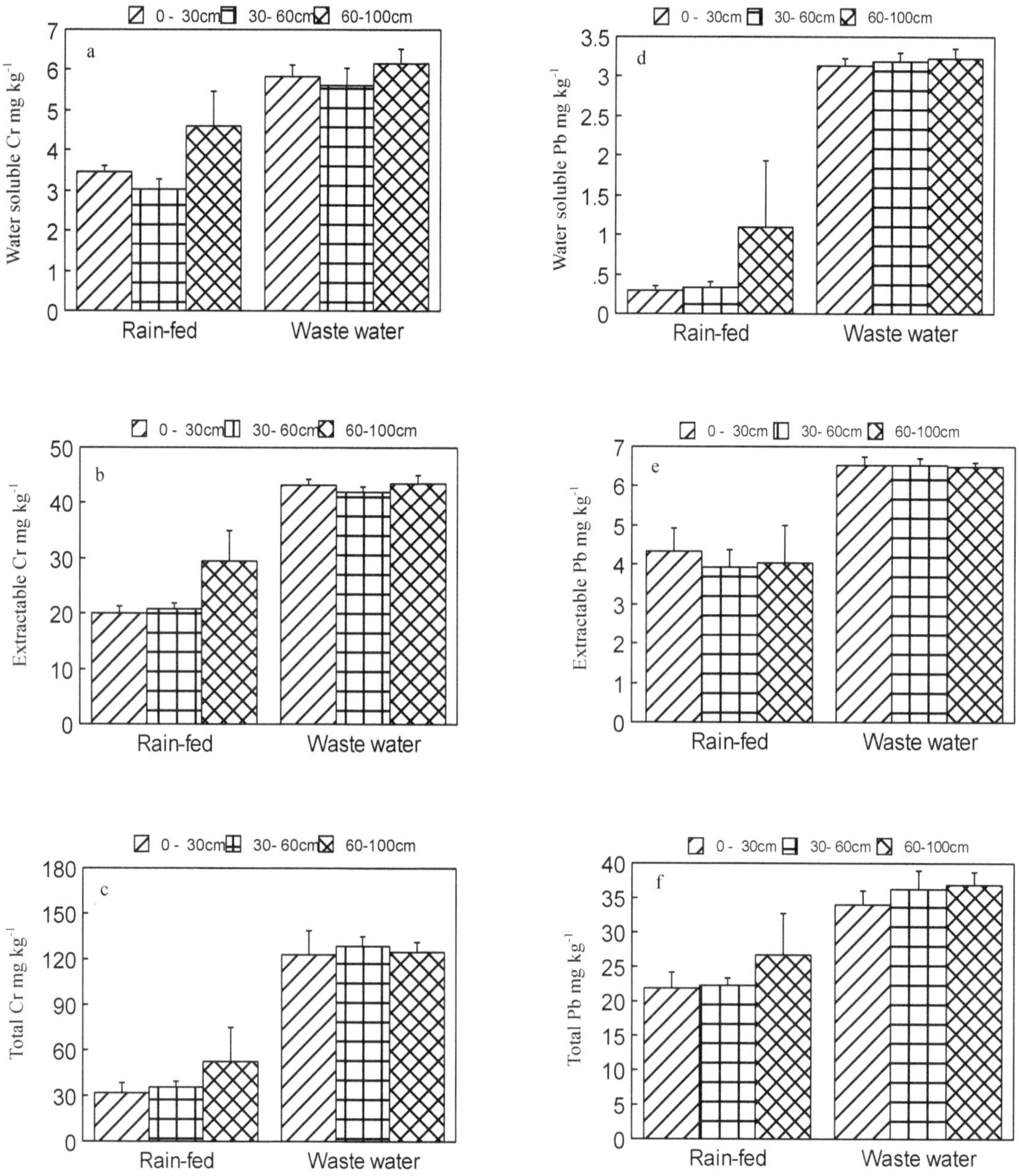

Figure 4. a, b and c show water soluble, extractable and total Cr (mg kg^{-1}) in rain fed and wastewater irrigated field. While d, e and f show water soluble, extractable and total Pb (mg kg^{-1}) in different soil layers in rain fed and wastewater irrigated field

Concentration of water soluble Pb was observed higher in the wastewater irrigated field as compared to the adjacent rain fed field as shown in Figure 4d. No significant differences in concentration of water soluble Pb were found in all three layers of wastewater irrigated field, however, significant variations were recorded for water soluble Pb in the adjacent rain fed soil. The amount of extractable Pb was observed higher in the wastewater irrigated field. Extractable Pb varied in the rain fed field in the order 0–30 > 60–100 cm > 30–60 cm (Figure 4e). Total Pb concentration was found higher in the wastewater irrigated field and observed lower in the rain fed field. Maximum amount of total Pb (36.83 mg kg^{-1}) was noted in the soil column (60–100 cm) in the wastewater irrigated field as shown in Figure 4f. Total Pb concentration was minimum the topsoil column (21.93 mg kg^{-1}) of rain fed soil.

3.2 Behaviour of Different Elements in Plant

Concentration of K found higher in the wastewater irrigated cabbage shoots as compared with the adjacent rain-fed field as shown in Figure 5a. Plant sampling was done from three random sites Site A, Site B and Site C in each field. Potassium contents were maximum in Site A (0.57 mg kg^{-1}) and minimum in Site B (0.50 mg kg^{-1}) Figure 5a. Amount of Ca and Mg in plant shoots cultivated in the wastewater irrigated field was noted higher than cabbage grown in rain fed field (Figures 5b and 5c). The concentration of Mg in the plant in Site A was found maximum with 0.74 mg kg^{-1} and noted same in Site B and Site C of the wastewater irrigated field (Figure 5b). On the other hand the amount of Mg in the cabbage shoots in rain fed field was almost same in Site A, Site B and Site C.

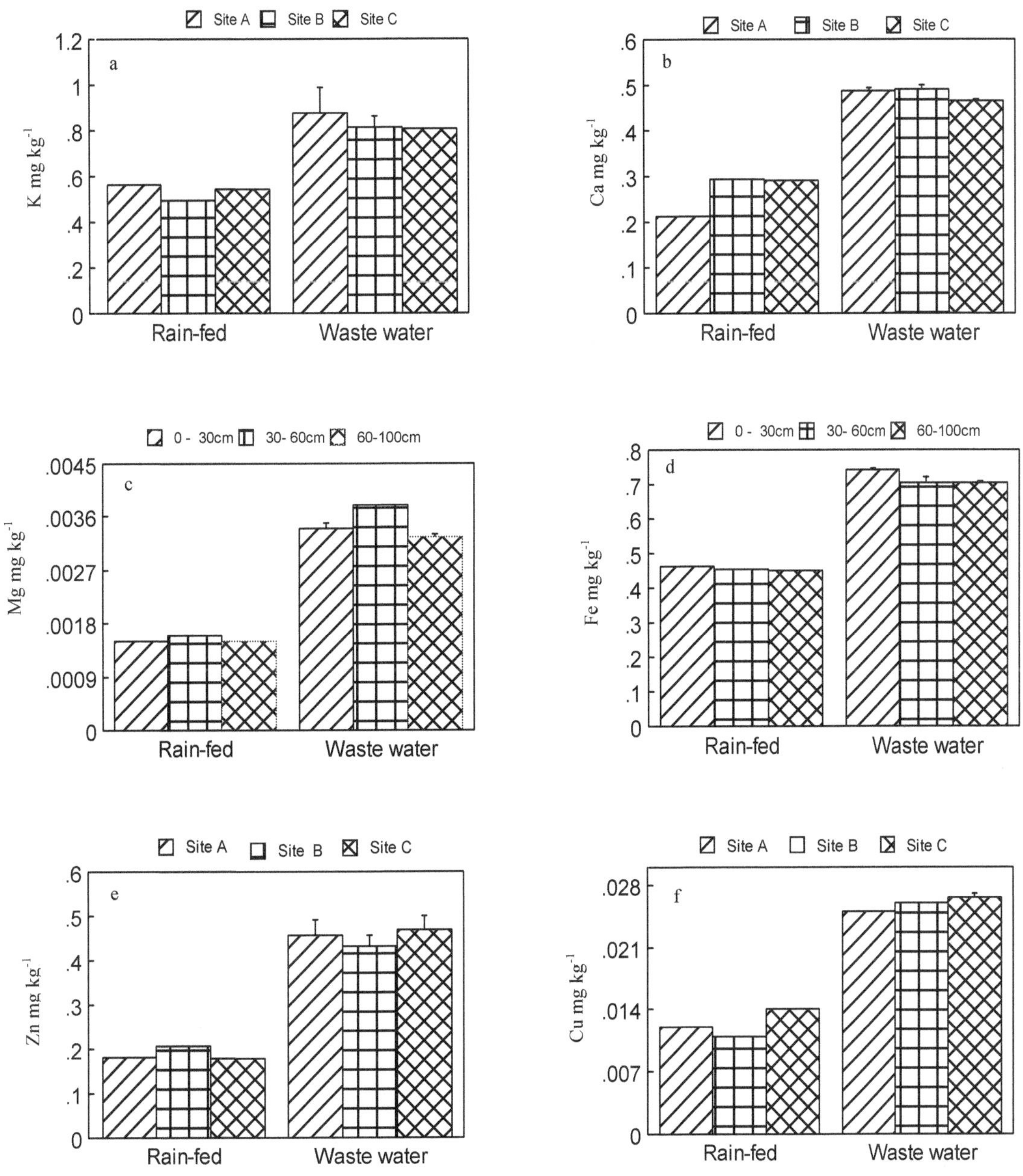

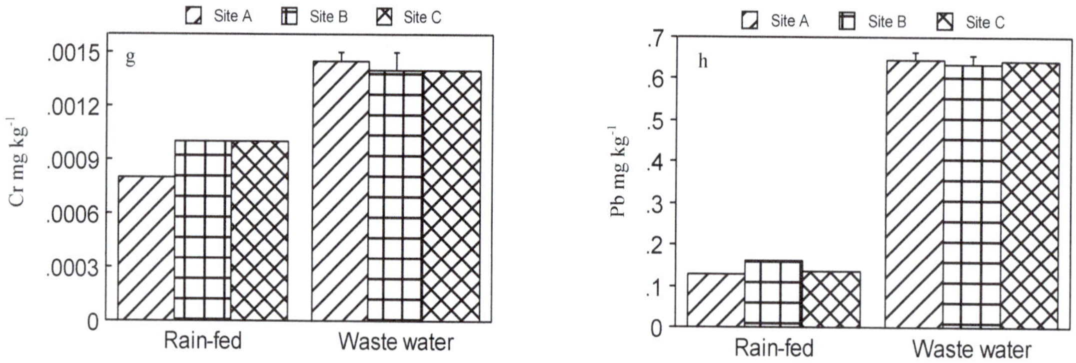

Figure 5. Concentration (mg kg^{-1}) of K (a), Ca (b), Mg (c), Fe (d), Zn (e), Cu (f), Cr (g) and Pb (h) in *Brassica Oleracea* L in wastewater irrigated and rainfed field

The Fe contents were found higher in the cabbage plant samples in all three sites of wastewater irrigated field and considerably decreased in the sample collected from Site A, B and C of the rain fed field (Figure 5d). Amount of Zn in the cabbage shoots of wastewater irrigated field were found significantly higher in all three sites A, B and C and observed lower in the Site A, B and C of the rain fed field (Figure 5e). In wastewater irrigated field the Zn contents varied in the order Site A > Site B > Site C (Table 10). Zn contents were found maximum (0.21 mg kg^{-1}) in the plant sample of Site B. No significant differences were observed in the plant samples collected from Site A and Site C of the rain fed field. Cu concentration was also noted much higher in the plant samples of the wastewater irrigated field as compared to the rain fed filed (Figure 5f). In the wastewater irrigated field Cu concentration found highest in the cabbage shoots collected from site A. Zn and Cu values were less than 60 mg kg^{-1} and 40 mg kg^{-1} in *Brassica oleracea* L. The concentration of Cu in the plant in all three sites were observed with marginal differences with variation in order Site C > Site B > Site A.

Cr concentration was observed higher in the plant shoots collected from the sites of wastewater irrigated field and lower concentration was observed in the samples collected from the rain fed field (Figure 5g). Cr was observed in the rain fed field with lesser concentration in Site A as compared with Site B and Site C. In the wastewater irrigated field concentration of Cr was same in Site B and Site C with maximum value (0.0014 mg kg^{-1}) in Site A. Higher amount of Pb content was found in the shoots of sample collected from the wastewater irrigated field as compared to the samples collected from adjacent rain fed field. There was no significant difference in the Pb content in the wastewater irrigated field in all three sites. In rain fed field the site wise concentration of Pb was Site B > Site C > Site A as shown in Figure 5h. Generally concentration of heavy metals Fe, Zn, Cu and Pb (except Cr) in cabbage (*Brassica oleracea* L.) was found significantly lesser in rain-fed field as compared with the adjacent wastewater irrigated field. Although total heavy metals in the soil are not available for plants and precise calculations cannot be done, still regression graph plotted between the total soil potassium and plant available potassium. Figure 6 shows increased concentration of plant available nutrient with increase in total soil contents. Simple regression analysis shows that with increase in the water soluble heavy metals the value of total heavy metals also increases. Same holds for Fe, Zn, Cr, Cu and Pb.

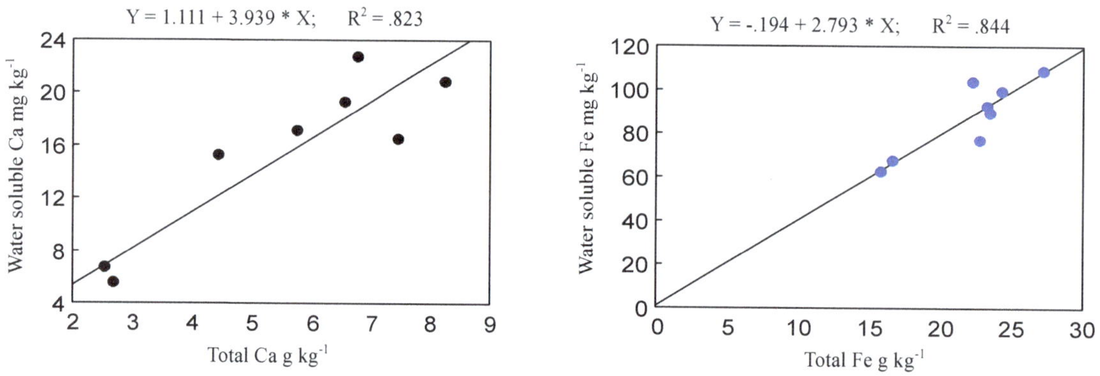

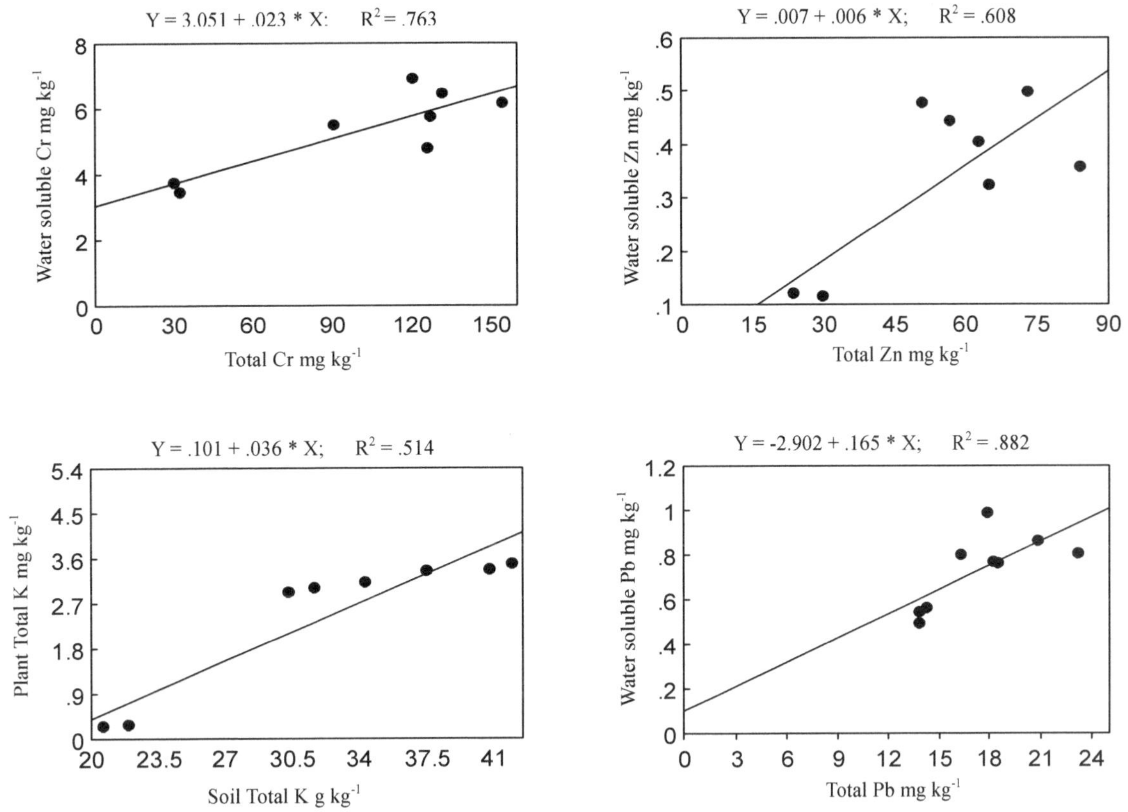

Figure 6. Simple regression analysis between the concentrations water soluble heavy metals the value of total heavy metals

4. Discussion

Main sources of water contamination are domestic waste products, paints, wrapping and packing materials, plumbing materials, vehicular emission, drainage from Abbottabad Maneshra road, limited scaled industry/hospital waste and auto workshops present in the vicinity of wastewater flow. For rain fed field concentration of water soluble K was minimum 19.97 mg kg^{-1} in top soil (0–30 cm) and noted maximum 23.14 mg kg^{-1} in sub soil having depth (60–100 cm). The logical reason may be continuous leaching of K due to rain in vertical column. The accumulation of heavy metals in agricultural soils has been remained an increasing concern for scientists from all over the world (Facchinelli et al., 2001; Abollino et al., 2002).

In case of heavy metals in soil results showed that concentrations of heavy metals in rain-fed fields were found within the permissible limits. However in the wastewater irrigated soil the concentrations of all heavy metals understudy were significantly higher. Human activities such as industrial and energy production, vehicle exhaust, waste disposal, coal and fuel combustion were the potential sources for heavy metals to enter agricultural soils (Chen et al., 2005; Lucho-Constantino et al., 2005). In wastewater irrigated field water soluble Fe varied in order of 30–60 cm > 60–100 cm > 0–30 cm whereas in rain fed field the concentration was 60–100 cm > 0–30 cm > 30–60 cm (Figure 4a). In the rain fed field it is in permissible limits (in the range of 6–8 mg kg^{-1}). The analysis of bioavailable metals is probably more significant than the analysis of total content, because the former allows prediction of the risk of metal uptake by plants and its mobility in the system (Bell et al., 1991). The toxicity of heavy metals in soil significantly varied with soil characteristics and the time elapsed after contamination (Doelman & Haanstra, 1984). Heavy metal concentrations in surface soil were significantly higher than those in the subsoil of cultivated field (Li et al., 1987). However, only soluble, exchangeable and chelated metal species in the soils were the labile fractions available for plants (Kabata-Pendias, 1993). At higher concentrations, some metals have strong toxic effects and are regarded as environmental pollutants (Nedelkoska & Doran, 2000; Chehregani et al., 2005).

Heavy metals after accumulating in soil get transferred to vegetables grown on these soils (Malla et al., 2007). Natural levels of lead in surface soils are usually below 50 ppm (Chaney et al., 1984; Reagan & Silbergeld,

1989).

Over the period of time population growth has resulted in increased domestic and industrial effluents. Wastewater of the cities and towns is thus a readily available and costless source of irrigation. People prefer wastewater as they get huge quantities of vegetable yields because of the presence of nutrients like Ca, Mg, K in abundance without knowing its effect on soils, plants and ultimately on consumers. The superiority of wastewater over rain is addition of nutrients present in it. Results in Figures 1-8 indicated that water soluble heavy metals varied in soil in the order of Fe > Cr > Pb > Zn > Cu irrespective of the soil depth and irrigation source. For extractable heavy metals the order was observed as Fe > Cr > Pb > Zn > Cu for rain fed field across all soil layers. In wastewater irrigated field similar concentrations were observed except the order of Zn and Cu was changed. For total heavy metals in all layers of soil the order was noted as Fe > Cu > Cr > Pb > Zn irrespective of the source of irrigation. On the other hand concentrations of water soluble essential elements under study varied as Mg > Ca > K for both rain fed and wastewater irrigated field. Heavy metal uptake by crops growing in contaminated soil is a potential hazard to human health because of transmission in the food chain (Gincchio et al., 2002; Friesl et al., 2006). There is also a concern with regard to heavy metal transmission through natural ecosystems (MacFarlane & Burchett, 2002; Walker et al., 2003). Parameters related to the heavy metal uptake have been used as sensitive indicators of heavy metal toxicity (Nannipieri et al., 1997; Wilke, 1991). Heavy metal behaviour in soils and plant uptake are strongly dependent on the nature of the metal, sludge/soil physico-chemical properties and plant species (McBride, 2003). While growing edible vegetables in contaminated soils, it is important to know the extent of accumulation in plants and their different parts. The toxicity of heavy metals in soil significantly varies with soil characteristics and the time elapsed after contamination (Doelman & Haanstra, 1984). Heavy metals are easily accumulated in the edible parts of the leafy vegetables. Vegetables take up heavy metals and accumulate them in their edible and in inedible parts in quantities high enough to cause clinical problems both to animals and human beings consuming these metal rich plants. Application of wastewater resulted in marked increase in cabbage (*Brassica Oleracea* L.) growth as compare to the rain fed field. Plant that received wastewater produced greater biomass then those received rain water. This superiority of wastewater over rain fed could possibly be due to the addition of nutrients present in the wastewater. Although sludge compost increase biomass of the crop but at the same time toxic metals can be transferred and concentrated into plant tissues from the soil (Alloway, 1995). Metals in water soluble fraction may be readily leachable and bioavailable in the environment. Contaminants in bulk soil might be transferred into soil solution, making them available to roots.

5. Conclusions

The study concluded that increased heavy metals were observed in the wastewater irrigated field as compared with the adjacent rain fed field. Total heavy metals were higher in wastewater irrigated field than rain fed field. For total heavy metals in all layers of soil vertical column the order was noted as Cr > Zn > Pb > Cu > Fe for wastewater irrigated field and Cr > Zn > Pb > Fe > Cu for rain fed field for all soil depths. The rain fed soil exhibited lower concentrations of water soluble heavy metals as compared to wastewater irrigated field which may reduce the risk of metal transfer from soil to plants. Fe, Cu and Cr were found higher in wastewater irrigated field whereas Pb and Zn showed their minimum concentrations. Mg and Ca were higher in both rain fed and wastewater irrigated soils whereas significantly decreased K was observed in both. We conclude that wastewater irrigation causes heavy metal accumulation in both soils and the plants. The use of wastewater for agriculture carefully rationalized before wastewater irrigation.

References

Abollino, O., Aceto, M., Malandrino, M., Mentasi, E., Sarzanini, C., & Petrella, F. (2002). Heavy metals in agricultural soils from Piedmont, Italy. Distribution, speciation and chemometric data treatment. *Chemosphere, 49*, 545-557. http://dx.doi.org/10.1016/S0045-6535(02)00352-1

Alloway, B. J. (1995). *Heavy Metals in Soils*. Chapman & Hall, London. http://dx.doi.org/10.1007/978-94-011-1344-1

American Public Health Association, Inc. (APHA). (2005). *Standard Methods for the Examination of Water and Wastewater* (21st ed.). New York, NY.

Bell, J. M., & Keith, M. O. (1991). A survey of variation in the chemical composition of commercial canola meal produced in Western Canadian crushing plant. *Can. J. Anim. Sci., 71*(2), 469-480. http://dx.doi.org/10.4141/cjas91-056

Burn, L. A., Maillet, J., Hinsinger, P., & Pepin, M. (2001). Evaluation of copper availability to plants in copper-contaminated vineyard soils. *Environ. Pollut., 111*, 293-302. http://dx.doi.org/10.1016/S0269-7491(00)00067-1

Chaney, R. L. (1984). *The potential for heavy metal exposure from urban gardens and soils* (pp. 37-84).

Chehregani, A., Malayeri, B., & Golmohammadi, R. (2005). Effect of heavy metals on the developmental stages of ovules and embryonic sac in Euphorbia cheirandenia. *Pak. J. Biol. Sci., 8*, 622-625. http://dx.doi.org/10.3923/pjbs.2005.622.625

Chen, T. B., Zheng, Y. M., Lei, M., Huang, Z. C., Wu, H. T., Chen, H., ... Tian, Q. Z. (2005) Assessment of heavy metal pollution in surface soils of urban parks in Beijing, China. *Chemosphere, 60*, 542-551. http://dx.doi.org/10.1016/j.chemosphere.2004.12.072

Doelman, P., & Haanstra, L. (1984). Short-term and long-term effects of cadmium, chromium, copper, nickel, lead and zinc on soil microbial respiration in relation to abiotic soil factors. *Plant Soil, 79*(3), 317-327. http://dx.doi.org/10.1007/BF02184325

Facchinelli, A., Sacchi, E., & Mallen, L. (2001). Multivariate statistical and GIS-based approach to identify heavy metal sources in soils. *Environment Pollution, 114*, 313-324. http://dx.doi.org/10.1016/S0269-7491(00)00243-8

Friesl, W., Friedl, J., Platzer, K., Horak, O., & Gerzabek, M. H. (2006). Remediation of contaminated agricultural soils near a former Pb/Zn smelter in Austria: batch, pot and field experiments. *Environ. Pollut., 144*, 40-50. http://dx.doi.org/10.1016/j.envpol.2006.01.012

Gee, G. W., & Bauder, J. W. (1986). Particle-size analysis. In A. Klute (Ed.), *Methods of soil analysis* (2nd ed., Part 1, Agron. Monogr. 9, pp. 383-411). ASA and SSSA, Madison, WI.

Gincchio, R., Rodriguez, P. H., Badilla-Ohlbaum, R., Allen, H. E., & Lagos, G. E. (2002). Effect of soil copper content and pH on copper uptake of selected vegetables grown under controlled conditions. *Environ. Toxicol. Chem., 21*, 1736-1744. http://dx.doi.org/10.1002/etc.5620210828

Haq, A. U., Bates, T. E., & Soon, Y. K. (1980). Comparison of extractants for plant-available zinc, cadmium, nickel, and copper in contaminated soils. *Soil Sci. Sot. Am. J., 44*, 772-777. http://dx.doi.org/10.2136/sssaj1980.03615995004400040023x

IBSRAM. (1994). *Methodological guidelines for IBSRAM's soil management networks* (2nd ed.). International Board for Soil Research and Management, Bangkok

Kabata-Pendias, A. (1993). Behavioural properties of trace metals in soils. *Applied Geochemistry, 2*, 3-9. http://dx.doi.org/10.1016/S0883-2927(09)80002-4

Li, Y. J., Ning, R. A., & Wang, D. Z. (1987). Mapping of soil background content map in 11 elements in aquic soils of the Yellow River downstream. *Agro-Environ., 6*,18-21.

Lindsay, W. L., & Norvell, W. A. (1978). Development of a DTPA soil test for zinc, iron, manganese and copper. *Soil Sci. Sot. Am. J., 42*, 421-428. http://dx.doi.org/10.2136/sssaj1978.03615995004200030009x

Lucho-Constantino, C. A., Álvarez-Suárez, M., Beltrán-Hernández, R. I., Prieto-Garcia, F., & Poggi-Varaldo, H. M. (2005). A multivariate analysis of the accumulation and fractionation of major and trace elements in agricultural soils in Hidalgo State, Mexico irrigated with raw wastewater. *Environment International, 31*, 313-323. http://dx.doi.org/10.1016/j.envint.2004.08.002

MacFarlane, G. R., & Burchett, M. D. (2002). Toxicity, growth and accumulation relationships of copper, lead and zinc in the grey mangrove *Avicennia marina* (Forsk.). *Vierh. Mar. Environ. Res., 54*, 65-84. http://dx.doi.org/10.1016/S0141-1136(02)00095-8

Malla, R., Tanaka, Y., Mori, K., & Totawat, K. L. (2007). Short term effect of sewage irrigation on chemical buildup in soil and vegetables. *The Agric. Engg. Int. The CIGR J., IX*, LW 07 006, 14.

Martens, D. C. (1968). Plant availability of extractable boron, copper, and zinc as related to selected soil properties. *Soil Sci., 106*, 23-28. http://dx.doi.org/10.1097/00010694-196807000-00004

McBride, M. B. (2003). Toxic metals in sewage sludge-amended soils: Has promotion of beneficial use discounted the risks? *Adv. Environ. Res., 8*, 5-19. http://dx.doi.org/10.1016/S1093-0191(02)00141-7

Mehlich, A. (1984) Mehlich 3 soil test extractant: A moditication of Mehlich 2 extractant. *Commun. Soil Sci. Plant Anal., 15*, 1409-1416. http://dx.doi.org/10.1080/00103628409367568

Nannipieri, P., Badalucco, L., Landi, L., & Pietramellara, G. (1997). Measurement in assessing the risk of chemicals to the soil ecosystem. In J. T. Zelikoff (Ed.), *Ecotoxicology: Responses and Risk Assessment, an OECD Workshop*. SOS Publications, Fair Haven, NJ.

Nedelkoska, T. V., & Doran, P. M. (2000). Characteristics of heavy metal uptake by plant species with potential for phytoremediation and phytomining. *Minerals Engineering, 13*, 549-561. http://dx.doi.org/10.1016/S0892-6875(00)00035-2

Nriagu, J. O. (1991). Human influence on the global cycling of trace metals. In J. G. Farmer (Ed.), *Heavy Metals in the Environment* (Vol. 1, pp. 1-5). CEP Consultants Ltd., Edinburgh, UK.

Oliver, M. A. (1997). Soil and human health: A review. *European Journal of Soil Science, 48*(4), 573-592. http://dx.doi.org/10.1046/j.1365-2389.1997.00124.x

Reagan, P. L., & Silbergeld, E. K. (1989). Establishing a health based standard for lead in residential soils. *Environ Geochem Health, 12*(Suppl.), 198-230.

SAS. (1999). *Stat View reference* (3rd ed.). SAS Institute Inc., Cary, NC, USA.

Speir, T. W., Kettles, H. A., Parshotam, A., Searle, P. L., & Vlaar, L. N. C. (1995). A simple kinetic approach to derive the ecological dose value, ED50, for the assessment of Cr(VI) toxicity to soil biological properties. *Soil Biol. Biochem., 27*(6), 801-810. http://dx.doi.org/10.1016/0038-0717(94)00231-O

Thomas, G. W. (1982). Exchangeable cations. In A. L. Page, R. H. Miller & D. R. Keeney (Eds.), *Methods of Soil Analysis* (Part 2, Agronomy series No. 9, pp. 159-165). Am. Soc. Agronomy and Soil Sci. Soc. Am. Inc., Publ. Madison W. I. USA.

Walker, D. J., Clemente, R., Roig, A., & Bernal, M. P. (2003). The effects of soil amendments on heavy metal bioavailability in two contaminated Mediterranean soils. *Environ. Pollut., 122*, 303-312. http://dx.doi.org/10.1016/S0269-7491(02)00287-7

Wilke, B. M. (1991). Effects of single and successive additions of cadmium, nickel and zinc on carbon dioxide evolution and dehydrogenase activity in a sandy luvisol. *Biol. Fert. Soil., 11*(1), 34-37. http://dx.doi.org/10.1007/BF00335831

Zulfiqar, A. B., Qaisar, M., Iftikhar, A. R., Amir, H. M., Naim, R., & Donglei, W. (2011). Integrated chemical treatment of municipal wastewater using waste hydrogen peroxide and ultraviolet light. *Physics and Chemistry of the Earth, 36*, 459-464. http://dx.doi.org/10.1016/j.pce.2010.03.024

21

Response of Two Wheat Cultivars to Supplemental Nitrogen under Different Salinity Stress

A. L. Saleh[1], A. A. Abd El-Kader[2] & A. K. Alva[3]

[1] Plant Nutrition Department, National Research Centre, Egypt

[2] Soils and Water Use Department, National Research Centre, Egypt

[3] Northern Plains Agricultural Research Laboratory, United States Department of Agriculture, Agricultural Research Service, Sidney, MT, USA

Correspondence: A. K. Alva, Northern Plains Agricultural Research Laboratory, United States Department of Agriculture, Agricultural Research Service, Sidney, MT 59270, USA. E-mail: ashok.alva@ars.usda.gov

Abstract

Effects of supplemental nitrogen (N), as either farmyard manure (FYM) or urea, on response of two wheat (*Triticum aestivum*) cultivars (a salt sensitive 'Sakha 69' and a salt tolerant 'Sakha 93') were investigated in a green house experiment under various salinity levels (control, 6, 9, or 12 dS m^{-1}). Grain and straw yields of both cultivars decreased with an increase in salinity levels. Supplamental N application, using FYM or urea, mitigated the adverse effects of salinity only at the low salinity level (6 dS m^{-1}). This effect was greater in a salt tolerant cultivar (Sakha 93) than that in a salt sensitive cultivar (Sakha 69). At the moderate and high salinity (9 and 12 dS m^{-1}) levels the supplemental N had no beneficial effects in mitigating the salinity stress of both cultivars. The mean grain yields, across all salinity levels and cultivars, of the plants received FYM and urea were greater by 11, and 8%, respectively, as compared to that of the plants received no supplemental N. The corresponding values for straw were 12 and 7%. The concentrations of N, P and K in the grain and straw significantly decreased with increasing salinity levels. Concentrations of Na, Cl, and Ca in the grain and straw were greater in salt sensitive cultivar than those in a salt tolerant cultivar. Concentrations of these elements significantly increased with an increase in salinity levels. This study demonstrated that supplemental N, as either FYM or as Urea, can mitigate negative effects of mild salinity stress, and that this beneficial effect was greater in a salt tolerant cultivar as compared to that in a salt sensitive cultivar.

Keywords: abiotic stress, salinity, nitrogen sources, farm yard manure, mineral nutrition, *Triticum aestivum*

1. Introduction

Soil salinity is a serious problem in Egypt with upto 33% of the cultivated land impacted by high salinity due to low precipitation (< 25 mm annual rainfall) and irrigation with saline water (El-Hendawy et al., 2005). Salinity is the major environmental factor limiting plant growth and productivity worldwide (Munns, 2002). The negative effects of salinity on plant growth may be associated to an increase in the osmotic potential of the soil solution that reduces plant available water, or increase in concentrations of certain ions that inhibits plant metabolism (Onani, 2005).

Wheat is the most important cereal crop in Egypt, however, domestic wheat production supplies only 40% of the demand. Increasing wheat production is a priority to maintain food security. Therefore, it is important to develop crop management alternatives to overcome the salinity stress, hance increase total wheat production.

Nitrogen (N) plays a vital nutritional and physiological role in plants. Plants can take up N as an anion (nitrate) or as cation (ammonium). Plant tolerance to high salt content in the growth medium may increase, decrease or remain unchanged with increasing rates of nutrients applications depending on the specific experimental conditions, including the level of salt stress (Irshad et al., 2002a). Therefore, the objective of this study was to evaluate the effects of different sources of supplemental N on the response of wheat cultivars to different levels of salinity.

2. Materials and Methods

Two wheat cultivars i.e. Sakha 93 (salt tolerant) and Sakha 69 (salt sensitive) were used. A pot experiment was conducted in a green house at the National Research Centre, Cairo, Egypt, using plastic pots of 30 cm height and 25 cm diameter with 8 kg clay loam soil (0-30 cm depth) per pot. Some physical and chemical properties of the soil used in the study are shown in Table 1.

Table 1. Some physical and chemical properties of the soil used in this experiment

Characteristics	Value
Particle size distribution	
Coarse sand (mg/g)	70.0
Fine sand (mg/g)	200.0
Silt (mg/g)	350.0
Clay (mg/g)	380.0
Textural class	Clay loam
pH (1:2.5)	7.4
ECe (dS m^{-1})	1.3
Soluble cations (meq/L)	
Calcium	4.9
Magnesium	3.5
Potassium	0.7
Sodium	4.7
Soluble anions (meq/L)	
Carbonate	0.0
Bicarbonate	8.1
Chloride	3.2
Sulphate	2.6
Organic matter (%)	1.8
Calcium Carbonate (%)	3.6
Available Nutrients	
Nitrogen (ppm)	76
phosphorus (ppm)	18
potassium (ppm)	160

Table 2. Some properties of the farmyard manure used in the study

pH (1:10)	EC (dS m^{-1}) (1:10)	OM (%)	OC (%)	C/N ratio	Concentrations of (%)			
					N	P	K	Ca
7.64	7.50	35.5	20.64	21.5	0.96	0.29	2.15	0.32

All pots received uniform rates of fertilizer application prior to sowing, i.e. 0.6, 0.8, and 0.4 g of urea, single superphosphate and potassium sulphate, respectively, which is equivalent to 180, 240, and 120 kg/ha N, P, and K, respectively. Ten seeds were planted per pot. N topdressing was done three weeks after sowing using urea at N rate equivalent to 180 kg/ha.

Treatments included three salinity levels of 6, 9, and 12 dS m^{-1} using 1:1 ratio of $NaCl:CaCl_2$. A control treatment with no added salinity was also included with an initial soil salinity of 1.3 dS m^{-1}. Subtreatments were supplemental N sources: i.e. (i) farmyard manure 81.63 g /pot before sowing (equivalent to 48 Mg/ha, at 0.96% N in FYM, total N application was 460 kg/ha); (ii) urea at 1.8 g/pot equivalent to 540 kg/ha applied in three doses, i.e. before sowing, and 3 and 5 weeks after sowing; (iii) no supplemental N, as a control. The experiment was conducted with four replications. Some properties of the farmyard manure used in this study are shown in Table 2.

Plants were harvested 150 days after sowing (maturity stage, late November 2011). Dry weights of straw and grain per pot were recorded. Concentrations of N, P, K, Na, Ca, and Cl in the straw and grian were analyzed following the procedure reported by Cottenie et al. (1982). Effects of treatments on plant response parameters were evaluated by analysis of variance (ANOVA) and mean separation test using SAS program (SAS, 1991).

3. Results and Discussion

3.1 Grain and Straw Yield

The grain and straw yields were significantly influenced by the cultivar, salinity levels as well as supplemental N treatments (Table 3). Only salinity X supplemental N interaction effect was significant on the grain yield. Straw yield, however, was significantly influenced by all two way interactions. The mean grain and straw yields, across both salinity and supplemental N treatments, were significantly greater for salt tolerant Sakha 93 cultivar as compared to those for a salt sensitive Sakha 69 cultivar (Table 4). The grain yields were 13.03, 12.31, 8.86, and 7.58 g/pot, while the straw yields were 14.95, 11.46, 9.52, and 7.41 g/pot for control and salinity levels of 6, 9 and 12 dS m^{-1}, respectively.

Table 3. Analysis of variance (ANOVA) for single factor and interaction effects of cultivars (C), salinity levels (S), and supplemental nitrogen (N) treatments on response parameters of wheat cultivars

Sources	Grain yield	Straw Yield	Na		Cl		Ca		N		P		K	
			Grain	Straw	Grain	Straw	Grain	Straw	Grain	Straw	Grain	Straw	Grain	Straw
Significance of F test														
C	0.26*	0.19*	0.007*	0.025*	0.014*	0.049*	NS	0.039*	0.024*	0.024*	0.015*	0.012*	0.016*	0.057*
S	0.37*	0.27*	0.01*	0.036*	0.02*	0.069*	0.020*	0.054*	0.068*	0.035*	0.020*	0.018*	0.023*	0.080*
N	0.32*	0.23*	NS	0.03*	0.016*	0.059*	NS	0.047*	0.059*	0.030*	NS	0.016*	NS	0.070*
CxS	NS	0.38*	0.016*	0.051*	0.027*	0.098*	0.03*	0.077*	0.096*	0.049*	0.028*	NS	NS	NS
SxN	0.63*	0.47*	NS	NS	NS	NS	NS	NS	NS	0.060*	0.015*	NS	NS	NS
CxN	NS	0.33*	NS	NS	NS	NS	NS	NS	NS	NS	NS	NS	NS	NS
CxSxN	NS	NS	NS	NS	NS	NS	NS	Ns	NS	NS	NS	NS	NS	NS

Note. NS = non significant; * = significant at P ≥ 95%.

Table 4. Effects of cultivars, salinity levels, and supplemental nitrogen sources on grain and straw yields of wheat cultivars and concentrations of selected mineral elements in grain and straw

Variables	Grain yield	Straw Yield	Na		Cl		Ca		N		P		K	
			Grain	Straw	Grain	Straw	Grain	Straw	Grain	Straw	Grain	Straw	Grain	Straw
	-------g/pot-------		----------Percent in DM (w/w)----------											
Cultivars:														
Sakha 93[#]	11.27a	11.86a	0.06b	0.69b	0.15b	2.45b	0.36a	2.24b	2.44b	0.55a	0.36a	0.21a	0.52a	2.90a
Sakha 69[$]	9.63b	9.81b	0.09a	0.91a	0.18a	2.94a	0.35a	2.32a	2.47a	0.47b	0.34b	0.19b	0.48b	2.83b
Salinity:														
Control[&]	13.03a	14.95a	0.04c	0.25d	0.12d	1.45d	0.28d	2.11d	2.89a	0.77a	0.42a	0.22b	0.56a	3.47a
6	12.31b	11.46b	0.05c	0.61c	0.15c	2.39c	0.33c	2.19c	2.45b	0.53b	0.39b	0.24a	0.54a	2.91b
9	8.86c	9.52c	0.08b	0.98b	0.18b	3.13b	0.38b	2.36b	2.34c	0.40c	0.31c	0.19b	0.47b	2.62c
12	7.58d	7.41d	0.12a	1.36a	0.21a	3.82a	0.43a	2.47a	2.16d	0.35d	0.28d	0.16c	0.44c	2.46d
N sources:														
Control[@]	9.76c	10.14c	0.072a	0.76c	0.18a	2.61c	0.35a	2.37a	2.41b	0.45c	0.34a	0.18c	0.50a	2.79c
FYM	10.99a	11.49a	0.075a	0.84a	0.16ab	2.87a	0.36a	2.15c	2.50a	0.56a	0.35a	0.23a	0.52a	2.94a
Urea	10.59b	10.87b	0.073a	0.80b	0.15b	2.71b	0.36a	2.32b	2.45ab	0.52b	0.34a	0.20b	0.50a	2.87b

Note. *Means followed by similar letters are not significantly different (P > 95%), by each column and by each variable; [#]Salt tolerant; [$]Salt sensitive; [&]No added salinity; [@]No Supplemental N.

Supplemental N, regardless of sources, increased the grain and straw yields of both cultivars only under no salinity or at the low salinity (6 dS m^{-1}) treatments (Figures 1 and 2). At the high salinity treatments (9 and 12 dS m^{-1}) supplemental N failed to provide any benificial effects of mitigating the negative effects of salinity. The magnitude of beneficial effects of supplemental N as compared to no supplemental N was greater in Sakha 93 than that in Sakha 69 cultivar. This is somewhat expected in line with the physiological difference among the cultivars, i.e. mitigation of negative effects of salinity by an external treatment is often greater for a salinity tolerant cultivar as compared to that for a sensitive cultivar. The grain yield increases were 11 and 8% For FYM and urea as compared to that for control, respectively, while the corresponding values for straw were 12 and 7%. The results are consistent with the findings of Sushila and Giri (2000), Uyanoz et al. (2006), and Bayu et al. (2006).

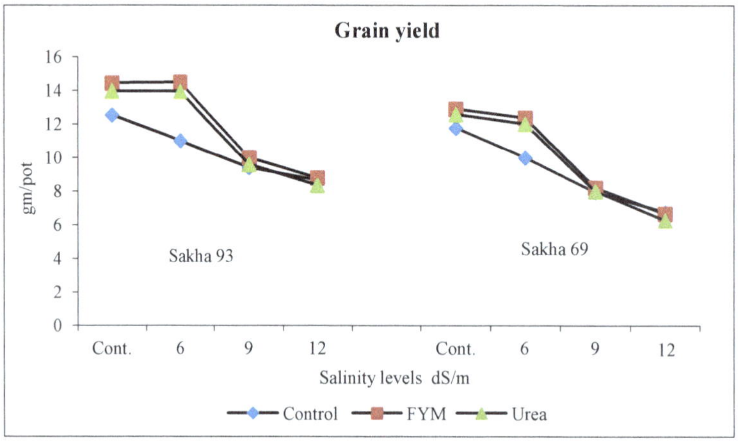

Figure 1. Effects of salinity levels and supplemental N treatments on grain yield of a salt tolerant (Sakha 93) and a salt sensitive (Sakha 69) wheat cultivars

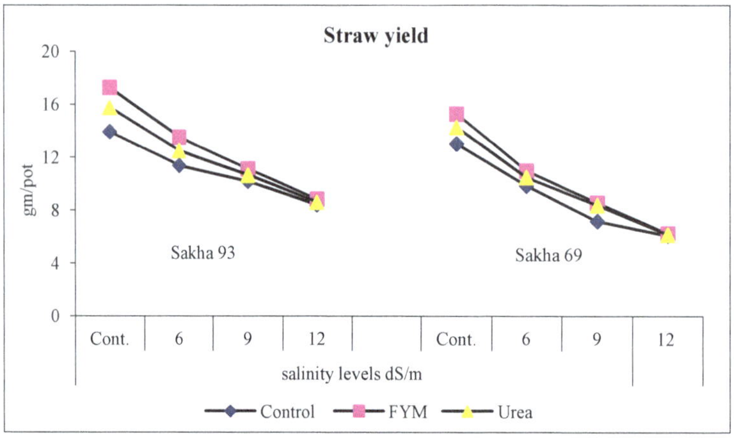

Figure 2. Effects of salinity levels and supplemental N treatments on straw yield of a salt tolerant (Sakha 93) and a salt sensitive (Sakha 69) wheat cultivars

The adverse effects of salinity were greater for Sakha 69 as compared to that for Sakha 93. The grain yield reductions at the highest salinity level as compared to that with the control were 47.2% and 36.9% for Sakha 69 and Sakha 93, respectively. The corresponding values for straw were 56.5% and 44.9%. Ahmed et al. (2005) reported 14% decrease in grain yield of salt tolerant wheat cultivar (LU 26S) with an increase in salinity level to 12.7 dS m^{-1} as compared to 52% decrease for a salt sensitive cultivar (SRC 3).

3.2 Concentrations of Na, Cl and Macronutrients in Grain and Straw

Concentrations of Na and Cl increased significantly in the grain and straw with an increase in salinity levels (Tables 3 and 4). The magnitude of increase was greater in the straw as compared to that in grain. This trend is in accordance with the finding of Ahmedi et al. (2009). Concentrations of Na and Cl, in both grain and straw were significantly greater in the salt sensitive Sakha 69 cultivar as compared to those in a tolerant cultivar Sakha 93 (Table 4). These results are in agreement with the findings of Schachtman and Munns (1992).

Increasing soil salinity levels significantly increased Ca concentrations in both grain and straw. Calcium concentrations in the grain were 0.28, 0.33, 0.38, and 0.43% with salinity treatments of control, 6, 9, and 12 dS m^{-1}, respectively, while the corresponding values in the straw were 2.11, 2.19, 2.36 and 2.47%. This could be attributable to increased Ca availability with increasing salinity levels that was achieved by using 1:1 ratio of $NaCl:CaCl_2$. Furthermore, the reduction in both grain and straw yields with an increase in salinity levels also contribute to an increase in Ca concentrations on dry matter basis. Ca concentration in the straw was significantly greater in Sakha 69 (2.32%) as compared to that in Sakha 93 (2.24%). On the other hand, there was no significant difference between the two cultivars with respect to Ca concentrations in the grain.

Concentrations of N, P, K in both grain and straw decreased with an increase in salinity levels from control to 12 dS m^{-1} (Table 4). The adverse effects of salinity on N concentration could be attributed to the presence of high amount of chloride which could decrease N uptake as reported by Saneoka, et al. (1999) and Uyanoz et al. (2006). Supplamental N application as FYM or Urea increased N concentration in the grain and straw. Concentrations of macronutrients in the grain and straw were greater in Sakha 93 as compared to those in Sakha 69, except the concentration of N in grain. The trend was reverse for the latter. El-Agroudi et al. (2005) reported greater N uptake by a salt tolerant wheat cultivar as compared to that by a salt sensitive cultivar.

Nitrogen concentration in the straw decreased significantly with increasing salinity levels, across both supplemental N and no supplemental N treatments. El-shafie et al. (2003) reported a significant increase in N uptake by wheat plants grown under saline condition with addition of FYM. Navarro et al. (2001) stated that phosphorus availability is reduced in saline soils because phosphate concentration in the soil solution is controlled by Ca-P interaction.

The concentrations of K in the grain and straw decreased with an increase in salinity levels (Table 4). This could be attributable to an antagonism for K uptake by an increase in Na and Ca availabity in the soil, since salinity increase was achieved by using NaCl and $CaCl_2$. Such antagonism for cation uptake was reported by Mostafa (2001) and Ahmedi et al. (2009).

Supplamental N application had no significant effects on K concentrations in the grain, but significantly increased K concentrations in the straw as compared to that of the plants with no supplemental N. This positive effect was greater with application of FYM than that with application of urea. These results are consistent with the findings of Hassan and Mostafa (2002), El-Shafie et al. (2003), and Irshad et al. (2002b).

References

Ahmadi, A., Eman, Y., & Pessarakli, M. (2009). Response of various cultivars of wheat and maize to salinity stress. *Journal food Agriculture and Environment, 7*, 123-128.

Ahmed, M., Niazi, B. H., Zaman, B., & Athar, M. (2005). Varietals differences in agronomic performance of six wheat varieties grown under saline field environment. *International Journal of Environmental Science and Technology, 2*, 49-57. http://dx.doi.org/10.1007/BF03325857

Bayu, W., Rethman, N. F. G., Hammes, P. S., & Alemu, G. (2006). Effects of Farmyard Manure and Inorganic Fertilizers on Sorghum Growth, Yield, and Nitrogen Use in a Semi-Arid Area of Ethiopia. *Journal of Plant Nutrition, 29*, 391-407. http://dx.doi.org/10.1080/01904160500320962

Cottenie, A., Verloo, L., Kiens, L., Velghe, G., & Camerlynch, R. (1982). *Chemical analysis of plant and soils.* Lab. of analytical and agro chemistry, State University of Ghent, Belgium.

El-Hendawy, S. E., Hu, Y., Yakout, G. M., Awad, A. M., Hafiz, S. E., & Schmidhalter, U. (2005). Evaluating salt tolerance of wheat genotypes using multiple parameters. *Europian Journal of Agronomy, 22*, 243-253. http://dx.doi.org/10.1016/j.eja.2004.03.002

El-Agroudi, M. W. M., Labeeb, G., Abd El-Hadi, M. R., & Abdou, T. A. M. (2005). Influence of soil salinity on nutrients uptake, yield and yield component of two wheat varieties. *Journal of Agricultural Science Mansoura University, 30*, 5689-5700.

El-shafie, F. S., & El-Shikha, S. A. (2003). Productivity and nutrients uptake of wheat and faba bean grown on cancerous soil as affected by water. *Minufiya Journal of Agricultural Research, 28*, 1025-1048.

Hassan, M. A. M., & Mostafa, M. M. (2002). Uptake of nutrients and heavy metals by barley plant grown on sandy and calcareous soils as affected by irrigation water salinity and sewage addition. *Zagazig Journal of Agricultural Research, 29*, 1929-1950.

Irshad, M., Honna, T., Eneji, A. E., & Yamamoto, S. (2002a). Wheat response to nitrogen source under saline conditions. *Journal of Plant Nutrition, 25*, 2603-2612. http://dx.doi.org/10.1081/PLN-120015525

Irshad, M., Yamamoto, S., Eneji, A. E., Endo, T., & Honna, T. (2002b). Urea and manure effect on growth and mineral contents of maize under saline conditions. *Journal of Plant Nutrition, 25*, 189-200. http://dx.doi.org/10.1081/PLN-100108790

Mostafa, M. M. (2001). Nutrient uptake and dry matter yield of barley as affected by salinity of irrigation water and addition of organic materials. *Zagazig Journal of Agricultural Research, 28*, 533-552.

Munns, R. (2002). Comparative physiology of salt and water stress. *Plant Cell and Environment, 25*, 239-250. http://dx.doi.org/10.1046/j.0016-8025.2001.00808.x

Navarro, J. M., Botella, M. A., Cerda, A., & Martinez, V. (2001). Phosphorus uptake and translocation in salt stressed melon plants. *Journal of plant physiology, 158*, 375-381. http://dx.doi.org/10.1078/0176-1617-00147

Onani, E. N. (2005). *Response of amaranth to salinity stress* (Ph.D. thesis, p. 119). Faculty of Agriculture, University of Pretoria.

Saneoka, H., Shiota, K., Kurban, H., Chaudhary, M. I., Premachandra, G. S., & Fujita, K. (1999). Effect of salinity on growth and solute accumulation in two wheat lines differing in salt tolerance. *Soil Science Plant Nutrition, 45*, 873-880. http://dx.doi.org/10.1080/00380768.1999.10414336

SAS Statistical Analysis System, Institute. (1991). *SAS Users guide* (Version 8) SAS Inst. Cary, N.C.

Sushila, R., & Giri, G. (2000). Influence of farmyard manure, nitrogen and biofertilizers on growth, yield attributes and yield of wheat (*Triticum aestivum*) under limited water supply. *Indian Journal of Agronomy, 45*, 590-595.

Schachtman, D. P., & Munns, R. (1992). Sodium accumulation in leaves for Triticum species that differ in salt tolerance. *Australian Journal of Plant Physiology, 19*, 331-340. http://dx.doi.org/10.1071/PP9920331

Uyanöz, R., Çetin, Ü., & Karaarslan, E. (2006). Effect of Organic Materials on Yields and Nutrient Accumulation of Wheat. *Journal of Plant Nutrition, 29*, 959-974. http://dx.doi.org/10.1080/01904160600651787

22

Effects of Drought Stress on Seed Germination and Seedling Growth of Different Maize Varieties

Mingli Liu[1], Meng Li[2], Kaichang Liu[2] & Na Sui[1]

[1] Key Laboratory of Plant Stress Research, College of Life Science, Shandong Normal University, Jinan, P. R. China

[2] Shandong Academy of Agricultural Sciences, Jinan, P. R. China

Correspondence: Kaichang Liu, Shandong Academy of Agricultural Sciences, Jinan 250100, P. R. China. E-mail: liukc@saas.ac.cn

Na Sui, Key Laboratory of Plant Stress Research, College of Life Science, Shandong Normal University, Jinan 250014, P. R. China. E-mail: suina800101@163.com

Abstract

Maize is susceptible to drought stress, especially during germination. The experiment of seed germination and seedling growth of two maize varieties under different concentration of mannitol solution was carried out. Four mannitol concentrations were set: 0, 200, 400 and 600 mM, and the corresponding solution osmotic potential was: 0, -0.49, -0.99, and -1.49 Mpa. Results indicated that Zhengdan958 germinated more quickly and in greater numbers than Liansheng15. The germination rate, germination potential, shoot and root fresh weight, shoot and root length of two maize varieties both decreased under drought stress, but they decreased more in Liansheng15 than in Zhengdan958. However, the germination and vigor index of Zhengdan958 were obviously higher than that of Liansheng15 under the stress of 200, 400 and 600 mM mannitol. These results showed that the drought resistances of Zhengdan958 was higher than that of Liansheng15 by having more powerful root system and higher osmotic adjustment ability.

Keywords: maize, mannitol stress, germination, drought resistance

1. Introduction

Abiotic stresses are governed by various factors, the most prominent is the elevated temperatures and scarcity of water (Pena & Hughes, 2007; Narusaka et al., 2003). As drought is one of the most important environmental factors affecting plant growth, development and crop yield, considerable progress has been made in understanding the changes in physiological processes caused by drought stress. It has been shown that the physiological responses of plants to drought stress are extremely complex and vary with plant species as well as with the degree and time of the exposure to drought (Evans et al., 1990, 1991; King, 2011). Differences in resistance to drought are known to exist within genotypes of plant species and were found in many studies, e.g. in maize (Martinielio & Lorenzoni, 1985; Lorens et al., 1987; Grzesiak, 1990; Grzesiak et al., 2012), wheat (Winter et al., 1988; Reynolds et al., 1998; Paknejad et al., 2007), rape seed (Richards & Thurling, 1978), oat (Larsson & Go'rny, 1988), coconut (Gomez et al., 2008) and triticale (Royo et al., 2000; Grzesiak et al., 2012). Unusual dry weather conditions adversely affect germination and seedling growth rates thus enhancing cell elongation sensitivity to damages induced under stressed conditions (Taylor et al., 1982; Delachiave & Pinho, 2003; Hamayun et al., 2010a). Deleterious effects of water stress have been reported in different crops such as tomato (Ragab et al., 2007), soybean (Sakthivelu et al., 2008; Hamayun et al., 2010b), corn (Khodarahmpour, 2011) and citrus (Ben-Hayyim, 1987).

Maize is one of the important crops, which serves as food and corn oil for human consumption, feed for livestock and poultry and raw material for agro-based industries. Expanding populations with greater food and energy needs are increasing demand for greater global maize production. Unfortunately, environmental limitations such as temperature and drought continue to restrain maize production levels as they have in earlier decades and in many areas, which is predicted to worsen with changing climates (Battisti & Naylor, 2009; Lobell et al., 2011; Mishra & Cherkauer, 2010). The increasing frequency of dry periods in many regions of the world result in the consecutive occurrence of drought on cultivated land (Hu & Schmidhalter, 2005). In China, maize planted in arid and semi arid area. At present, China's annual maize planting area is around 24 million hm^2, of which nearly 2/3 area is dry-land maize. Drought has become an important limiting factors of agricultural production in the world, especially in arid agricultural production. As dry-land crops, corn not only need more water, but also are more sensitive to water stress. Drought is the major cause of our country and around the world maize production low and unstable, generally can be reduced by 20%-30%. Therefore, the breeding of new maize varieties with strong drought resistance is one of effective ways for high and stable yield. However, the accurate identification of drought resistance of maize varieties is the basis for breeding drought resistance maize varieties.

In recent years, many scholars have studied the drought mechanism of maize, identification methods and indicators of drought resistance. Wang et al. have studied the germination characters of maize varieties with a hypertonic solution PEG-6000 simulated drought stress. Xu et al. have studied the germination potential, germination rate and utilization efficiency of storage material, amylase activity. Hou et al. have studied the drought resistance in the germination period of maize by hypertonic solution germination, TTC quantitative measurement, seed water absorption detection method with mannitol solution simulated water stress. These methods and indicators reflected the drought resistance of maize varieties from different angles and extent, also see scattered reports on bud seedling stage of maize.

Mannitol is a kind of macromolecule osmotic regulation substances that can't through the cell wall, small toxicity to cells, an ideal material to simulate the soil drought. This experiment with mannitol hypertonic solution simulated water stress and studied the difference of drought resistance on germination and seedling growth of two maize varieties to filter out strong drought resistance varieties, in order to provide theoretical guidance for drought resistance breeding and promotion of maize varieties.

2. Materials and Methods

2.1 Plant Culture and Treatments

Maize seeds were provided by the shandong academy of agricultural sciences. Dry seeds were stored in a refrigerator at −4 °C before being used.

Selecting the grain full, same size, no worm holes maize seeds, 70% ethanol solution disinfected for 2 min and then rinsed with running water. Each variety seeds germinated in different germination boxes which containing 20 seeds. Each variety sets four concentration: 0, 200, 400 and 600 mM mannitol, three times repeated and the corresponding solution osmotic potential were: 0, -0.49, -0.99, and -1.49 Mpa. New mannitol solution of the same amount and normal water were replaced every day. The seeds germinated in a glasshouse in the light condition of 14 h light/10 h dark. The temperature in the glasshouse was 25 °C during the day and 20 °C at night.

2.2 Germination Rate (%)

Observing germination situation of maize seeds every day and the germination rate of seeds was measured by the national seed inspection standards. Computing the final germination rate after germination for seven days. Germination rate was calculated as follows: Germination rate = (Germinated seed number/Test seed number) × 100%

2.3 Germination Potential

Germination potential is the index to evaluate the seed germination rate and germination uniformity. Germination potential was calculated as follows: Germination potential = (Germinated seed number at germination peak/Test seed number) × 100%

2.4 Germination and Vigor Index of Drought Resistant

Germination and vigor index of drought resistant were calculated as follows: Seed germination index of drought resistance = seed promptness index under water stress (PIS)/controlled seed promptness index (PIC); Promptness index PI = nd_2 (1.00) + nd_4 (0.75) + nd_6 (0.50) + nd_8 (0.25); nd_x = number of germinated seeds by the xth day of measurement (Grzesiak et al., 2013).

Seed vigor index of drought resistance = Seed vigor index under water stress (VIS)/Controlled seed vigor index (VIC); Vigor index VI = PI × Sx; Sx is the average length of bud for seventh days.

2.5 Shoot and Root Fresh Weight and Root-Shoot Ratio

After germination for seven days, seedlings grew consistently were chosen to be cut the root and shoot with a shear and weighed respectively. Root-shoot ratio was calculated as follows: Root-shoot ratio = Root fresh weight/Shoot fresh weight.

2.6 Shoot and Root Length

After germination of maize seeds for seven days, the shoot and root length were measured with a ruler.

2.7 Drought Resistance Coefficient

Drought resistance coefficient can reflect the sensitivity of plants to drought and it was calculated as follows: Drought resistance coefficient = (Index value under drought stress/Index value under normal moisture) × 100%

3. Results

3.1 Germination Rate during Drought Stress

The germination of Zhengdan958 was better than Liansheng15 under different mannitol treatment (Figure 1). The germination rate of two maize varieties declined obviously under drought stress. Under 200 mM mannitol treatment, germination percentage of Zhengdan958 and Liansheng15 decreased 1.75% and 13.46%, respectively. Under 400 mM mannitol treatment, germination percentage of Zhengdan958 and Liansheng15 decreased 28.07 % and 50.00%, respectively. Under 600 mM mannitol treatment, germination percentage of Zhengdan958 and Liansheng15 decreased 50.88% and 75.00%, respectively (Figure 2).

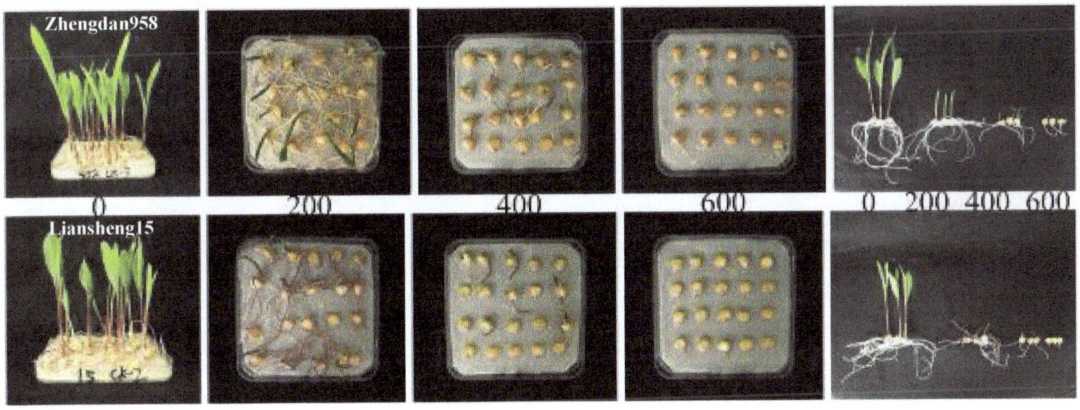

Figure 1. Photos of germination of Zhengdan958 and Liansheng15 under different concentrations of mannitol

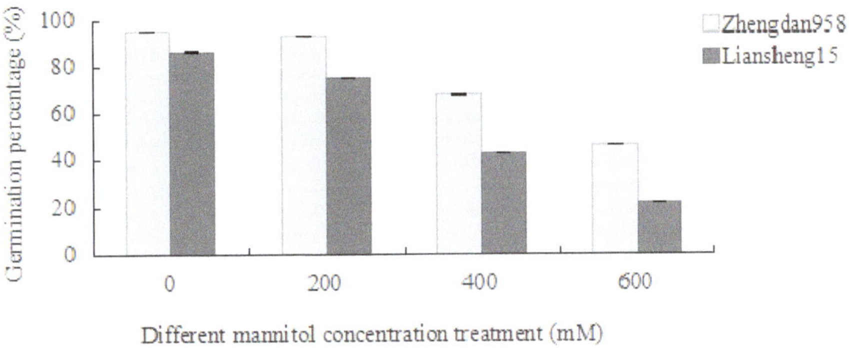

Figure 2. Germination rate of different maize varieties in response to different concentrations of mannitol. Each point represents the means ± SD of five measurements on each of five plants (n = 5)

3.2 Germination Potential during Drought Stress

The germination potential of two maize varieties declined significantly under mannitol stress and there was significant difference between different varieties. Under 200 mM mannitol treatment, there was nearly no decrease in germination potential of Zhengdan958, but it decreased 34.62% in Liansheng15. Under the stress of 600 mM mannitol, the germination potential of Zhengdan958 decreased 67.92% and the germination potential of Liansheng15 was 0 (Figure 3).

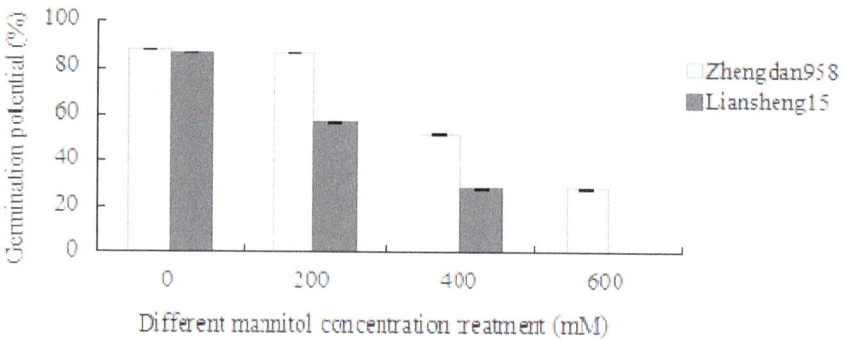

Figure 3. Germination potential of different maize varieties in response to different concentrations of mannitol. Values are means ±SD of five measurements for each of five plants (n = 5)

3.3 Germination and Vigor Index of Drought Resistance

The germination index (PIS/PIC) and vigor index (VIS/VIC) of drought resistance of the two maize varieties both decreased with the increase of mannitol concentration. The PIS/PIC of Zhengdan958 was higher than Liansheng15 under the stress of 200, 400 and 600 mM mannitol. The VIS/VIC of Zhengdan958 was also higher than that of Liansheng15 under 200 and 400 mM mannitol and there was no obvious difference under the stress of 600 mM mannitol (Figures 4 and 5).

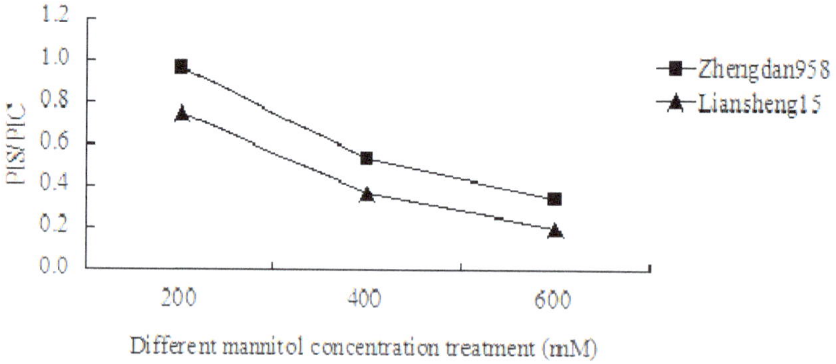

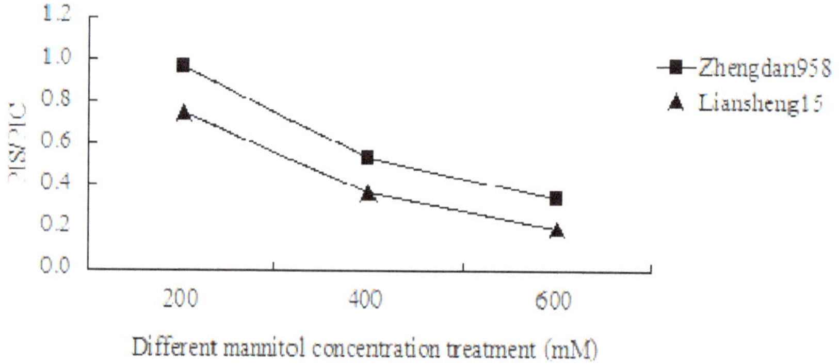

Figure 4. Germination index of drought resistance of different maize varieties in response to different concentrations of mannitol. Data are means of 5 replicates (n = 5) ± SD

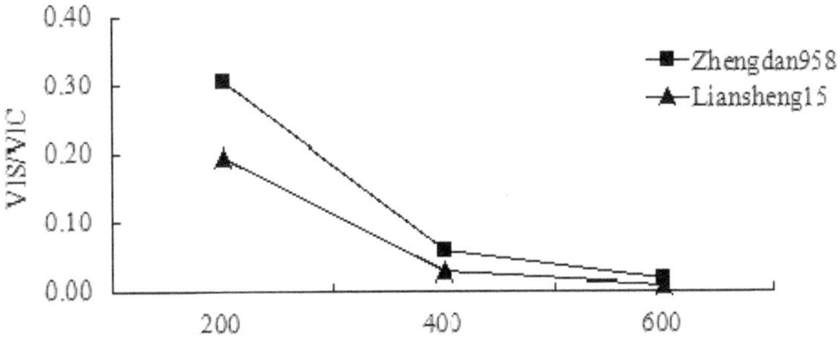

Figure 5. Vigor index of drought resistance of different maize varieties in response to different concentrations of mannitol. Each point represents the means ± SD of five measurements on each of five plants (n = 5)

3.4 Shoot and Root Fresh Weight and Root-Shoot Ratio during Drought Stress

Shoot and root fresh weight of two maize varieties under drought stress decreased obviously with the increase of mannitol concentration. Under 200 mM mannitol treatment, shoot fresh weight of Zhengdan958 and Liansheng15 decreased 71.13% and 79.51%, respectively (Figure 6). Under 400 mM mannitol treatment, shoot fresh weight of Zhengdan958 and Liansheng15 decreased 90.84% and 95.55%, respectively. Under 600 mM mannitol treatment, shoot fresh weight of Zhengdan958 and Liansheng15 decreased 95.49% and 98.36%, respectively. Root fresh weight of Zhengdan958 and Liansheng15 under 200, 400 and 600 mM mannitol treatment decreased 43.73% and 46.30%, 70.28% and 87.27%, 85.87% and 96.00%, respectively (Figure 7).

The root-shoot ratio of Zhengdan958 increased during mannitol treatment. It increased before 400 mM mannitol treatment and then decreased in Liansheng15 (Figure 8). Root-shoot ratio of Zhengdan958 increased from 0.82 under the control condition to 3.11 under 600 mM mannitol treatment, and the ratio of Liansheng15 increased from 1.04 under the control condition to 2.15 under 600 mM mannitol treatment.

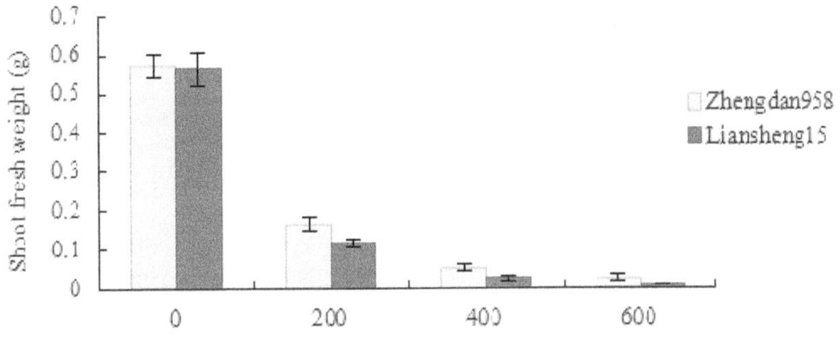

Figure 6. Shoot fresh weight of different maize varieties in response to different concentrations of mannitol. Values are means ±SD of five measurements for each of five plants (n = 5)

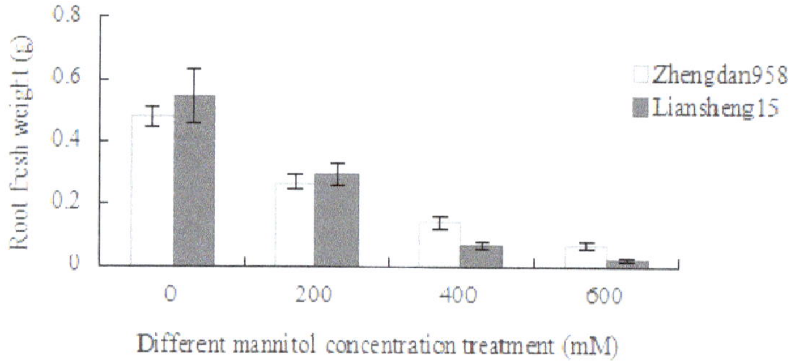

Figure 7. Root fresh weight of different maize varieties in response to different concentrations of mannitol. Data are means of 5 replicates (n = 5) ± SD

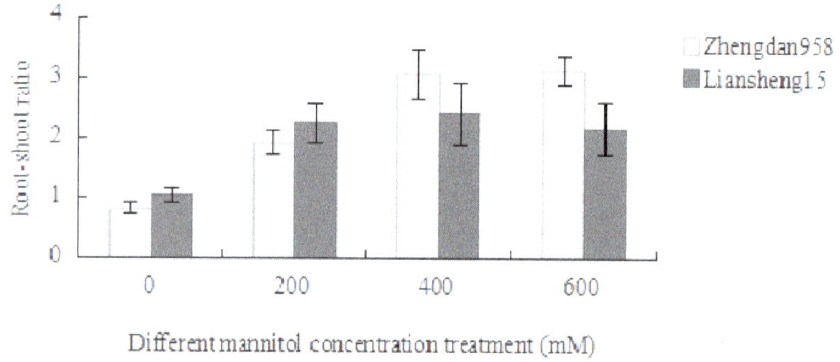

Figure 8. Root-shoot ratio of different maize varieties in response to different concentrations of mannitol. Each point represents the means ± SD of five measurements on each of five plants (n = 5)

3.5 Changes of Shoot and Root Length during Drought Stress

Shoot and root length of two maize varieties decreased obviously under the stress of 200, 400 and 600 mM mannitol (Figures 9 and 10). Under the stress of 200, 400 and 600 mM mannitol, shoot length of Zhengdan958 decreased 68.17%, 88.66% and 94.33%, respectively. While shoot length of Liansheng15 under the stress of 200, 400 and 600 mM mannitol decreased 75.20%, 92.32% and 96.80%, respectively (Figure 9). Root length of Zhengdan958 under the stress of 200, 400 and 600 mM mannitol decreased 37.73%, 64.53% and 84.83%, respectively. And root length of Liansheng15 under the stress of 200, 400 and 600 mM mannitol decreased 49.65%, 82.39% and 92.74%, respectively (Figure 10).

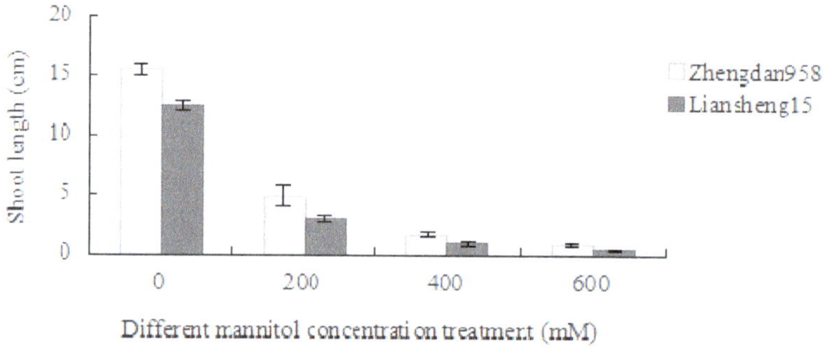

Figure 9. Shoot length of different maize varieties in response to different concentrations of mannitol. Values are means ±SD of five measurements for each of five plants (n = 5)

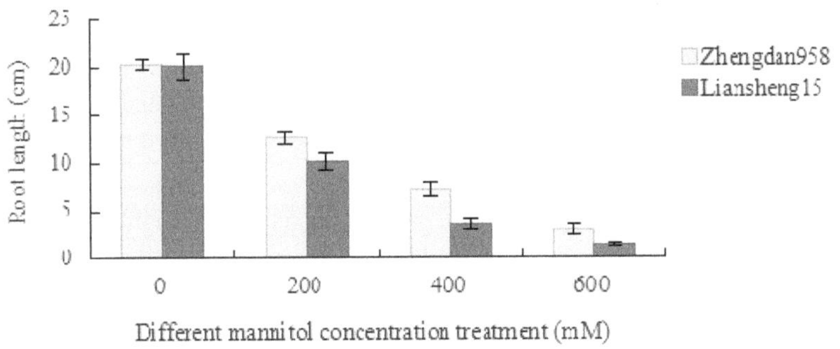

Figure 10. Root length of different maize varieties in response to different concentrations of mannitol. Data are means of 5 replicates (n = 5) ± SD

3.6 Drought Resistance Coefficient

The drought resistance coefficients of shoot and root fresh weight and length of two maize varieties under drought stress decreased obviously with the increase of mannitol concentration. But the drought resistance coefficients of Zhengdan958 were higher than that of Liansheng15. The drought resistance coefficient of root-shoot ratio of Zhengdan958 increased during mannitol treatment. While the ratio of Liansheng15 increased first and then decreased (Table 1). This showed that the drought resistances of Zhengdan958 was higher than that of Liansheng15 on seed germination condition.

Table 1. Drought resistance coefficient of two maize varieties under 200, 400 and 600 mM mannitol treatment. Each point represents the means ± SD of five measurements on each of five plants (n = 5)

Variety	Mannitol concentration (mM)	Shoot fresh weight	Root fresh weight	Root-shoot ratio	Shoot length	Root length
Zhengdan958	200	0.29^{a}	0.56^{a}	2.34^{c}	0.32^{a}	0.62^{a}
	400	0.09^{b}	0.30^{b}	3.74^{ab}	0.11^{b}	0.34^{b}
	600	0.05^{c}	0.14^{c}	3.82^{a}	0.06^{c}	0.15^{c}
Liansheng15	200	0.24^{a}	0.54^{a}	2.16^{b}	0.25^{a}	0.50^{a}
	400	0.06^{b}	0.13^{b}	2.33^{a}	0.08^{b}	0.18^{b}
	600	0.02^{c}	0.04^{c}	2.07^{c}	0.03^{c}	0.07^{c}

4. Discussion

Soil drought is one of the most important environmental factors responsible for the limit of maize productivity in many areas of the world. Maize appears very sensitive to drought stress during seedling emergence. Some researches (Grzesiak, 1990; Lorens et al., 1987a, 1987b; Martiniello & Lorenzoni, 1985; Ristic & Cass, 1991) confirmed the existence in maize of a wide range of genotypic variability of response to drought. Plants develop different morphological, physiological and biochemical mechanisms which inhibit or remove the harmful effects of drought stresses (Boyer, 1982; Larsson & Go'rny, 1988; Chaves et al., 2002; Reynolds et al., 1998; Asharaf, 2010). Relatively little research was carried out on the response to drought stress of some plant species. Intragenotypic variability was studied in maize (Martiniello & Lorenzoni, 1985; Lorens et al., 1987a, 1987b; Evans et al., 1990; Grzesiak, 1990; Ristic & Cass, 1991). Developing cultivars of maize that can perform well under heat and drought stress is an important goal throughout the world. However, progress in breeding of drought resistant maize genotypes requires future study into physiological mechanisms underlying the responses of maize to water stress (Larsson & Go'rny, 1988; Richards, 1991). In maize, similarly to other species, genotypic differences in response to soil water deficit were identified with respect to plant physiological characteristics. In the present study, we investigated the effects of drought stress on seeds germination and seedlings growth of two maize varieties.

The germination rate under drought stress can show the germination ability of drought resistance of various varieties. In this experiment, we preliminary judged the strength of drought resistance according to the germination rate of two maize varieties under the simulated drought stress with mannitol. The germination rate of two maize varieties declined obviously under drought stress, and there was no significant difference between the two varieties. Under the stress of 200, 400 and 600 mM mannitol, germination percentage of Zhengdan958 was still higher than that of Liansheng15, suggesting that Zhengdan958 had a stronger drought resistance (Figure 2).

Germination potential is a index to evaluate the seed germination rate and germination uniformity. It has been shown that the germination potential of two maize varieties both declined under drought stress. But Liansheng15 decreased more than that of Zhengdan958 (Figure 3). This suggested that Zhengdan958 germinated more quickly than Liansheng15, showing a stronger drought resistance of Zhengdan958. In addition, the effect of drought stress on the germination potential was greater than germination rate, explaining the greater influence of drought stress on early seed germination. This may be due to the bigger seeds suction influence of Zhengdan958 under drought stress on the early stage of germination.

Vigor index of drought resistance can better reflect the drought resistance and germination characteristics. The germination and vigor index of drought resistance of two maize varieties decreased with the increase of mannitol concentration, but Zhengdan958 decreased less than that of Liansheng15 under the stress of 200, 400 and 600 mM mannitol (Figures 4 and 5). The seedlings of Zhengdan958 grew better under osmotic stress, which could also be reflected by the seedling fresh weight (Figures 6 and 7) and length (Figures 9 and 10).

Osmotic adjustment is one of main strategies by which plants ensured water uptake during drought stress conditions. Drought resistance coefficient reflects the sensitive degree of plants to drought stress. From our results, the drought resistance coefficients of Zhengdan958 were all higher than that of Liansheng15 (Table 1). Roots is an important absorption organ of plants, developed level of roots is one of the main indexes of drought resistance. From the physiological point of view, a great root-shoot ratio is beneficial to alleviate drought stress. From the experimental results, Zhengdan958 had higher root-shoot ratio under drought stress (Figure 8), suggesting that Zhengdan958 developed powerful root system to fully use of water in deep soil under drought condition, thus showing stronger drought resistance.

5. Conclusion

In conclusion, under the same drought stress, the stronger drought resistance varieties which have higher germination and vigor index of drought resistance can still maintain higher germination potential and germination rate. In this experiment, Zhengdan958 germinated in greater numbers and more quickly than Liansheng15. And Zhengdan958 also had more powerful root system and stronger ability of osmotic adjustment to ensure seedling establishment in their drought environments, which might be the important reason for higher resistance to drought stress of Zhengdan958.

Acknowledgements

We are grateful for financial support from the Supporting Plan of National Science and Technology of China (2013BAD07B06), the NSFC (National Natural Science Research Foundation of China (31300205), Natural Science Research Foundation of Shandong (ZR2013CQ009), the Science and Technology Development Projects of Shandong Province (2014GNC113005) and the Program for Scientific research innovation team in colleges and universities of Shandong Province.

References

Asharaf, M. (2010). Inducing drought tolerance in plants: Recent advances. *Biotech Adv, 28*, 199-238. http://dx.doi.org/10.1016/j.biotechadv.2009.11.005

Battisti, D. S., & Naylor, R. L. (2009). Historical warnings of future food insecurity with unprecedented seasonal heat. *Science, 323*, 240-244. http://dx.doi.org/10.1126/science.1164363

Ben-Hayyim, G. (1987). Relationship between salt tolerance and resistance to polyethylene glycol-induced water stress in cultured citrus. *Plant Physiol, 85*, 430-433. http://dx.doi.org/10.1104/pp.85.2.430

Boyer, J. S. (1982). Plant productivity and environment. *Science, 218*, 443-448. http://dx.doi.org/10.1126/science.218.4571.443

Chaves, M. M., Pereira, J. S., Maroco, J., Rodrigues, M. L., Ricardo, C. P. P., Osorio, L. M., ... Pinheiro, C. (2002). How plants cope with water stress in the field? Photosynthesis and growth. *Ann Bot, 89*, 907-916. http://dx.doi.org/10.1093/aob/mcf105

Delachiave, M. E. A., & Pinho, S. Z. D. (2003). Germination of Sennaoccidentalis link: seed at different osmotic potential levels. *Braz Arch Tech, 46*, 163-166. http://dx.doi.org/10.1590/S1516-89132003000200004

Evans, R. O., Skagss, R. W., & Sneed, R. E. (1990). Normalized crop susceptibility factors for corn and soybean to excess water stress. *Trans ASAE, 33*, 1153-1161. http://dx.doi.org/10.13031/2013.31452

Evans, R. O., Skagss, R. W., & Sneed, R. E. (1991). Stress day index models to predict corn and soybean relative yield under high water table conditions. *Trans ASAE, 5*, 1997-2005. http://dx.doi.org/10.13031/2013.31829

Gomez, F., Oliva, M. A., Mielke, M. S., Almeida, A. A. F., Leita, H. G., & Aquino, L. A. (2008). Photosynthetic limitations in leaves of young Brazilian Green Dwarf coconut (Cocos macifera L) "nana" palm under well-watered conditions or recovering from drought stress. *Environ Exp Bot, 62*, 195-204. http://dx.doi.org/10.1016/j.envexpbot.2007.08.006

Grzesiak, M. T., Marcin'ska, I., Janowiak, F., Rzepka, A., & Hura, T. (2012). The relationship between seedling growth and grain yield under drought conditions in maize and triticale genotypes. *Acta Physiol Plantarum, 34*, 1757-1764. http://dx.doi.org/10.1007/s11738-012-0973-3

Grzesiak, M. T., Waligorski, P., Janowiak, F., Marcinska, I., Hura, K., Szczyrek, P., & Głab, T. (2013). The relations between drought susceptibility index based on grain yield (DSIGY) and key physiological seedling traits in maize and triticale genotypes. *Acta Physiol Plant, 35*, 549-565. http://dx.doi.org/10.1007/s11738-012-1097-5

Grzesiak, S. (1990). Reaction to drought of inbreds and hybrids of maize (*Zea mays* L.) as evaluated in field and greenhouse experiments. *Maydica, 35*, 303-311. Retrieved from http://www.cabdirect.org/abstracts/19901617370.html

Hamayun, M., Khan, S. A., Shinwari, Z. K., Khan, A. L., Ahmad, N., & Lee, I. (2010b). Effect of polyethylene glycol induced drought stress on physio-hormonal attributes of soybean. *Pak J Bot, 42*, 977-986. Retrieved from https://scholar.google.com.hk/scholar?q=+Effect+of+polyethylene+glycol+induced+drought+stress+on+physio-hormonal+attributes+of+soybean.&btnG=&hl=zh-CN&as_sdt=0%2C5

Hamayun, M., Sohn, E. Y., Khan, S. A., Shinwari, Z. K., Khan, A. L., & Lee, I. J. (2010a). Silicon alleviates the adverse effects of salinity and drought stress on growth and endogenous plant growth hormones of soybean (*Glycine max* L.). *Pak J Bot, 42*(3), 1713-1722. Retrieved from https://scholar.google.com.hk/scholar?q=Silicon+alleviates+the+adverse+effects+of+salinity+and+drought+stress+on+growth+and+endogenous+plant+growth+hormones+of+soybean+%28Glycine+max+L%29&btnG=&hl=zh-CN&as_sdt=0%2C5

Hu, Y., & Schmidhalter, U. (2005). Drought and salinity: A comparison of their effects on mineral nutrition of plants. *J Plant Nutr Soil Sci, 168*, 541-549. http://dx.doi.org/10.1002/jpln.200420516

Khodarahmpour, Z. (2011). Effect of drought stress induced by polyethylene glycol (PEG) on germination indices in corn (*Zea mays* L.) hybrids. *Afr J Biotech, 10*, 18222-18227. Retrieved from http://www.ajol.info/index.php/ajb/article/view/98591

Larsson, S., & Go'rny, A. G. (1988). Grain yield and drought resistance indices of oat cultivars in field rain shelter and laboratory experiments. *J Agron Crop Sci, 161*, 277-286. http://dx.doi.org/10.1111/j.1439-037X.1988.tb00668.x

Lobell, D. B., Schlenker, W., & Costa-Roberts, J. (2011). Climate trends and global crop production since 1980. *Science, 333*, 616-620. http://dx.doi.org/10.1126/science.1204531

Lorens, G. F., Bennett, J. M., & Loggale, L. B. (1987). Differences in drought resistance between two corn hybrids. I. Water relations and root length density. *Agron J, 79*, 802-807. http://dx.doi.org/10.2134/agronj1987.00021962007900050009x

Lorens, G. F., Bennett, J. M., & Loggale, L. B. (1987b). Differences in drought resistance between two corn hybrids. II. Component analysis and growth rate. *Agron J, 79*(5), 808-813. http://dx.doi.org/10.2134/agronj1987.00021962007900050009x

Martinielio, P., & Lorenzoni, C. (1985). Response of maize genotypes to drought tolerance tests. *Maydica, 30*, 361-370. Retrieved from http://agris.fao.org/agris-search/search.do?recordID=IT876013788

Mishra, V., & Cherkauer, K. A. (2010). Retrospective droughts in the crop growing season: Implications to corn and soybean yield in the midwestern united states. *Agric For Meteorol, 150*, 1030-1045. http://dx.doi.org/10.1016/j.agrformet.2010.04.002

Narusaka, Y., Nakashima, K., Shinwari, Z. K., Sakuma, Y., Furihata, T., Abe, H., ... Shinozaki, K. Y. (2003). Interaction between two cis-acting elements, ABRE and DRE, in ABA-dependent expression of

Arabidopsis rd29A gene in response to dehydration and high salinity stresses. *The Plant Journal, 34*(2), 137-149. http://dx.doi.org/10.1046/j.1365-313X.2003.01708.x

Paknejad, F., Nasri, M., Moghadam, H. R. T., Zahedi, H., & Alahmadi, M. F. (2007). Effects of drought stress on chlorophyll fluorescence parameters, chlorophyll content and grain yield of wheat cultivars. *J Biol Sci, 7*, 841-847. http://dx.doi.org/10.3923/jbs.2007.841.847

Pena, R. D. L., & Hughes, J. (2007). Improving vegetable productivity in a variable and changing climate. *SAT eJournal, 4*, 1-22. Retrieved from http://agropedia.iitk.ac.in/openaccess/sites/default/files/sp1.pdf

Ragab, A. R., Abdel-Raheem, A. T., Kasem, Z. A., Omar, F. D., & Samera, A. M. (2007). Evaluation of R1 tomato somaclone plants selected under poly ethylene glycol (PEG) treatments. *Afr Crop Sci Soc, 8*, 2017-2025. Retrieved from http://www.acss.ws/upload/xml/research/468.pdf

Reynolds, M. P., Singh, R. P., Ibrahim, A., Agech, O. A. A., Larque-Saavedra, A., & Quick, J. S. (1998). Evaluation physiological traits to complement empirical selection for wheat in warm environments. *Euphytica, 100*, 84-95. http://dx.doi.org/10.1023/A:1018355906553

Richards, R. A., & Thurling, N. (1978). Variation between and within species of rapeseed (Brasica campestris and B napus) in response to drought stress. I. Sensitivity at different stages of development. *Aust J Agric Res, 29*, 469-477. http://dx.doi.org/10.1071/AR9780491

Ristic, Z., & Cass, D. D. (1991). Leaf anatomy of Zea mays L in response to water shortage and high temperature: a comparison of drought-resistant and drought-sensitive lines. *Bot Gaz, 152*(2), 173-185. http://dx.doi.org/10.1086/337878

Royo, C., Abaza, M., Bianco, R., & Moral, L. F. G. (2000). Triticale grain growth and morphometry as affected by drought stress, late sowing and simulated drought stress. *Aust J Physiol, 27*, 1051-1059. http://dx.doi.org/10.1071/PP99113

Sakthivelu, G., Devi, M. K. A., Giridhar, P., Rajasekaran, T., Ravishankar, G. A., Nedev, T., & Kosturkova, G. (2008). Drought induced alterations in growth, osmotic potential and in vitro regeneration of soybean cultivars. *Genet Appl Plant Physiol, 34*, 103-112. Retrieved from http://www.bio21.bas.bg/ippg/bg/wp-content/uploads/2011/06/08_pisa_1-2_103-112.pdf

Taylor, A. G., Moles, J. E., & Kirkham, N. B. (1982). Germination and seedling growth characteristics of three tomato species affected by water deficits. *J Am Soc Hort, 107*, 282-285. Retrieved from http://agris.fao.org/agris-search/search.do?recordID=US19830891444

Winter, S. R., Musick, J. T., & Porter, K. B. (1988). Evaluation of screening techniques for breeding drought resistant winter wheat. *Crop Sci, 28*, 512-516. http://dx.doi.org/10.2135/cropsci1988.0011183X002800030018x

23

Inoculation of Arbuscular Mycorrhizal Fungi Increase the Growth of Cocoa and Coffee Seedling Applied with Ayamaru Phosphate Rock

Antonius Suparno[1], Saraswati Prabawardani[2], Sudirman Yahya[3] & Novita A. Taroreh[4]

[1] Department of Agriculture, Faculty of Agriculture and Technology, The University of Papua (UNIPA), Manokwari, Indonesia

[2] Agronomy Laboratory, Faculty of Agriculture and Technology, The University of Papua (UNIPA), Manokwari, Indonesia

[3] Department of Agronomy, Faculty of Agriculture, Bogor Agriculture Institute (IPB), Bogor, Indonesia

[4] Department of English Literature, Faculty of Letters The University of Papua (UNIPA), Manokwari Papua Barat, Indonesia

Correspondence: Antonius Suparno, Department of Agriculture, Faculty of Agriculture and Technology, The University of Papua (UNIPA), Manokwari, Indonesia. E-mail: anton.sprn@gmail.com

Abstract

Researh was carried out aimed to study the response of cocoa and coffee seedlings inoculated by Arbuscular Mycorrhizal Fungi (AMF) and applied with Ayamaru Phosphate Rock (APR). The research was cunducted at Bogor Agriculture Institute (IPB) and Papua University (Unipa) Manokwari Indonesia. Cocoa seed (Upper Amazone Hybrid F-1) and Arabica coffee seed (*Coffea arabica*) were collected from the Indonesian Coffee and Cocoa Research Institute (ICCI) Jember Indonesia. Cocoa dan coffee seedling were planted in polybag with acid soil medium, under 60% shading net. Cocoa and coffee seedlings were grown for 16 weeks after planting. AMF inoculum used Mycofer that was collected from IPB consisted of four species i.e *Gigaspora margarita* (INVAM-105), *Glomus etunicatum* (NPI-126), (*Acaulospora tuberculata* (INDO-2), *Glomus manihotis* (INDO-1) and indigenous AMF Manokwari consisted of three species i.e *A. tuberculata* Janos & Trappe, *A. scrobiculata* Trappe, and *G. aggregatum* Schenck & Smith. Ayamaru Phosphate Rock (APR) was collected from Ayamaru District, West Papua, Indonesia. The result showed that the growth of cocoa seedling applied with APR up to 4 g $P_2O_5{\cdot}seedling^{-1}$ and the growth of coffee seedling applied with APR up to 8 g $P_2O_5{\cdot}seedling^{-1}$ increased linearly. Compared to control (without AMF), inoculation of AMF Mycofer increased the cocoa seedling's height up to 50.58% and shoot dry-weight 127.55%, whereas height of coffee seedling increased by 27.29% and shoot dry-weight by 121.21%. For all APR dosage, the growth of cocoa and coffee seedling inoculated with AMF Mycofer inoculum was better than inoculated by indigenous AMF and without AMF (control).

Keywords: cocoa, coffee, mycorrhiza, phosphate rock, seedling

1. Introduction

Arbuscular Mycorrhiza Fungi (AMF) is widely known to have an important role in improving growth, yield, and quality of agriculture plants, horticulture, plantation, forestry, recovering land degradation and stabilizing upland ecosystem (Vosátka & Albrechtová, 2009; Gianinazzi et al., 2010).

Association between AMF and plant root was formed by the obligate fungi system with approximately 80 – 90% of upland plants (Wang & Qui, 2006) or approximately 73% of flowering plant family (Brundrett, 2009), both those that have root and without root (Read et al., 2000; Pressel et al., 2010), and they also have mutualism characteristic which covering 83% of dicotyledone and 79% of monocotyledone (Swift, 2004).

Swift (2004) states that one of the several significant responses of AMF infection on host plant is the increase of P uptake that mainly due to the capability of AMF to absorb P from soil and transfer it to the host plant roots. When the plant is in deficiency of P or N, this symbiotic association will be advantage and improve plant growth (Morgan et al., 2005).

Cocoa and coffee are perennial plants that are able to symbiotic with AMF to improve their plant growth.

Miyakasa and Habte (2001) state that cocoa and coffee are depend highly on symbiotic with AMF. Mutualism symbiotic between AMF and plants is beneficial for plants in enhancing growth and yield, improving plant vitality, and increasing minerals absorbtion area by root plants, due to smaller diameter of AMF hyphae than hair diameter of plant root (Sylvia, 2005).

Generally, great number of high quality seedlings are expected by seedling producer. High quality seedlings produced by breeding or genetic selection, culture technic, planting method, and accurate plant management will determine the success of seedling growth in the field (van den Driessche, 1994; South, 2000; Carneiro et al., 2007), particularly at marginal lands. Marginal lands are generally acid soil which P is bounded by Fe and Al. This causes unavailability of P to plant and reduces plant growth. Cocoa and coffee are perennial plants, therefore the effectiveness of AMF inoculation can be more optimal if the inoculation process was done in the seedling period and as a result when transferred to the field, plants will grow and yield better.

Cacao and coffee are exported commodities which generate high economic income, and these commodities have been broadly cultivated. Therefore, researh was carried out to study the growth of cocoa and coffee seedlings in response to the inoculation of Arbuscular Mycorrhizal Fungi (AMF) and applied with Ayamaru Phosphate Rock (APR). Effective AMF will produce good quality of cacao and coffee seedlings and this will increase fertilizer efficiency.

2. Materials and Methods

The experiment was carried out in 2009 at Cikabayan University Farm, IPB Bogor, Indonesia and at Unipa's Green House in 2012. The plant material was Hybrid cocoa seedling F-1 Upper Amazone Hibrid (UAH) and Arabica coffee collected from the Indonesian Coffee and Cocoa Research Institute (ICCRI) in Jember, East Java. Growing media was the acid soil with Al_{dd} 17.03 $cmol{\cdot}kg^{-1}$. APR which was obtained from Ayamaru District, West Papua Indonesia.

The plantlets were grown for four months under 60 % shading net. The experiment was set up in a Completely Randomized Design. For cocoa seedling trial, the first factor was AMF inoculation, consisted of none AMF (mo), indigenous AMF from cocoa rhizosphere Manokwari (m_1), and Mycofer from IPB Bogor (m_2); the second factor was five dosages of APR 0, 0.5, 1.0, 1.5, and 2.0 g $P_2O_5{\cdot}seedling^{-1}$, with 2.0 g P_2O_5 SP-36$\cdot seedling^{-1}$ was as a comparison. For coffee seedling trial, the first factor was AMF inoculation, consisted of none AMF (mo), indigenous AMF from cocoa rhizosphere Manokwari (m_1), and Mycofer from IPB Bogor (m_2); the second factor was five dosages of APR 0, 2, 4, 6, and 8 g $P_2O_5{\cdot}seedling^{-1}$, with 2.0 g P_2O_5 SP-18.seedling^{-1} was as a comparison.

The experiment was conducted three times by using 20 cm × 30 cm sized polybag which was filled with 3 kg soil. Each seed was grown in each polybag with three polybags for each experimental unit.

The inoculum production was done by *Sorghum bicolor* band zeolit as a growing media. The inoculum consisted of propagule which was a mixture of spore, infected root, hyphal and growing_media. After Most Probable Number test, each plant was inoculated by 10 grams inoculum.

The observed variables were growth of seedlings, shoot dry-weight, leaf number, seedling height, stem diameter, root dry-weight, shoot-root ratio, shoot P uptake, root colonization, and the effectiveness of AMF inoculum.

Root colonization is measured by the following formula:

$$Root\ colonization\ (\%) = \frac{number\ of\ infected\ roots}{number\ of\ observed\ roots} \times 100\% \qquad (1)$$

The effectiveness of AMF inoculum is measured by formula:

$$The\ AMF\ effectiveness\ (\%) = \frac{Mycorrhyza\ Plant - Non\ Mycorrhyza\ Plant}{Mycorrhyza\ Plant} \times 100\% \qquad (2)$$

P uptake was measured by:

$$Shoot\ P\ uptake = Shoot\ P\ content\ \times\ shoot\ dry\ weight \qquad (3)$$

The data of the experimental result was analysed by using Analysis of Variance, and followed by LSD test if the treatment showed a significant effect. LSD test was used to compare the effect of AMF between APR and SP-36 application. The effect of APR dosages levels to the growth of cocoa and coffea seedling was analysed by regression correlation analysis. All the data were analysed using SAS v 9.0 software.

3. Result and Discussion

Cocoa and coffee belong to perennial plant that depend highly on symbiotic with AMF. AMF inoculation and APR application increased growth of both of cocoa and coffee seedlings. As the application level of APR dosage increased the seedling growth also increased at all APR dosage levels. AMF inoculation produced better response on the growth of cocoa and coffee seedling (Tables 1 and 3). AMF Mycofer was more effective to increase the growth of cocoa and coffee seedling if compared to seedling without AMF and seedling inoculated by indigenous AMF (Figures 1 and 3).

Table 1. Seedling height, stem diameter, roots dry-weight, and shoot-root ratio of cocoa seedling resulted from AMF inoculation in several levels of APR dosages

Treatment	Seedling Height (cm)	Stem Diameter (mm)	Root Dry-Weight (g·seedling^{-1})	Shoot-root Ratio
APR dosage (g P_2O_5·seedling^{-1}):				
Control:				
0	32.48 b	6.15 c	1.32 c	2.57 d
0.5	34.83 ab	6.41 bc	1.43 bc	3.05 bc
1.0	36.39 a	6.58 ab	1.59 abc	3.01 c
1.5	37.00 a	6.85 a	1.75 a	3.07 bc
2.0	38.56 a	6.92 a	1.64 ab	3.47 b
SP-36: 2.0 g P_2O_5·seedling^{-1}	38.06 a	6.97 a	1.68 ab	3.90 a
LSD Test (95%)	*3.93*	*0.49*	*0.29*	*0.42*
AMF Inoculum:				
m_0	27.21 C	5.69 C	1.01 C	3.01 A
m_1	38.00 B	6.93 B	1.69 B	3.21 A
m_2	43.15 A	7.26 A	2.01 A	3.31 A
Effectiveness (%) :				
m_1 vs m_0	39.65	21.79	67.33	6.64
m_2 vs m_1	13.55	4.76	18.93	3.12
m_2 vs m_0	58.58	27.59	99.01	9.96

Note. The number in one column followed by the same letter is not significantly different in LSD Test for 95%. m_0: non AMF; m_1: Manokwari indigenous AMF; m_2: Mycofer AMF; vs: versus.

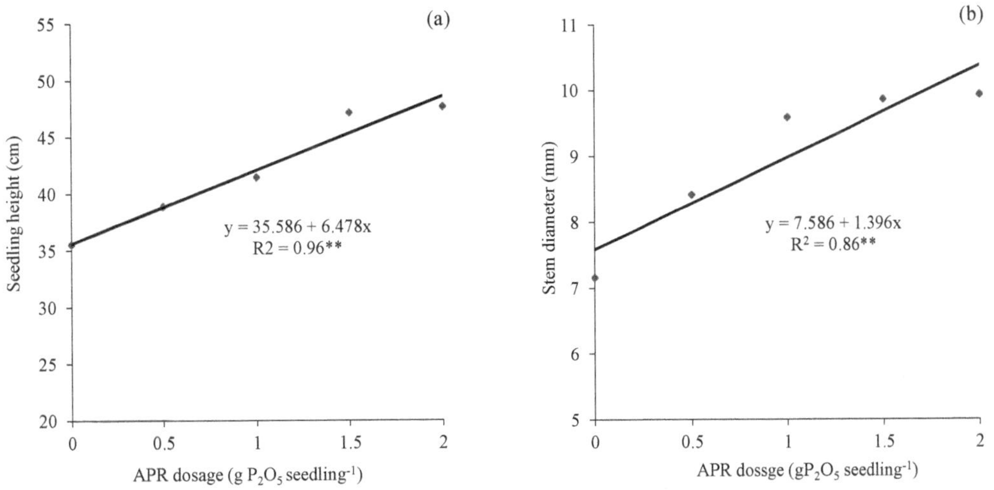

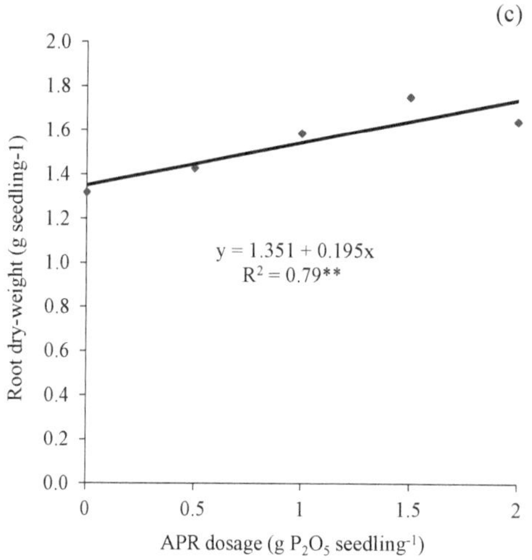

Figure 1. The relationship of seedling height (a), stem diameter (b), and root dry-wight (c) of cocoa seedlings to APR dosage levels

Note. **: highly significant; *: significant.

Table 2. Response of shoot dry-weight response, shoot P uptake, root colonization, and acid phosphatase activity of cocoa seedling to APR dosage at different AMF inoculums

Treatment	APR dosage (g P_2O_5·seedling^{-1})					SP-36	Curve Response
	0	0.5	1.0	1.5	2.0		
	Shoot dry-weight (g·seedling^{-1})						
m_0	1.68	2.74	2.93	3.40	3.23	3.88	Linear
m_1	3.62	4.70	5.24	4.70	6.33	7.82	Linear
m_2	4.98	5.55	6.33	7.09	7.35	7.85	Linear
Effectiveness (%):							
m_1 vs m_0	115.48	71.53	78.84	38.23	95.97	101.55	
m_2 vs m_0	196.43	102.55	116.04	108.53	127.55	102.32	
m_2 vs m_1	37.57	18.09	20.80	50.85	16.11	0.0030	
	P uptake (mg·shoot^{-1})						
m_0	3.50	4.30	5.85	6.85	7.75	8.20	Linear
m_1	5.50	6.50	7.05	8.35	9.40	13.60	Linear
m_2	7.60	8.20	9.90	10.85	11.25	13.70	Linear
Effectiveness (%):							
m_1 vs m_0	57.14	51.16	20.51	21.90	21.29	65.85	
m_2 vs m_0	117.14	90.70	69.23	58.39	45.16	67.07	
m_2 vs m_1	38.18	26.15	40.43	29.94	19.68	0.007	
	Root colonization (%)						
m_0	10.00	13.33	6.67	10.00	6.67	3.33	Quadratic
m_1	73.33	73.33	86.67	93.33	96.67	13.33	Linear
m_2	86.67	90.00	93.33	100.00	100.00	16.67	Linear
Effectiveness (%):							
m_1 vs m_0	633.30	450.11	1199.40	833.30	1349.33	300.30	
m_2 vs m_0	766.70	575.17	1299.25	900.00	1399.25	400.60	
m_2 vs m_1	18.19	22.73	7.68	7.15	3.44	25.06	

	--------------- Acid phosphatase activity ($\mu g \cdot g^{-1} \cdot hour^{-1}$) ---------------						
m_0	1.44	3.47	4.12	3.30	3.34	2.33	Quadratic
m_1	3.49	5.45	4.84	5.45	3.69	3.56	Linear
m_2	3.59	5.52	6.54	5.46	5.68	5.16	Linear
Effectiveness (%):							
m_1 vs m_0	142.36	57.06	17.48	65.15	10.48	52.79	
m_2 vs m_0	149.31	59.08	58.73	65.45	70.06	121.46	
m_2 vs m_1	2.87	1.28	35.12	0.18	53.93	44.94	

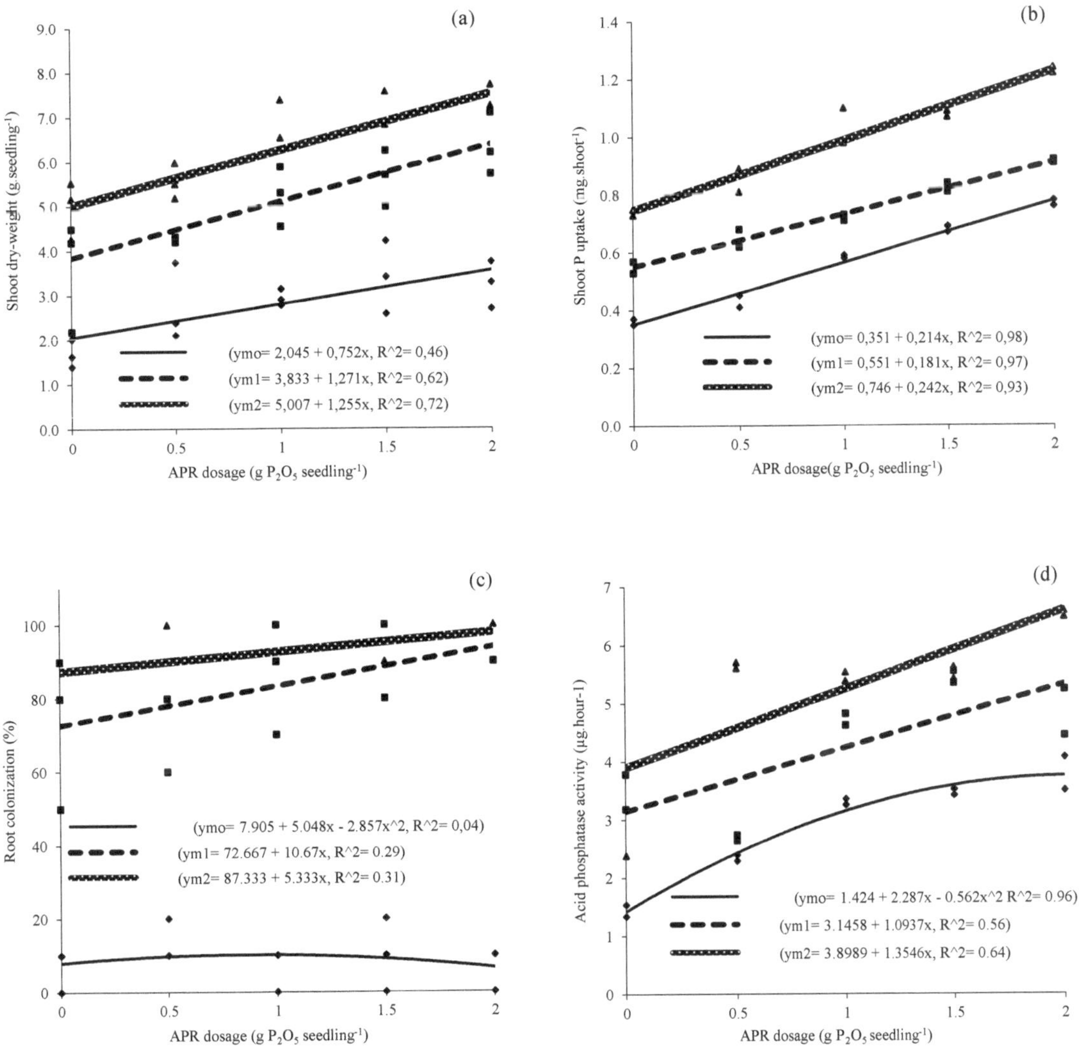

Figure 2. The relationship of Shoot dry-weight (a), shoot P uptake (b), root colonization (c), and acid phosphatase activity (d) of cocoa seedlings to APR dosage levels

Note. m_0 = non AMF, m_1 = Manokwari indigenous AMF, m_2 = Mycofer AMF.

The interaction between AMF inoculation and APR dosage level affected leaf number, shoot dry-weight, P uptake, and root AMF colonization of cocoa seedling, whereas it affected on root AMF colonization only in the coffee seedling (Table 3; Figures 3 and 4). However, when APR dosage level was increased to 8 g $P_2O_5 \cdot seedling^{-1}$ on coffee seedling, the level of root AMF colonization, acid phosphatase activity, leaves number, stem diameter, seedling height, shoot dry weight decreased, while P uptake had still increased. However, AMF Mycofer always produced good results in all dosage levels both on cocoa and coffee seedlings. This shows that effective AMF inoculation is needed to produce high quality seedlings.

Table 3. Effect of APR dosage and AMF Inoculum on leaves number, seedling height, stem diameter, shoot dry weight, root dry weight, and root-shoot ratio of coffee seedling

Treatment	Leaves number	Seedling height (cm)	Stem Diameter (mm)	Shoot dry weight (g)	Root dry weight (g)	Root-shoot ratio
APR dosage (g·seedling^{-1}):						
Control:	11.16 b	13.38 b	2.41 b	3.74 b	1.52 b	344.79 a
2	12.47 a	16.28 a	3.19 a	5.36 a	1.96 ab	317.94 a
4	12.73 a	17.36 a	3.07 a	6.07 a	2.86 a	242.69 a
6	12.52 a	16.03 a	3.09 a	5.28 a	2.28 ab	278.80 a
8	12.48 a	17.50 a	3.13 a	5.98 a	2.56 ab	276.45 a
SP-18 (2 g·seedling^{-1})	12.22 a	15.80 a	3.08 a	5.84 a	2.80 a	229.83 a
LSD Test (95%)	*0.86*	*1.93*	*0.51*	*1.29*	*1.12*	*120.33*
AMF inoculum:						
m_0	11.52 C	14.00 C	2.58 B	3.44 C	1.41 C	304.70 A
m_1	12.32 B	16.34 B	3.05 A	5.39 B	2.24 B	293.85 A
m_2	12.96 A	17.82 A	3.36 A	7.30 A	3.34 C	246.69 A
Effectiveness (%):						
m_0 vs m_1	6.94	16.71	18.22	56.69	58.88	3.56
m_0 vs m_2	12.50	27.29	30.23	112.21	136.88	19.04
m_1 vs m_2	5.19	10.57	10.16	35.44	49.11	16.05

Note. The number in each column followed by the same letter are not significantly different in LSD Test for 95%. m_0 = without AMF, m_1 = Indigenous AMF, m_2 = AMF Mycofer, SP = Super Phosphate, vs = versus.

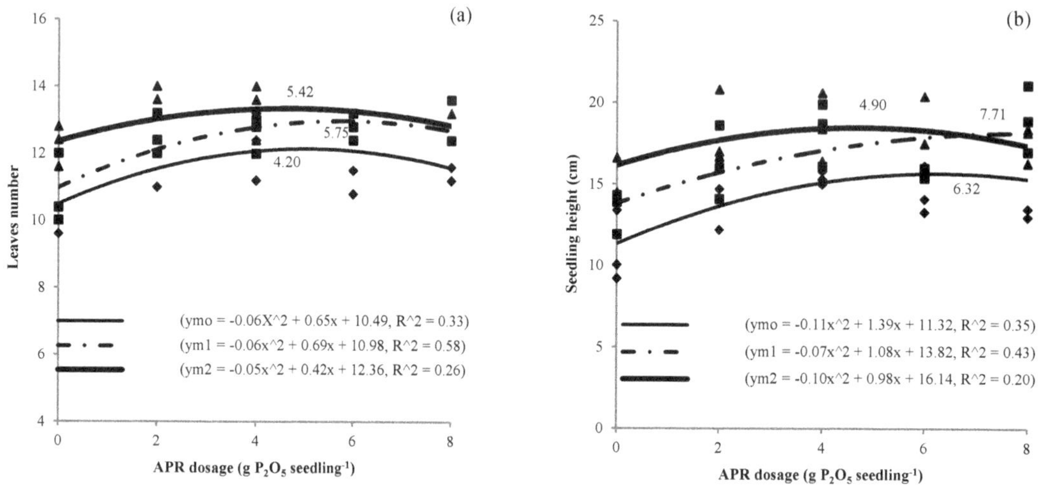

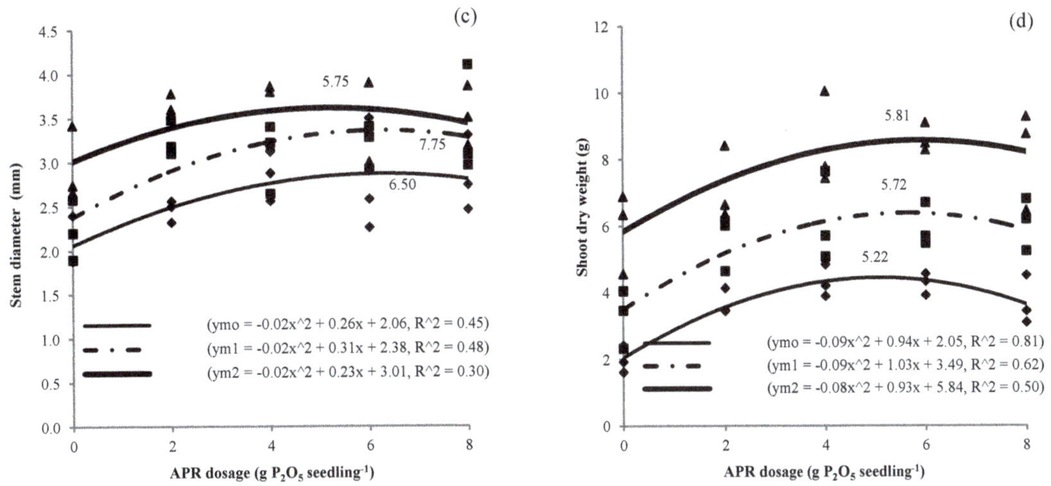

Figure 3. The interaction curve response of leave number (a), seedling height (b), stem diameter (c) shoot dry weight (d) as resulted from APR dosage levels with different AMF inoculum on coffee seedling

Note. m_0 = non AMF, m_1 = Indigenous AMF, m_2 = Mycofer AMF.

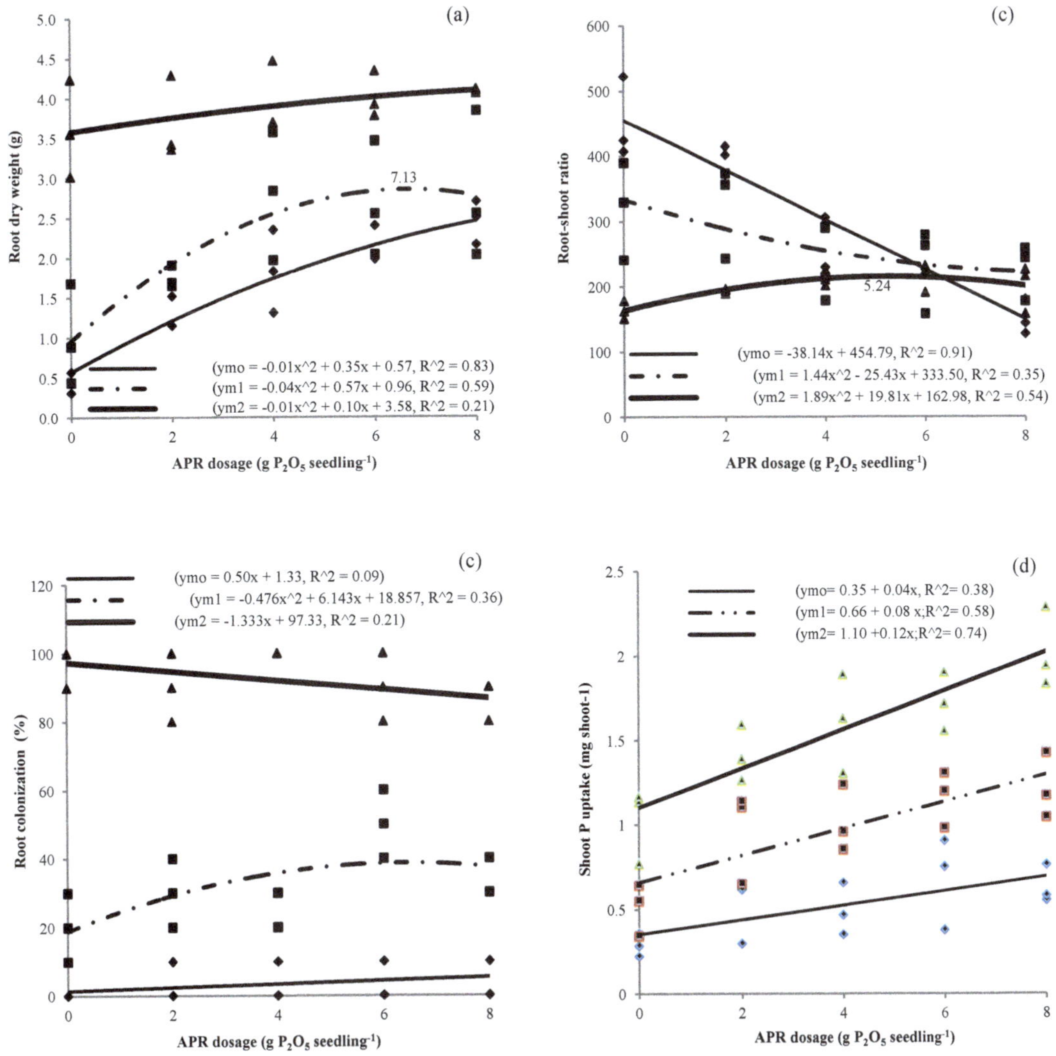

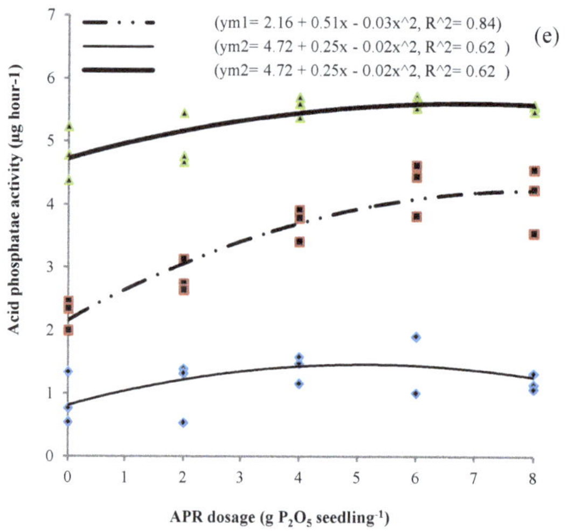

Figure 4. The interaction curve response of root dry-weight (a) root-shoot ratio (b) root colonization (c) shoot P uptake (d) acid phosphatase activity (e) as resulted from APR dosage levels with different AMF inoculums on coffee seedling

Note. mo = non AMF, m_1 = Indigenous AMF, m_2 = Mycofer AMF.

The utilization of low release P such as APR and bones powder even though in high dosage application, was effective to maintain AMF and to increase plant growth (Cardoso, 1996; Nikolaou et al., 2002). P content was reported to negatively correlated (Amijee et al., 1989; Koide & Li, 1990), no correlation (Tawaraya et al., 1996), or positively correlated (Carrenho et al., 2001) with AMF, depends on AMF species, plant species, P source and its solubility (Cardoso, 1996; Tawaraya et al., 1996; Nikolaou et al., 2002; Bhadalung et al., 2005). Maximum AMF symbiotic was achieved if soil P content was not more than 50 $mg{\cdot}kg^{-1}$ (50 ppm) (Ishii, 2004).

Cocoa and coffee seedling which are inoculated with AMF grew well because the AMF infected the root, intraradical hyphae (IH) and extraradical hyphae (EH) developed and spread, hence it produced positive effect on plants. The spreading of EH in soil was soon formed after root colonization. EH produce multiple function such as the transport of water and essential mineral, spore production, soil aggregation, and host plant protection from pathogen. The role of EH in essential mineral transport especially organic P, is important because EH is able to reach out the depletion zone, that is unreachable and unavailable to plant roots (Zhu et al., 2001).

The contribution of AMF on water and organic P uptake was affected by AMF species, plant, and environment which indicated that the functional suitability between AMF and plant/crop is not always be related to its colonization (Pearson & Jakobsen, 1993; Burleigh et al., 2002). AMF inoculation was done one time only, which was in planting time, but the result will be in a long time effect. This method could reduce the investment on fertilizer and pesticide supply, and therefore farmer able to produce safe agriculture products with high quality and lower in cost.

The use of AMF inoculation is proven in improving the resistance of plant to biotic and abiotic stress that result to reduce fertilizer supply and pesticide investment (Herrera et al., 1993; Douds & Johnson, 2007; Vosatka & Albrechtova, 2009; Koltai & Kapulnik, 2010). The application of AMF inoculant could produce more productive plants and from stable ecosystem (van der Heijden et al., 1998; Siddiqui et al., 2008; Cameron, 2010), especially in restoring the degradation land (Herrera et al., 1993). AMF mycofer inoculant shows better effect to the growth of cocoa and coffee seedling. This is presumably because of AMF Mycofer consists of four species, and hence the benefit of each species can support one another.

Mycofer AMF is more effective because it consists of four AMF species i.e *Gigaspora margarita*, *Acaulospora crobiculata*, *Glomus manihotis*, and *Glomus etunicatum*. Positive effect of *G. etunicatum* could be nutrient functionality such as increasing in macro-nutrient uptake (Govindarajulu et al., 2005; Li et al., 2006), micro-nutrient (Purakayastha et al., 2001; Nogueira & Cardoso, 2002; Andrade et al., 2010), and non-nutrient functionality such as secondary metabolic production (Yao et al., 2003; Fester & Hause, 2007).

AMF has been reported engaged in rhizodeposition process to enrich the soil with carbon substrate with the result of the increasing of the soil biological activity (Rillig et al., 2001), and soil aggregation (Rillig et al., 2002;

Rillig & Mummey, 2006). Symbiosis of *G. etunicatum* has the impact to plant root morphology as well (Bressan & Vasconcellos, 2002), hence it speeds up the early growing of vascular plant seedling planted at post mining area (Flores-Aylas et al., 2003; Santos et al., 2008).

Generally, plants that are symbiotic with AMF have lower shoot-root ratio than plants without AMF (Piccini et al., 1988; Khalil et al., 1994). Shoot-root ratio is the description of carbon translocation from top to the root of plants. The shoot-root ratio value get smaller due to the changing on rhizosphere dynamic that is reflected by carbon translocation to roots. And this shows one of the plant mechanisms to find out the limited nutrition. Carbon translocation to roots was showed by an increasing of root biomass, therefore it increases the root surface colonized by AMF.

The success of AMF inoculation was commonly observed based on plant indicators such as the increasing of plant dry-weight and yield dry-weight, nutrition uptake-Phosphor in particular (Rossiana & Supriatun, 2003; Simarmata & Herdiani, 2003). In fact, cocoa and coffee seedling inoculated with AMF always produce better growth in all of APR dosage levels.

AMF is widely known producing various benefits so the plant is capable to cope with biotic and abiotic stress (Smith & Read, 2008; Gianinazzi et al., 2010; Smith et al., 2010). The real advantage of AMF inoculation on seedling production is in an increasing growth of seedlings when transplanted to the field. The growth of seedlings without AMF are often dwarf and leaf suffer from necrosis (Landis & Amaranthus, 2009).

The diameter of AMF hypha is smaller than root hair, therefore it was able to reach out soil water and nutrient which were difficult to get to by plant roots. *Glomus etunicatum* has greater IH diameter (2.2–18.0 μm), greater number of hypha and spreads all over the root, however EH developed slowly and non-extensive. *A. tuberculata's* has smaller IH diameter (1.2–9.1 μm), fever hypha number but in developed faster and spread extensively on root surface (Dodd et al., 2000; Smith & Read, 2008). Greater hypha diameter will enables greater carbon volume transported to the rhizosphere for EH formation and extraradical spore.

Better nutrient availability determine better growth of host plant and better establishment of AMF intra-structure (IH, vesicle, and arbuscul) and extra-structure (EH and spore) (Olsson et al., 2010). The increasing of nutrient uptake by host plants ensured to vigorous plant in order to continually supply carbon to rhizosphere (Kaschuk et al., 2010) which is needed for the formation of EH and AMF sporulation.

4. Conclusion

Mycofer inoculum is more effective to increase the potency of APR as fertilizer and the growth of coffee and cocoa seedling. Cocoa seedling applied with APR up to 4 g P_2O_5·seedling^{-1} and coffee seedling applied with APR up to 8 g P_2O_5·seedling^{-1} linearly increased the growth of both cocoa and coffee seedling. Compared to control (without AMF), inoculation of AMF Mycofer increased the cocoa seedling's height up to 50.58% and shoot dry-weight 127.55%, whereas height of coffee seedling increased by 27.29% and shoot dry-weight by 121.21%. For all APR dosage, the growth of cocoa and coffee seedling inoculated with AMF Mycofer inoculum was better than inoculated by indigenous AMF and without AMF (control). This showed that AMF inoculation in both cocoa and coffee seedling increased the APR's effectiveness.

Acknowledgements

Our great gratitude goes to the Agroclimate Laboratory, Agriculture Faculty of Papua University Manokwari-Indonesia, Laboratory of Biotechnology on Forest and Environment, Research Center of Natural Resource and Biotechnology IPB, Bogor-Indonesia, the Indonesian Coffee and Cocoa Research Institute (ICCRI) Jember, Directurate General of Higher Education, and everyone who supported this research.

References

Amijee, F., Tinker, P. B., & Stribley, D. P. (1989). The development of endomycorrhizal root systems. VII. A detailed study of effects of soil phosphorus on colonization. *New Phytol., 111*, 435-446. http://dx.doi.org/10.1111/j.1469-8137.1989.tb00706.x

Andrade, S. A. L., Gratão, P. L., Azevedo, R. A., Silveira, A. P. D., Schiavinato, M. A., & Mazzafera, P. (2010). Biochemical and physiological changes in jack bean under mycorrhizal symbiosis growing in soil with increasing Cu concentrations. *Environm. Exp. Bot., 68*, 198-207. http://dx.doi.org/10.1016/j.envexpbot.2009.11.009

Bhadalung, N. N., Suwanarit, A., Dell, B., Nopamornbodi, O., Thamchaipenet, A., & Rungchuang, J. (2005). Effects of long–term NP–fertilization on abundance and diversity of arbuscular mycorrhizal fungi under a maize cropping system. *Plant Soil, 270*, 371-382. http://dx.doi.org/10.1007/s11104-004-1829-4

Bressan, W., & Vasconcellos, C. A. (2002). Morphological alterations on root system of maize induced by mycorrhizal fungi (*Glomus etunicatum* and *Glomus clarum*) and phosphorus. *Pesq. Agrope.c Brasileira, 37*, 509-517. http://dx.doi.org/10.1590/S0100-204X2002000400013

Brundrett, M. C. (2009). Mycorrhizal associations and other means of nutrition of vascular plants: understanding the global diversity of host plants by resolving conflicting information and developing reliable means of diagnosis. *Plant Soil, 320*, 37-77. http://dx.doi.org/10.1007/s11104-008-9877-9

Burleigh, S. H., Cavagnaro, T. R., & Jakobsen, I. (2002). Functional diversity of arbuscular mycorrhizas extends to the expression of plant genes involved in P nutrition. *J. Exp. Bot., 53*, 1593-1601. http://dx.doi.org/10.1093/jxb/erf013

Cameron, D. D. (2010). Arbuscular mycorrhizal fungi as (agro)ecosystem engineers. *Plant Soil., 333*, 1-5. http://dx.doi.org/10.1007/s11104-010-0361-y

Cardoso, E. J. B. N. (1996). Interaction of mycorrhiza, phosphate and manganese in soybean. In C. Azcon-Aguilar & J. M. Barea (Eds.), *Mycorrhizas in Integrated Systems from Genes to Plant Development* (pp. 304-306). Proc. of the 4th European Symposium on Mycorrhiza. Brussels, Luxembourg: Directorat General XII (Science, Research and Development), European Commision. http://dx.doi.org/10.1007/s00572-011-0375-1

Carneiro, J. G. A., Barroso, D. G., & da Silva-Soares, L. M. (2007). Growth of bare root *Pinus taeda*, L. seedlings cultivated under five densities in nursery. *Sci. Agric. (Piracicaba, Braz.), 64*, 23-29. http://dx.doi.org/10.1590/S0103-90162007000100004

Carrenho, R., Silva, E. S., Trufem, S. F. B., & Bononi, V. L. R. (2001). Succesive cultivation of maize and agricultural practices on root colonization, number of spores and species of AM fungi. *Braz. J. Microbiol., 32*, 262-270. http://dx.doi.org/10.1590/S1517-83822001000400002

Dodd, J. C., Boddington, C. L., Rodriguez, A., Gonzalez-Chavez, C., & Mansur, I. (2000). Mycelium of arbuscular mycorrhizal fungi (AMF) from different genera: form, function and detection. *Plant Soil., 226*, 131-151. http://dx.doi.org/10.1023/A:1026574828169

Douds, D. D., & Johnson, Jr. N. C. (2007). Contributions of arbuscular mycorrhizas to soil biological fertility. In L. K. Abbott, & D. V. Murphy (Eds.), *Soil Biological Fertility - A Key to Sustainable Land Use in Agriculture* (pp. 129-162). New York: Springer Science+Business Media.

Fester, T., & Hause, B. (2007). Drought and symbiosis – Why is abscisic acid necessary for arbuscular mycorrhiza? *New Phytol., 175*, 383-386. http://dx.doi.org/10.1111/j.1469-8137.2007.02171.x

Flores-Aylas, W. W., Saggin, O. J., Siqueira, J. O., & Davide, A. C. (2003). Effects of *Glomus etunicatum* and phosphorus on initial growth of woody species at direct seeding. *Pesq. Agropec. Brasileira., 38*, 257-266. http://dx.doi.org/10.1590/S0100-204X2003000200013

Gianinazzi, S., Gollotte, A., Binet, M. N., van Tuinen, D., Redecker, D., & Wipf, D. (2010). Agroecology: The key role of arbuscular mycorrhizas in ecosystem services. *Mycorrhiza., 20*, 519-530. http://dx.doi.org/10.1007/s00572-010-0333-3

Govindarajulu, M., Pfeffer, P. E., Jin, H. R., Abubaker, J., Douds, D. D., Allen, J. W., ... Shachar-Hill, Y. (2005). Nitrogen transfer in the arbuscular mycorrhizal symbiosis. *Nature, 435*, 819-823. http://dx.doi.org/10.1038/nature03610

Herrera, M. A., Salamanca, P., & Barea, J. M. (1993). Inoculation of woody legumes with selected arbuscular mycorrhizal fungi and rhizobia to recover desertified Mediterranean ecosystems. *Appl. Environm. Microbiol., 59*, 129-133.

Ishii, T. (2004). Vesicular–arbuscular (VA) mycorrhizae. Retrieved August 18, 2005, from http://www.bio.kpu.ac.jp/ pomlab/VAMinf.html

Kaschuk, G., Leffelaar, P. A., Giller, K. E., Alberton, O., Hungria, M., & Kuyper, T. W. (2010). Responses of legumes to rhizobia and arbuscular mycorrhizal fungi: A meta–analysis of potential photosynthate limitation of symbioses. *Soil Biol. Biochem., 42*, 125-127. http://dx.doi.org/10.1016/j.soilbio.2009.10.017

Khalil, S., Loynachan, T. E., & Tabatabai, M. A. (1994). Mycorrhizal dependency and nutrient uptake by improved and unimproved corn and soybean cultivars. *Agron. J., 86*, 949-958. http://dx.doi.org/10.2134/agronj1994.00021962008600060005x

Koide, R. T., & Li, M. (1990). On host regulation of the vesicular–arbuscular mycorrhizal symbiosis. *New*

Phytol., 114, 59-65. http://dx.doi.org/10.1111/j.1469-8137.1990.tb00373.x

Koltai, H., & Kapulnik, Y. (2010). *Arbuscular Mycorrhizas: Physiology and Function.* Heildelberg: Springer Science-Business Media B.V. http://dx.doi.org/10.1007/978-90-481-9489-610

Landis, T. D., & Amaranthus, M. A. (2009). Inoculate with mycorrhizae, rebuild your soil, and help stop global warming. *Forest Nursery Note,* 13-16.

Li, H., Smith, S. E., Holloway, R. E., Zhu, Y., & Smith, F. A. (2006). Arbuscular mycorrhizal fungi contribute to phosphorus-fixing soil even in the absence of positive growth responses. *New Phytol., 172*, 536-543. http://dx.doi.org/10.1111/j.1469-8137.2006.01846.x

Miyakasa, S. C., & Habte, M. (2001). Plant mechanisms and mycorrhizal symbioses to increase phosphorus uptake efficiency. *Commun. Soil Sci. Plant Anal., 32*, 1101-1147. http://dx.doi.org/10.1081/CSS-100104105

Morgan, J. A. W., Bending, G. D., & White, P. J. (2005). Biological costs and benefits to plant-microbe interactions in the rhizosphere. *J. Exp. Bot., 56*, 1729-1739. http://dx.doi.org/10.1093/jxb/eri205

Nikolaou, N., Karagiannidis, N., Koundouras, S., & Fysarakis, I. (2002). Effects of different P sources in soil on increasing growth and mineral uptake of mycorrhizal *Vitis vinifera* L. (cv Victoria) vines. *J. Int. Sci. Vigne Vin., 36*, 195-204.

Nogueira, M. A., & Cardoso, E. J. B. N. (2002). Microbial interactions on manganese availability and uptake by soybean. *Pesq. Agropec. Brasil., 37*, 1605-1612. http://dx.doi.org/10.1590/S0100-204X2002001100012

Olsson, P. A., Rahm, J., & Aliasgharzadeh, N. (2010). Carbon dynamics in mycorrhizal symbioses is linked to carbon costs and phosphorus benefits. *FEMS Microbiol. Ecol., 72*, 123-131. http://dx.doi.org/10.1111/j.1574-6941.2009.00833.x

Pearson, J. N., & Jakobsen, I. (1993). Symbiotic exchange of carbon and phosphorus between cucumber and three arbuscular mycorrhizal fungi. *New Phytol., 124*, 481-488. http://dx.doi.org/10.1111/j.1469-8137.1993.tb03839.x

Piccini, D., Ocampa, J. A., & Bedmar, E. J. (1988). Possible influence of *Rhizobium* on VA mycorrhiza metabolic activity in double symbiosis of alfalfa plants (*Medicago sativa* L.) grown in a pot experiment. *Biol. Fertil. Soils., 6*, 65-67.

Pressel, S., Bidartondo, M. I., Ligrone, R., & Duckett, J. G. (2010). Fungal symbioses in bryophytes: new insights in the twenty first century. *Phytotaxa., 9*, 238-253. http://dx.doi.org/10.11646/phytotaxa.9.1.13

Purakayastha, T. J., & Chhonkar, P. K. (2001). Influence of vesicular–arbuscular mycor–rhizal fungi (*Glomus etunicatum* L.) on mobilization of zinc in wetland rice (*Oryza sativa* L.). *Biol. Fertil. Soils., 33*, 323-327. http://dx.doi.org/10.1007/s003740000330

Read, D. J., Duckett, J. G., Francis, R., Ligrone, R., & Russell, A. (2000). Symbiotic fungal associations in 'lower' land plants. *Phil. Transact. Royal Soc. London (B), Biol. Sci., 355*, 815-830. http://dx.doi.org/10.1098/rstb.2000.0617

Rillig, M. C., Wright, S. F., & Eviner, V. T. (2002). The role of arbuscular mycorrhizal fungi and glomalin in soil aggregation: comparing effects of five plant species. *Plant Soil., 238*, 325-333. http://dx.doi.org/10.1023/A:1014483303813

Rillig, M. C., Wright, S. F., Nichols, K. A., Schmidt, W. F., & Torn, M. S. (2001). Large contribution of arbuscular mycorrhizal fungi to soil carbon pools in tropical forest soils. *Plant Soil., 233*, 167-177. http://dx.doi.org/10.1023/A:1010364221169

Rillig, M. R., & Mummey, D. L. (2006). Mycorrhizas and soil structure. Tansley Review. *New Phytol., 171*, 41-53. http://dx.doi.org/10.1111/j.1469-8137.2006.01750.x

Rossiana, N., & Supriatun, T. (2003). Fitoremediasi lumpur minyak bumi dengan tanaman sengon (*Paraserianthes falcataria* L. Nielsen) yang diinokulasi mikoriza. In T. Simarmata, D. H. Arief, Y. Sumarni, R. Hindersah, A. Azirin & A. M. Kalay (Eds.), *Teknologi Produksi dan Pemanfaatan Inokulan Endo–Ektomikoriza untuk Pertanian, Perkebunan, dan Kehutanan* (pp. 71-77). Prosiding Seminar Mikoriza, Bandung 16 September 2003. Bandung: Asosiasi Mikoriza Indonesia dan UNPAD.

Santos, J. G. D., Siqueira, J. O., & Moreira, F. M. D. (2008). Efficiency of arbuscular mycorrhizal fungi isolated from bauxite mine spoils on seedling growth of native woody species. *Rev Bras Ciencia do Solo., 32*, 141-150. http://dx.doi.org/10.1590/S0100-06832008000100014

Siddiqui, Z. A., Akhtar, M. S., & Futai, K. (2008). *Myorrhizae: Sustainable Agriculture and Forestry*. Heildelberg: Springer Science-Business Media B.V.

Simarmata, T., & Herdiani, E. (2003). Efek pemberian inokulan CMA dan pupuk kandang terhadap P tersedia, retensi P dalam tanah dan hasil bawang merah (*Allium ascalonicum* L.) pada Andisols. In T. Simarmata, D. H. Arief, Y. Sumarni, R. Hindersah, A. Azirin, & A. M. Kalay (Eds.), *Teknologi Produksi dan Pemanfaatan Inokulan Endo–Ektomikoriza untuk Pertanian, Perkebunan, dan Kehutanan* (pp. 71-77). Prosiding Seminar Mikoriza, Bandung 16 September 2003. Bandung: Asosiasi Mikoriza Indonesia dan UNPAD.

Smith, S. E., Facelli, E., Pope, S., & Smith, A. (2010). Plant performance in stressful environments: interpreting new and established knowledge of the roles of arbuscular mycorrhizas. *Plant Soil., 326*, 3-20.

Smith, S. E., & Read, D. J. (2008). *Mycorrhizal Symbiosis* (3rd ed.). San Diego: Academic Press.

South, D. B. (2000). Planting morphologically improved pine seedlings to increase survival and growth. *Forestry Wildlife Series., 1*, 1-14.

Swift, C. E. (2004). *Mycorrhiza and soil phosphorus levels* (pp. 1-4). Colorado State University, Cooperation Extention. Retrieved December 19, 2005, from http://www.colostate.edu/Depts/CoopExt/TRA/PLANTS/mycorhiza.html?

Sylvia, D. M. (2005). Mycorrhizal Symbioses. *Principles and Applications of Soil Microbiology* (2nd ed., pp. 263-282). New Jersey: Pearson Prentice Hall, Upper Saddle River.

Tawaraya, K., Watanabe, S., Yoshida, E., & Wagatsuma, T. (1996). Effect of onion root exudates on the hyphal growth of *Gigaspora margarita*. *Mycorrhiza., 6*, 57-59. http://dx.doi.org/10.1007/s005720050106

van den Driessche, R. (1994). Relationship between spacing in the nursery in relation to growth yield, and performance of the stock. *Forest Chronicle., 60*, 345-355.

van der Heijden, M. G. A., Klironomos, J. N., Ursic, M., Moutoglis, P., Streitwolf-Engel, R., Boller, T., ... Sanders, I. R. (1998). Mycorrhizal fungal diversity determines plant biodiversity, ecosystem variability and productivity. *Nature, 396*, 69-72. http://dx.doi.org/10.1038/23932

Vosátka, M., & Albrechtová, J. (2009). Benefits of Arbuscular Mycorrhizal Fungi to Sustainable Crop Production. In M. S. Khan, A. Zaidi & J. Mussarat (Eds.), *Microbial Strategies for Crop Improvement* (pp. 205-224). Berlin: Springer-Verlag. http://dx.doi.org/10.1007/978-3-642-01979-1_9

Wang, B., & Qui, Y. L. (2006). Phylogenetic distribution and evolution of mycorrhizas in land plants. *Mycorrhiza, 16*, 299-363.

Yao, M. K., Desilets, H., Charles, M. T., Boulanger, R., & Tweddell, R. J. (2003). Effect of mycorrhization on the accumulation of rishitin and solavetivone in potato plantlets challenged with *Rhizoctonia solani*. *Mycorrhiza, 13*, 333-336. http://dx.doi.org/10.1007/s00572-003-0267-0

Zhu, Y. G., Cavagnaro, T. R., Smith, S. E., & Dickson, S. (2001). Backseat driving? Accessing phosphate beyond the rhizosphere–depletion zone. *Trends Plant Sci., 6*, 194-195. http://dx.doi.org/10.1016/S1360-1385(01)01957-4

Permissions

All chapters in this book were first published in JAS, by Canadian Center of Science and Education; hereby published with permission under the Creative Commons Attribution License or equivalent. Every chapter published in this book has been scrutinized by our experts. Their significance has been extensively debated. The topics covered herein carry significant findings which will fuel the growth of the discipline. They may even be implemented as practical applications or may be referred to as a beginning point for another development.

The contributors of this book come from diverse backgrounds, making this book a truly international effort. This book will bring forth new frontiers with its revolutionizing research information and detailed analysis of the nascent developments around the world.

We would like to thank all the contributing authors for lending their expertise to make the book truly unique. They have played a crucial role in the development of this book. Without their invaluable contributions this book wouldn't have been possible. They have made vital efforts to compile up to date information on the varied aspects of this subject to make this book a valuable addition to the collection of many professionals and students.

This book was conceptualized with the vision of imparting up-to-date information and advanced data in this field. To ensure the same, a matchless editorial board was set up. Every individual on the board went through rigorous rounds of assessment to prove their worth. After which they invested a large part of their time researching and compiling the most relevant data for our readers.

The editorial board has been involved in producing this book since its inception. They have spent rigorous hours researching and exploring the diverse topics which have resulted in the successful publishing of this book. They have passed on their knowledge of decades through this book. To expedite this challenging task, the publisher supported the team at every step. A small team of assistant editors was also appointed to further simplify the editing procedure and attain best results for the readers.

Apart from the editorial board, the designing team has also invested a significant amount of their time in understanding the subject and creating the most relevant covers. They scrutinized every image to scout for the most suitable representation of the subject and create an appropriate cover for the book.

The publishing team has been an ardent support to the editorial, designing and production team. Their endless efforts to recruit the best for this project, has resulted in the accomplishment of this book. They are a veteran in the field of academics and their pool of knowledge is as vast as their experience in printing. Their expertise and guidance has proved useful at every step. Their uncompromising quality standards have made this book an exceptional effort. Their encouragement from time to time has been an inspiration for everyone.

The publisher and the editorial board hope that this book will prove to be a valuable piece of knowledge for researchers, students, practitioners and scholars across the globe.

List of Contributors

H. N. Subudhi
Central Rice Research Institute, Cuttack, India

S. P. Panda
P. G. Department of Botany, Utkal University, Bhubaneswar, India

P. K. Behera
Panchayat College, Baragarh, Orissa, India

Chitra Patnaik
P. N. College, Khurda, Orissa, India

Vera Alves de Sá
Programa de Pós-Graduação em Entomologia e Conservação da Biodiversidade, Universidade Federal da Grande Dourados (UFGD), Dourados, MS, Brazil

Marcos Gino Fernandes
Programa de Pós-Graduação em Entomologia e Conservação da Biodiversidade, Universidade Federal da Grande Dourados (UFGD), Dourados, MS, Brazil

Andrea Battiato
Agroscope, Institute for Sustainability Sciences (ISS), Ettenhausen, Switzerland
Department of Land, Environment, Agriculture and Forestry, University of Padua, Legnaro, Italy

Abdallah Alaoui
Centre for Development and Environment (CDE), University of Bern, Bern, Switzerland

Etienne Diserens
Agroscope, Institute for Sustainability Sciences (ISS), Ettenhausen, Switzerland

Saad Drissi
Plant Production, Protection and Biotechnology Department, Hassan II Agronomy and Veterinary Institute, Rabat, Morocco

Abdelhadi Aït Houssa
Department of Agronomy and Plant Breeding, National Agricultural School of Meknes, Meknes, Morocco

Ahmed Bamouh
Plant Production, Protection and Biotechnology Department, Hassan II Agronomy and Veterinary Institute, Rabat, Morocco

Mohamed Benbella
Department of Agronomy and Plant Breeding, National Agricultural School of Meknes, Meknes, Morocco

Homayoun Kanouni
Seed and Plant Unit, Agriculture and Natural Resources Research Center of Kurdistan, Sanandaj, Iran

Yadolah Farayedi
Food Legume Research Department, Dryland Agricultural Research Institute (DARI), Maragheh, Iran

Ali Saeid
Seed and Plant Unit, Agriculture and Natural Resources Research Center of West-Azerbaijan, Urmia, Iran

Sayyed Hossein Sabaghpour
Seed and Plant Unit, Agriculture and Natural Resources Research Center of Hamedan, Hamedan, Iran

Luciana Faldoni
Centro APTA Citros Sylvio Moreira, Cordeirópolis-SP., Brazil
Universidade Federal de São Carlos, Araras-SP, Brazil

Mariângela Cristofani-Yaly
Centro APTA Citros Sylvio Moreira, Cordeirópolis-SP., Brazil

Leonardo Pires Boava
Centro APTA Citros Sylvio Moreira, Cordeirópolis-SP., Brazil

Evandro Henrique Schinor
Centro APTA Citros Sylvio Moreira, Cordeirópolis-SP., Brazil

Kátia Cristina Kupper
Centro APTA Citros Sylvio Moreira, Cordeirópolis-SP., Brazil
Universidade Federal de São Carlos, Araras-SP, Brazil

Brian T. Campbell
Virginia Tech Southern Piedmont AREC, Blackstone, VA, USA

John C. Waller
Department of Animal Science, The University of Tennessee, Knoxville, TN, USA

William M. Backus
Research and Extension Center at Ames Plantation, The University of Tennessee, Grand Junction, TN, USA

Arnold M. Saxton
Department of Animal Science, The University of Tennessee, Knoxville, TN, USA

Deborah K. Joines
The Tennessee Soil, Plan, and Pest Center, Nashville, TN, USA

P. K. Maina
School of Biological Sciences, University of Nairobi, Nairobi, Kenya

P. M. Wachira
School of Biological Sciences, University of Nairobi, Nairobi, Kenya

S. A. Okoth
School of Biological Sciences, University of Nairobi, Nairobi, Kenya

J. W. Kimenju
Department of Plant Science and Crop Protection, University of Nairobi, Nairobi, Kenya

M. Otipa
Kenya Agricultural and Livestock Research Organization, Nairobi, Kenya

J. W. Kiarie
School of Biological Sciences, University of Nairobi, Nairobi, Kenya

Kwame D. Ansah
Department of General Agriculture, School of Applied Science & Technology, Sunyani Polytechnic, Sunyani, Ghana

Henry O. Sintim
Department of General Agriculture, School of Applied Science & Technology, Sunyani Polytechnic, Sunyani, Ghana

Samuel Awuah
Department of General Agriculture, School of Applied Science & Technology, Sunyani Polytechnic, Sunyani, Ghana

Joseph E. Ali
Department of General Agriculture, School of Applied Science & Technology, Sunyani Polytechnic, Sunyani, Ghana

George Oteng
Department of General Agriculture, School of Applied Science & Technology, Sunyani Polytechnic, Sunyani, Ghana

Abraha Gebrekiros Assefa
Department of Land Resources Management and Environmental Protection, Mekelle University, Mekelle, Ethiopia

Solomon Habtu Mesgina
Department of Land Resources Management and Environmental Protection, Mekelle University, Mekelle, Ethiopia

Yirga Weldu Abrha
Department of Land Resources Management and Environmental Protection, Mekelle University, Mekelle, Ethiopia

Lempie K. Ekandjo
Leibniz-Institute of Vegetable- and Ornamental Crops, Großbeeren/Erfurt e.V., Theodor-Echtermeyer Weg 1, Grossbeeren, Germany

Silke Ruppel
Leibniz-Institute of Vegetable- and Ornamental Crops, Großbeeren/Erfurt e.V., Theodor-Echtermeyer Weg 1, Grossbeeren, Germany

Anamika
Department of Agriculture, CAFF, Fiji National University, Fiji Island

Juan Valente Megchún-García
Colegio de Postgraduados, Campus Veracruz, Manlio Fabio Altamirano, Veracruz, México

Cesáreo Landeros-Sánchez
Colegio de Postgraduados, Campus Veracruz, Manlio Fabio Altamirano, Veracruz, México

Alejandra Soto-Estrada
Colegio de Postgraduados, Campus Veracruz, Manlio Fabio Altamirano, Veracruz, México

María del Refugio Castañeda-Chávez
Colegio de Postgraduados, Campus Montecillo, Montecillo, Texcoco, México

Juan Pablo Martínez-Dávila
Colegio de Postgraduados, Campus Veracruz, Manlio Fabio Altamirano, Veracruz, México

Iouri Nikolskii-Gavrilov
Colegio de Postgraduados, Campus Montecillo, Montecillo, Texcoco, México

Itzel Galaviz-Villa
Instituto Tecnológico de Boca del Río, Carretera Veracruz, Boca del Río, Veracruz, México

Fabiola Lango-Reynoso
Instituto Tecnológico de Boca del Río, Carretera Veracruz, Boca del Río, Veracruz, México

E. Carrillo-Ávila
Colegio de Postgraduados Campus Campeche, Carretera Haltunchén-Edzná, Sihochac, Champotón, Campeche, México

C. García-Acedo
Secretaria de Desarrollo Rural, Calle Ricardo Castillo Oliver Lote 12, Barrio de San Francisco, Campeche, México

J. Arreola-Enríquez
Colegio de Postgraduados Campus Campeche, Carretera Haltunchén-Edzná, Sihochac, Champotón, Campeche, México

C. Landeros-Sánchez
Colegio de Postgraduados, Campus Veracruz, Carretera Federal Xalapa-Veracruz, vía Paso de Ovejas entre Puente Jula y Paso San Juan, Tepetates, Veracruz, México

M. L. Osnaya-González
Colegio de Postgraduados Campus Campeche, Carretera Haltunchén-Edzná, Sihochac, Champotón, Campeche, México

C. Castillo Aguilar
Colegio de Postgraduados Campus Campeche, Carretera Haltunchén-Edzná, Sihochac, Champotón, Campeche, México

Truong Tuan Lin
Laboratory of Agricultural and Farm Management, Department of Agricultural and Resource Economics, Kyushu University, Higashiku, Fukuoka, Japan
Thai Nguyen University of Information and Communication Technology, Thai Nguyen, Viet Nam

Teruaki Nanseki
Faculty of Agriculture, Kyushu University, Higashiku, Fukuoka, Japan

Yosuke Chomei
Faculty of Agriculture, Kyushu University, Higashiku, Fukuoka, Japan

Sylvester Mpandeli
University of Venda, School of Environmental Sciences, Department of Geography and Geo-Information Sciences, Thohoyandou, South Africa
Water Research Commission of South Africa, Gezina, South Africa

Edward Nesamvuni
University of Venda, Office the Academic Registrar, Thohoyandou, South Africa

Phokele Maponya
Agricultural Research Council, Vegetable and Ornamental Plant Institute, Pretoria, South Africa

Juan Damian Marques Fong
Faculty of Life and Environmental Science, Shimane University, Matsue, Japan

Tsugiyuki Masunaga
Faculty of Life and Environmental Science, Shimane University, Matsue, Japan

Kuniaki Sato
Faculty of Life and Environmental Science, Shimane University, Matsue, Japan

H. Kang
Department of Crop Science and Biotechnology Chonbuk National University, Jeonju, Korea
National Institute of Crop Science, RDA, Suwon, Korea

C. S. Kang
National Institute of Crop Science, RDA, Suwon, Korea

H. Kim
National Institute of Crop Science, RDA, Suwon, Korea

C. S. Park
Department of Crop Science and Biotechnology Chonbuk National University, Jeonju, Korea
Institute of Agricultural Sciences and Technology, Chonbuk National University, Jeonju, Korea

Hamid Raeisi Vanani
Irrigation and Drainage engineering, Isfahan University of Technology, Isfahan, Iran

Ali Reza Soltani Todeshki
Water Engineering Department, Isfahan University of Technology, Isfahan, Iran

Kaveh Ostad Ali Askari
Civil Engineering, Water Engineering, Lecturer of Islamic Azad University of Najafabad, Iran

Mohammad Shayannejad
Water Engineering Department, Isfahan University of Technology, Isfahan, Iran

Faridullah
Department of Environmental Sceinces, COMSATS Institute of Information Technology Abbottabad, Pakistan

Farid Ul Haque
Department of Environmental Sceinces, COMSATS Institute of Information Technology Abbottabad, Pakistan

Alias Bin Abdullah
Faculty of Arts and Social Sciences, Department of East Asian Studies, University of Malaya, Kuala Lumpur, Malaysia

Muhammad Irshad
Department of Environmental Sceinces, COMSATS Institute of Information Technology Abbottabad, Pakistan

Arif Alam
Department of Development Studies, COMSATS Institute of Information Technology Abbottabad, Pakistan

Akhtar Iqbal
Department of Environmental Sceinces, COMSATS Institute of Information Technology Abbottabad, Pakistan

A. L. Saleh
Plant Nutrition Department, National Research Centre, Egypt

A. A. Abd El-Kader
Soils and Water Use Department, National Research Centre, Egypt

A. K. Alva
Northern Plains Agricultural Research Laboratory, United States Department of Agriculture, Agricultural Research Service, Sidney, MT, USA

Mingli Liu
Key Laboratory of Plant Stress Research, College of Life Science, Shandong Normal University, Jinan, P. R. China

Meng Li
Shandong Academy of Agricultural Sciences, Jinan, P. R. China

Kaichang Liu
Shandong Academy of Agricultural Sciences, Jinan, P. R. China

Na Sui
Key Laboratory of Plant Stress Research, College of Life Science, Shandong Normal University, Jinan, P. R. China

Antonius Suparno
Department of Agriculture, Faculty of Agriculture and Technology, The University of Papua (UNIPA), Manokwari, Indonesia

Saraswati Prabawardani
Agronomy Laboratory, Faculty of Agriculture and Technology, The University of Papua (UNIPA), Manokwari, Indonesia

Sudirman Yahya
Department of Agronomy, Faculty of Agriculture, Bogor Agriculture Institute (IPB), Bogor, Indonesia

Novita A. Taroreh
Department of English Literature, Faculty of Letters The University of Papua (UNIPA), Manokwari Papua Barat, Indonesia